Perspectives on Modern Optics and Imaging

With Practical Examples Using Zemax® OpticStudio®

RONIAN SIEW

ISBN: 978-1-5499-9369-5

To Babs.
And to my Mom, Dad, and bros.

Foreword

Optical design is evolving rapidly, and this book is an excellent review of the state of the art in several specific application areas. Being anchored firmly in both theory and modern optical design methods, this book helps the reader successfully straddle both aspects of product design. Optical design theory is of course covered extensively in the literature, but modern design methods are usually jealously guarded: this is where Ronian Siew has been very generous.

With examples from machine vision, Google's Glass™ wearable optics, and autostereoscopic 3D displays in mobile phones (and many others), Ronian presents a compelling vision of both fundamental theory and practical application in a modern product-engineering environment of optical systems that are relevant to today's burgeoning optical product development field.

Ronian's ability to explain design in deep theoretical detail and in a practical manner, and keen ability to evaluate optical designs in relation to the rest of the system, make this guide essential reading for today's practicing optical product designer.

Mark Nicholson, Ph.D.
CEO of Zemax LLC,
Kirkland, WA

Preface

Optical engineers and designers who have participated in technical group discussions on social media and professional networking sites can often relate to the continual need to seek answers to questions about many modern applications that may not be covered elsewhere. For example, in a blog on Linkedin®, I described an automated approach to improve the correction of secondary chromatic aberration in machine vision lenses with extended field of view, and I also included Zemax® OpticStudio® (OS) lens files and prescriptions for a specific design example. I felt thankful and gratified to receive many "shares" and "likes" on that blog. The number of visits to that blog also seemed to have helped increase the number of downloads of the original article I had published on that topic in a research journal [R. Siew, "Practical automated glass selection and the design of apochromats with large field of view," *App. Opt.* **55**(32), pp. 9232 – 9236 (2016)]. That experience motivated me to write this book, which I take as an opportunity to share a little more of my perspective on optics and imaging. In particular, I provide a brief review of some key optics principles, and I discuss their role in a selection of modern applications. Additionally, I provide some practical examples with lens prescriptions using Zemax OS.

This book assumes that readers possess basic optics knowledge (e.g., Snell's law), and for most cases, basic knowledge of optical system design using Zemax OS. Sec. 2 provides a focused review of optics fundamentals, highlighting only the most essential principles that are relevant to understand the topics covered in Sec. 3, which discusses a selection of modern visible light imaging applications, accompanied by a number of practical examples using Zemax OS (version 16). Topics covering some of the research that I have published in peer-refereed journals and conference proceedings are also provided in some sections of this book. They include the automated color correction method mentioned above (Sec. 3.2.4.4), nonimaging characteristics of imaging lenses (Sec. 2.3.5), the concept of "irradiance magnification" and the connection between Fourier optics and radiometry (Sec. 2.3.4), the f-number for lenses with non-circular pupils (Sec. 2.3.6), and the impact of "differential distortion" on relative illumination (Secs. 2.3.7 and 3.2.3.4).

Evidently, this book is not an introduction to optics, nor is it a textbook on optical design. It is also not an introduction to the use of Zemax OS, nor does it focus solely on Zemax OS modeling examples. Rather, this book is a technical resource offering my perspective on various optics and imaging fundamentals and their relationship to a select number of modern applications. Some of those applications have been relevant to my career experiences, while others have been among my research interests. Optical engineers and designers in industry are often too busy to write books, which is a pity for many fresh graduates and students who could benefit from being exposed to the rich variety of insights and perspectives that experienced engineers would have gained in their work. This book is an effort on my part to hopefully fill some of that gap.

For a long time, I struggled in deciding between having a less complex life, or to stop procrastinating and get even busier by writing a book. I guess I chose the latter. I have found in life that pushing one's limits and keeping busy is a good thing. Doing what we live for is the reason that we are alive. See, I'm just a guy who hasn't stopped searching for answers. So that's what I live for – to seek answers. And I enjoy writing about what I've found, what I still don't understand, and I do sometimes enjoy caricaturizing them (I've included cartoon sketches in some parts of this book). As we write, we also learn and acquire profound humility, for it is through writing that we really find out how much we do *not* know about something. As an author, therefore, I am a student of my own book. And I hope that you will benefit from it as much as I had fun and pain writing it.

Ronian Siew
Vancouver, British Columbia, Canada

Acknowledgments

When I was in middle-school, my social studies teacher once said that the greatest gift a parent could give a child is an education. I have found this to be true. I am thankful to my parents for supporting my college education. My Mom, a retired middle-school math teacher, provided much support and academic guidance in my early years in school. My Dad, a retired human resources specialist, supported my early interests in astronomy and showed me the stars. I also thank my brothers. My elder brother, Ivan, is a man of few words, with a big heart. When I was in high school, and he had already started university, he bought me a scientific calculator, a book on lasers, a book on Richard Feynman, a winter coat, a comforter, and more. In later years, he got me the complete Rocky movie collection (love that show), and a book on patents. My younger brother, Jeremy, is a man with virtually unlimited talent. His artwork and encouragement provided much inspiration for starting my own doodles. I thank my college professors at the University of Rochester for putting up with my constant questions, my bad grades, and giving me a chance to redeem myself. I thank Mark Nicholson and Alison Yates at Zemax LLC for their support and encouragement. And, saving the best for last, I thank my wife, Claire, aka Babs, aka Babsie, for so many things. Where do I even start? When we first met, naturally, she'd ask what I did for a living. I tried explaining optical engineering as well as I could. Instead of the usual "eyeglasses" response, she exclaimed, "You're a rocket scientist!" I was in love. And I believe she felt the same. She had to put up with my long hours at work, and my long hours at home working. Working on my papers. Working on my ideas. Working on my book. Working on everything. I thank her, from the bottom of my heart and soul, for her patience, support, care, great home-cooked food, and for loving me despite myself. And for saying "yes" to our first date, despite what I wore that day: a disheveled shirt, broken flip-flops, and torn jeans. I have improved much since.

Book updates, errata, etc.

This is the book's first edition. Over time, perhaps some errors (typographical, conceptual, or any other error) may be found in the book, and there should be a convenient way to share them with readers. Also, in some instances, perhaps I may develop new ideas, or perhaps I would find a better way to explain something that I had originally attempted to do in the book, which I would like to share with readers openly. Therefore, I have decided that perhaps the best way to do all of these is to have a webpage dedicated to book updates, which would include information about errata, possible new ideas, and reader feedback. So, to stay informed, please visit:

http://www.inopticalsolutions.com/p/the-book.html

Ronian Siew
ronian@inopticalsolutions.com

Contents

FOREWORD I

PREFACE II

ACKNOWLEDGMENTS IV

BOOK UPDATES, ERRATA, ETC. V

1. INTRODUCTION 1

  1.1 THE NATURE OF LIGHT 1

    1.1.1 What was light? 4

    1.1.2 What is light today? 5

    1.1.3 What light is in this book 7

    1.1.4 Who would benefit from reading this book 7

  1.2 THE NATURE OF OPTICS AND IMAGING 8

    1.2.1 Optical engineers, lens designers: What are we? 8

    1.2.2 Where do optics and imaging fit in school and the world? 12

  1.3 THE FUTURE OF OPTICS AND IMAGING 13

    1.3.1 Where do we go from here, and what does the future hold? 13

    1.3.2 Who will be the quantum optical engineers? 15

    1.3.3 Why geometrical optics is well and alive and probably forever 18

REFERENCES 19

2. OPTICS AND IMAGING IN THEORY 22

  2.1 GEOMETRICAL OPTICS: THE BACKBONE OF OPTICAL ENGINEERING 22

    2.1.1 What a lens is, and the "imaging principle" 23

    2.1.2 The paraxial thin lens model (PTLM) 31

    2.1.3 Aperture stops, pupils, and f-number 33

    2.1.4 Relative illumination (Part One): Vignetting 36

    2.1.5 Aberrations and "apparitions" of imaging systems 37

    2.1.6 Depth of field and focus 44

  2.2 PHYSICAL OPTICS: THE BACKBONE OF GEOMETRICAL OPTICS 54

    2.2.1 Diffraction and interference in imaging: A brief review of Fourier optics principles 54

    2.2.2 Physical optical imaging systems: Some examples 60

      2.2.2.1 Diffractive optics 61

      2.2.2.2 The approximately diffraction-free lens: Apodization 66

      2.2.2.3 Zemax OS example: Apodization and resolution enhancement 69

      2.2.2.4 A short note about coherence 75

  2.3 RADIOMETRY: THE BACKBONE OF ILLUMINATION AND NONIMAGING OPTICS 76

    2.3.1 Radiance, etendue, intensity, and all that: A brief review 76

    2.3.2 Radiance theorem for real and "fake" but useful lenses 84

2.3.3   Invariance of image brightness with distance................................86
2.3.4   Impact of diffraction and aberrations on image irradiance..........90
2.3.4.1   Zemax OS example: Aberrations and image irradiance 99
2.3.5   Nonimaging characteristics of imaging lenses..............................105
2.3.5.1   Zemax OS example: Axial nonimaging property of lenses 105
2.3.6   F-number of a lens with non-circular pupils................................112
2.3.7   Relative illumination (Part Two): Effect of angles, image distortion and pupil size..116

REFERENCES **122**

3.   OPTICS AND IMAGING IN MODERN APPLICATIONS **126**

3.1   MODERN OPTICAL ENGINEERING: PUSHING THE LIMITS OF THE THREE BACKBONES...127
3.2   MACHINE VISION IMAGING PRINCIPLES AND PRACTICES 128
3.2.1   The world of machine vision optics 128
3.2.2   Optical characteristics of "stock" lenses 129
3.2.2.1   What exactly is a lens's "field of view"? 129
3.2.2.2   What exactly is an "image circle"?...137
3.2.2.3   What exactly is "working distance"? 142
3.2.2.4   Determining principal plane locations in real life 143
3.2.2.5   The chief ray angle in image space 144
3.2.2.6   Sensor formats, image circles, and resolution 145
3.2.2.7   Illumination in machine vision 150
3.2.3   Customization and optical design for machine vision 155
3.2.3.1   Finding a starting point and scaling a lens 155
3.2.3.2   Optimizing in steps 156
3.2.3.3   Telecentricity does not imply uniform relative illumination........161
3.2.3.4   Zemax OS example: Relative illumination and distortion 163
3.2.4   Chromatic aberration in machine vision 178
3.2.4.1   Achromatism 179
3.2.4.2   Apochromatism 181
3.2.4.3   Zemax OS example: Automated apochromat glass selection 183
3.2.4.4   Zemax OS example: Apochromatic machine vision lens with extended field of view 190
3.2.5   Curved image sensors 202
3.2.5.1   Where to put the best focus..202
3.2.5.2   Distortion on a curved surface..205
3.2.6   Focusing a lens in machine vision...206
3.2.6.1   Shifting the image plane versus shifting a lens 206
3.2.6.2   How to maintain the working distance..208
3.2.6.3   Liquid lenses in machine vision...210
3.2.6.4   Zemax OS example: Position of a liquid lens in a lens 211
3.2.7   Depth sensing with machine vision lenses 218
3.2.7.1   Triangulation with an imaging lens..218
3.2.7.2   Impact of entrance and exit pupils on depth sensing...................219
3.2.7.3   Impact of lens symmetry on depth sensing 221

3.2.7.4   Zemax OS example: Symmetric lens for depth sensing 222
3.2.7.5   Depth determination by focus variation 228
3.2.7.6   Extended depth of field imaging 228
3.2.8   The limits of miniaturization *231*
3.3   MOBILE PHONE IMAGING ATTACHMENTS 234
3.3.1   Lens attachments and the meaning of magnification *234*
3.3.2   Significance of entrance and exit pupils *239*
3.3.3   The OOWA ^{TM} lens from Dynaoptics *245*
3.3.4   Spectrometer mobile phone attachments *249*
3.3.4.1   Basic concepts for spectrometer attachments 249
3.3.4.2   Zemax OS example: Simple spectrometer phone attachment 254
3.4   GOOGLE'S GLASS™ WEARABLE COMPUTING DEVICE 262
3.4.1   Basic imaging principle of the Glass ^{TM} eyepiece *262*
3.4.2   Zemax OS example: Glass ^{TM} as a fundus imager *264*
3.4.3   Use of adaptive optics *274*
3.5   STEREO IMAGING SYSTEMS 275
3.5.1   Stereoscopy vs. Autostereoscopy vs. Holography *276*
3.5.2   Stereo imaging with a single stationary lens *279*
3.4.2.1   Zemax OS example: Single lens stereoscopy 280
3.5.3   Autostereoscopic 3D displays in mobile phones *283*
3.4.3.1   Zemax OS example: Time-multiplexed autostereoscopic 3D display optical system for mobile phone 284
3.6   IMAGING USING THE GRATING LIGHT VALVE™ 292
3.6.1   Operating principle of the GLV *292*
3.6.2   Zemax OS example: On/off switching of GLV pixels in a simple GLV-based imaging system *301*
3.7   FLUORESCENCE DETECTION AND DNA ANALYSIS 310
3.7.1   The polymerase chain reaction (PCR) *312*
3.7.2   Real-time PCR *314*
3.7.3   Zemax OS Example: Optimizing detection from replicate wells using a lens' POP characteristic *316*

REFERENCES **323**

INDEX **330**

1. Introduction

1.1 The nature of light

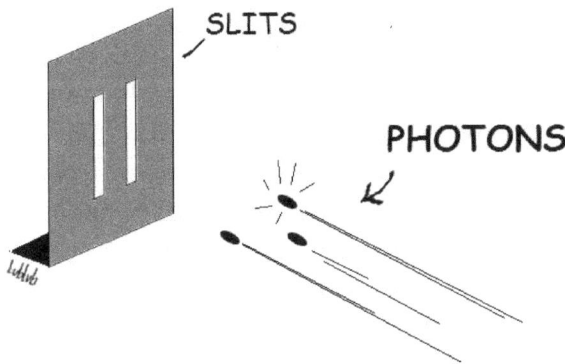

"Hey guys. Is it against the rules if we just went **AROUND** them?"

W hy would optical engineers – who mostly design instruments rather than perform fundamental research – concern themselves with questions regarding the nature of light? The reasons are plenty. Consider the following: If the optical engineer is designing an interferometer, then light is (mostly) a wave. If the optical engineer is developing an image sensor, then light might perhaps be particles, but it could also still be a wave, because it is sufficient to treat the sensor's atoms as quantum systems that absorb light in discrete energies. If the optical engineer is designing a lens, then light is (mostly) a "ray", whose path through lenses is determined by the propagation of wavefronts. Equivalently, the path that rays take through lenses may also be considered a consequence of Fermat's principle (i.e., light takes a path that, approximately speaking, takes the least time).

Besides the reasons stated above, questions regarding the nature of light may also serve as inspiration for the aspiring optical engineer.

For instance, my path to optical design and engineering started from asking not just questions about how telescopes work, but also about what light is, how it travels through glass, and how it reflects off mirrors. When I was in high school, I had a question about light that plagued me until I entered university, where I had found some satisfactory answers. That question was, "If light were a stream of particles, then why do they bounce off a polished mirror at equal angles?" That is, why is the angle of reflection equal to the angle of incidence (Fig. 1.1a)? The answer is not as obvious as it might seem. One might say that, by the conservation of momentum, a particle bouncing off a smooth surface should behave like a ball bouncing off the walls on a pool table. Yet if a light particle (let's call it a "photon") were "point-like" and therefore has no size, and if atoms were tiny particles, then on the scale of atoms comprising the surface of a mirror, one would expect photons to bounce off in all directions in a manner similar to throwing a tennis ball onto a bed of round stones (Fig. 1.1b).

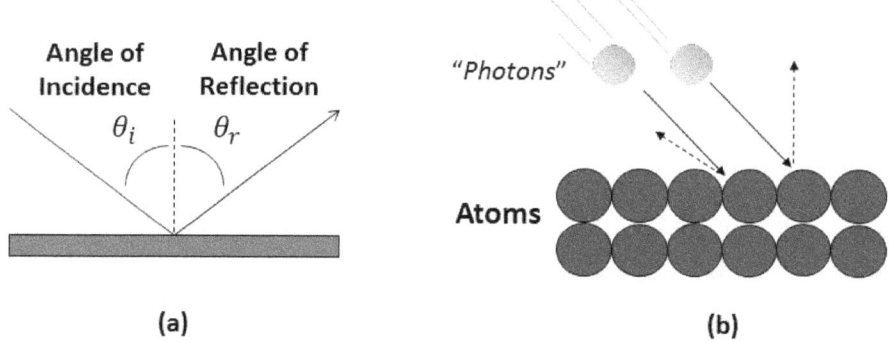

(a)   (b)

FIGURE 1.1 Reflection of light. (a) At the scale of humans, it seems natural that the angles of incidence and reflection should be equal. (b) At the scale of atoms, things are not so obvious.

Yet that isn't how it all happens [e.g., 1 – 8]. Whatever photons really are physically, it turns out that they do not just bounce off surfaces. In the case of a collection of closely packed molecules comprising a non-metallic material such as a slab of glass, photons effectively shake the electrons that are bound like springs to atoms comprising the material. Then, they are spat back out towards one of two possible directions: The direction of the reflected light (in accordance with the law of reflection), or the direction of the refracted light (in accordance with Snell's law). For glasses with a refractive index of 1.5, there is about a 4% chance of the photon being spat out in the reflected direction, and a 96% chance that it would refract through the glass. Since, to some extent, these outcomes may be considered consequences of the shaking of electric dipoles and interfering electromagnetic waves within a solid, then even the behavior of photons appear to be determined by laws that govern waves. And yet in some other cases, wave theory alone cannot explain the full behavior of the interaction of photons with matter [e.g., 9].

The mysteries do not end there. What happens when a photon is incident on a single atom or molecule rather than a collection of closely packed molecules? In this case, can one define an angle of incidence? Further, if an isolated molecule absorbs a photon, then in which direction would this photon go when it is spat back out by the molecule? Even Einstein had considered such questions, where, according to him, after a photon has been absorbed by a molecule, "There is no radiation in spherical waves. In a spontaneous emission process the molecule suffers a recoil of magnitude $h\nu/c$ in a direction that in the present state of the theory is determined only by 'chance'" [10].

Actually, the concept that a "photon" is like a "particle" is rather over-simplified. An accurate scientific theory of the photon is highly mathematical, somewhat more abstract, and yet also rather physical. For instance, the "photon" picture of light particles arises from the concept of lumps of light energy called "light quanta", and such quantized light energy arises from the energy states of standing wave modes of the electromagnetic field in a cavity. But we digress. This book concerns the workings of optical imaging systems, whose functions depend on theories and models of light that best suit those functions (we shall return to this point in Sec. 1.1.3). Still, in order to get readers to come on board with the spirit of this book, it is instructive to very briefly revisit those models of light that have helped to shape many of the imaging systems we know of today.

1.1.1 What was light?

Up until the late 18th century, the development of most of the geometrical aspects of optical imaging instruments (such as the magnification and focal position of an image) appears to have been driven sufficiently by a ray model of light, in which a "ray" quite simply represents the trajectory of a thin light beam that obeys the laws of reflection and refraction, and can spread further into rays of different colors upon refraction. For example, the magnifying glass (earlier than 65 A.D.), eyeglasses (earlier than the 1500s), the compound microscope and Galileo's and Kepler's refracting telescopes (during the 1600s), Newton's reflecting telescope (17th century), and Dolland's achromatic lens (during the 1700s) [11]. However, throughout that period, the analysis of aberrations and the manner in which rays travel through lenses also required the concept of "wavefronts", largely formulated by P. de. Fermat (1667), E. Malus (1808), C. Dupin (1816), W. R. Hamilton (1820 – 1830), P. L. von Seidel (1856), and perhaps others [12, 13].

Starting around the 18th to 19th centuries, Ernst Abbe (1840 – 1905) and Lord Rayleigh (1842 – 1919) laid the foundations for a wave theory of imaging that accounted for diffraction phenomena, which we now know quite generally as Fourier optics [14, 15]. However, the modern formulation of applied Fourier optics imaging principles took place around the mid-20th century, with contributions from P. M. Duffieux [16], O. Schade [17], and H. H. Hopkins [18]. Of course, it should also be noted that diffractive wave theory had its roots from the work of C. Huygens (1690), T. Young (1801), A. Fresnel (1866), and G. R. Kirchhoff (1891) [11, 12]. Fourier optics compliments the methods of geometrical optics in that the precise contrast of images formed by aberration-free optical imaging systems is shown to be limited by diffraction, interference, and coherence, which has contributed, for instance, to our understanding of the role of condensers in microscope image resolution. Additionally, the wave theory of imaging is the foundation for holography, invented in 1948 by Dennis Gabor [19], who won the 1971 Nobel Prize in physics for his invention. Thus, one can say that throughout much of the history of the development of image forming optical systems, it had been sufficient to model light as either rays or waves.

1.1.2 What is light today?

Today, most optical imaging devices in modern industrial applications still apply much of the theory of rays and waves that were highlighted in the previous section. However, their functions involve applying the ray-wave model and clever techniques to achieve such things as miniaturizing lenses, using liquid lenses for machine vision, imaging beyond the classical limits of resolution, extending the depth of field of a lens system, and seeing clearly through the Earth's turbulent atmosphere. These modern imaging techniques and others are the subjects of discussion in Sec. 3.

It is in the sense of applying clever techniques within the confines of the ray and wave models of light to address product requirements and to push the limits of imaging systems that, in my view, makes these imaging systems "modern". Hence, within the context of this book, "modern imaging" means "modern classical imaging", which does not necessarily require the application of the quantum theory of light. At most, they may involve the application of a so-called "semi-classical" theory of light-matter interactions, which treats light as electromagnetic waves, and matter (i.e., atoms, electrons...etc.) as quantum systems [20]. Within the realm of such semi-classical theories, much of the function of lasers and light absorption processes (such as the photoelectric effect) actually do not require a model of light being comprised of photons [20, 21]. Accordingly, since the process of light detection by charged coupled devices (CCDs) and complimentary metal-oxide semiconductor (CMOS) sensors (which are both present in many modern imaging systems) rely on the so-called internal photoelectric effect [22], one may say that the functions of these detectors in most practical cases would also not require a model of light being comprised of photons. However, in the research laboratory, aspects of the quantum nature of light have indeed been applied to imaging systems, which has come to be known as Quantum Imaging [for an overview, see, e.g., 23 – 25].

As far as fundamental physics is concerned, questions concerning the foundations of quantum optics, such as "what is light?" and "what is a photon?", continue to be investigated [e.g., 26 – 29]. If you are confused, you are not alone. In fact, the late physicist and co-winner of the 1955 Nobel Prize in Physics, Willis E. Lamb, Jr. (1913 – 2008), once wrote, "there is no such thing as a photon. Only a comedy of errors and historical accidents led to its popularity among physicists and optical scientists," [30]. Those who have studied physics are aware that much of the mysteries concerning the nature of light may be appreciated from observing that an interference pattern (which in theory is ordinarily associated with wave phenomena) on a screen placed behind a pair of slits develop over time even under conditions where only a single photon is allowed to pass through one or the other slit (hence, the cartoon at the beginning of the introduction). Thus, even today, the search for answers continues, which makes the subject of optics ever more exciting.

But for the rest of this book, fundamental questions need not concern us any more than what has been already highlighted. We are mainly concerned with optics and imaging within the context of modern industrial applications (i.e., applications that either involve concept prototypes that people are developing, or products that people are already using). To that end, one might ask, "But if the ray and wave models of light have been applied to develop imaging systems, how does one know when to use which model?" This is the subject of the next section.

1.1.3 What light is in this book

We shall take the following approach: From a practical standpoint, light is whatever it needs to be in order to design and analyze optical imaging systems, within mostly the confines of the ray and wave theories. However, we shall be reminded in Sec. 2.1 that these two models of light are not necessarily distinct, for the former is actually derived from the latter (we had already encountered this idea briefly in Sec. 1.1.1 when it was mentioned that the analysis of ray trajectories through lenses applies the concept of "wavefronts"). Hence, in the development of imaging systems, it is not really a question of deciding which model or theory of light to apply. Rather, it is a matter of taking the appropriate limiting representation of light that best suits the application. In many cases, for example, the ray model may be taken as an approximation in the limit that the sizes of optical components are much greater than the wavelength of light.

1.1.4 Who would benefit from reading this book

Although this book would perhaps mainly benefit optics students, optical engineers, and scientists, newcomers to optics and imaging who have strong scientific and engineering backgrounds may also find the book comprehensible. For the serious newcomers, this book could serve as a companion to many of the fine optics textbooks that universities use, and it may also serve to "put the pieces together", so to speak, in their self-study of imaging. This book could also serve as an "imaging counterpart" to the very practical book by Hobbs [31], which presents material with a spirit similar to that in this book, and covers topics as diverse as electro-optical systems, signal processing, and best practices in the lab.

Practitioners of optical design and engineering could perhaps refer to this book for some ideas and alternative perspectives. Additionally, those optical designers who routinely use Zemax's OpticStudio (OS) program (version 16) would certainly benefit from the design and modeling examples provided in this book. However, because lens prescriptions are also provided, optical designers who use other programs may also try out those examples for themselves by entering lens prescription data into any program of their choice.

The level of mathematics in this book spans a wide range, from elementary trigonometry and algebra to multidimensional calculus, but mostly does not go beyond methods covered in undergraduate and introductory graduate engineering courses. The focus of course is always on the concepts, but much math is unavoidable, for it is the language of science (in some cases, it *is* the science). Still, I think that the math used here would be accessible to anyone who has some handy calculus textbooks nearby. As the reader proceeds from the introduction, the reader will notice a gradual increase in the level and rate at which concepts are presented. And from Sec. 2.2 onwards, the material covered is rather advanced. In summary, I hope that students, fresh graduates, scientists, technologists, technopreneurs, geeks smarter than me, and essentially anyone who would like to know a little more about the theory and practice of optical imaging would benefit from reading this book.

1.2 The nature of optics and imaging

1.2.1 Optical engineers, lens designers: What are we?

They or *we* can sometimes be a weird bunch. I believe it is rather accurate to say that anyone who has ever worked at a large technology company that hires and uses the skills of optical engineers can relate to experiences in which one finds a "corner" of the company where a certain group of people (let's call them optical engineers) sit, stare at incomprehensible data, and often speak of some incomprehensible things that contradict a layperson's experience, such as "laser TEM_{00} beams follow a hyperboloidal trajectory and so do not travel in straight lines neither within nor beyond the Rayleigh range". When asked about the cost of goods for the build of material for a product, the optical engineer's list of prices quite often amounts to being among the highest and with the longest lead times. If you manage the hardware engineering group in such a company, you may know what I mean. If you're an optical engineer, you may also know what I mean.

But for those new to managing a group comprised of a variety of engineers, some whom are optical people, you might wonder what is the difference between, say, an optical engineer and an optical designer. You may also wonder if a lens designer is some other type of creature. If you are wondering these, then perhaps the illustration in Fig. 1.2 might be helpful. Traditionally, lens designers are those who focus a significant portion of their time designing a lens. This lens could be a modern single lens reflex (SLR) camera lens, or a mobile phone's lens, or any lens system, and it must meet a defined list of optical specifications such as image contrast, distortion, modulation transfer function (MTF), field curvature, astigmatism, relative illumination, and more. Additionally, the lens must meet certain environmental requirements, such as shock, vibration, immersion in water…etc.

FIGURE 1.2 Lens designers and optical engineers.

The work of a lens designer is tedious, to say the least. They are forced within their job responsibilities to be meticulous, and they must concentrate fully on tracing light rays through lenses, prisms, reflection off mirrors…etc., all day, every day. When one manages such a group, it pays to be somewhat patient, and to allow lens designers their time at their desk (or lab). If we should bother them, then we could be the reason that the lens designer's concentration has been broken, and therefore, he or she has to re-start the process of "balancing aberrations" all over (because the process is rather iterative), thereby delaying product development and time-to-market. Hence, resist the temptation to call them into irrelevant meetings, yet do also continue to encourage communication, for it usually leads to good and better things. Finally, though they might seem strange at times, lens designers are not really another type of creature. They are indeed, people, and they enjoy their work. They also enjoy movies. And popcorn.

Nowadays, the term "optical designer" is used synonymously with "lens designer", but the work of an optical designer may consist not just of tracing rays through lenses to develop an image forming optical system. It could also involve the design of illumination systems. In some cases, optical designers may be called "illumination engineers" if their work and expertise focuses on the application of radiometry and nonimaging principles.

What are optical engineers? Optical engineers on the whole may not necessarily possess the expertise of designing lenses. A physics graduate, for example, can be an optical engineer.[*] But more often, optics graduates who did not specialize in lens design would become optical engineers (I personally did not specialize in lens design in my university studies, and I had actually hated that aspect of work as a student. But eventually, I acquired an appetite for lens design, and proceeded to learn it on the job as well as through self-study, with loads of practice).

Optical engineers, for example, apply the principles of optics to develop models that aid in the conceptualization, design, and integration of an entire optical instrument, such as the fluorescence detection system shown in Fig. 1.3 [32]. They also specify optical

---

[*] Of course, anyone can be an optical engineer. It only takes interest, maybe even some passion, and certainly much discipline, self-study, asking questions, and perhaps about "10,000" or more hours of practice [33].

components for the purpose of procurement, construct breadboards or prototypes on an optics lab bench, test optical components and instruments, and discuss requirements with optical parts vendors. Nowadays, the term "optical system engineers" is used synonymously with "optical engineers". But the work and responsibilities of both optical designers and optical engineers are not limited to the above, for anything is possible. For example, when I worked as an optical designer at a company in Singapore, I was often required to travel with business managers to various parts of the world to support technical discussions with customers. I was also asked to present technical marketing material at road shows. Business managers and the marketing department often appreciate the technical support provided by optical designers and engineers. So, it seems that the more skills one possesses, the more responsibilities one might acquire, and that may not be a bad thing from the point of view of gaining experience and opening one's eyes to the big wide world in which we live.

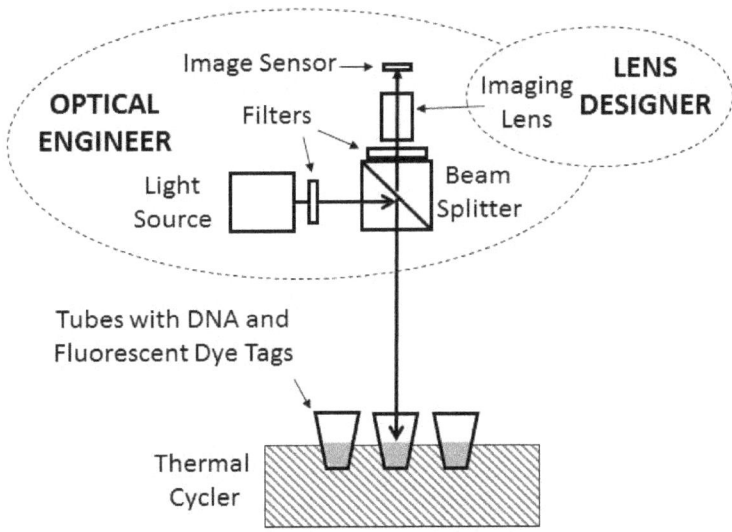

FIGURE 1.3 Typical optical design responsibilities for an integrated optical system to produce a complete product, in this case, a fluorescence detection system for DNA analysis [32].

1.2.2 Where do optics and imaging fit in school and the world?

If you know of a high school student who is considering an optics education, or if you are a technopreneur new to optics and imaging, then you may wish to know how optical engineers study this material at a university. In American colleges or universities that offer specialization on optics and optical engineering (such as the University of Rochester, University of Arizona, University of Alabama in Huntsville, University of Central Florida, and others), a typical curriculum would include a slew of math and physics courses, for those are the essential tools of an optical engineer. Within the physics or optics department, three major or core topics are mostly covered: geometrical optics, physical optics, and radiometry. Other more advanced subjects could be lens design, optical metrology, lasers, color science, vision, and perhaps quantum optics. There are usually plenty of other advanced course offerings in university. In this book, brief refreshers covering the most essential principles of geometrical optics, physical optics, and radiometry (Secs. 2.1, 2.2., and 2.3 respectively) are provided, with the purpose of setting the stage for understanding a selected number of modern imaging applications.

The question of where optics fits in the world is generally not obvious to everyone. There are, perhaps, the more common examples: Mobile phone cameras, digital camcorders, digital cinema projectors, and of course, spectacle lenses or eyeglasses. Of these, perhaps eyeglasses are the first pieces of technology any person thinks of when an optics graduate says "I'm an optical engineer". In reality, essentially any product that has to do with light would have likely been the result of work done on it by an optical designer or optical engineer. Moreover, anything in nature that has to do with any segment of the electromagnetic spectrum is of interest to the optical engineer (but this book's focus is on the visible spectrum). Gamma-rays, x-rays, ultraviolet rays, visible light rays, infrared rays, microwaves, and radio waves – they are all light. They are all electromagnetic radiation[*], but at different frequencies and wavelengths. Therefore, those so-called "thermal imaging cameras" (of the microbolometer type) actually detect infrared electromagnetic radiation, not heat, which is the random shaking motion of atoms and molecules.

Evidently, optics and imaging plays a significant role in technology and the world. Our understanding of light not only helps us design and build optical products, but it also helps us rationalize much of the optically related natural phenomena around us. If the phenomena are visible to our eyes, we need only observe. If not, then they are probably happening at some other electromagnetic frequency, and we need only use an image sensor that is sensitive to that frequency.

1.3 The future of optics and imaging

1.3.1 Where do we go from here, and what does the future hold?

The short answer is that nobody knows. At least, not with absolute certainty. For instance, field curvature and astigmatism have both been known for a long time, yet only today is the prospect of practical curved image sensors possibly realizable [34] (this is discussed further in Sec. 3.2.5.) Technology evolves unpredictably, and it is mostly reported in research journals, company websites, and patent publications. However, research and development in companies (and perhaps the military) is not always fully disclosed, for obvious reasons. Therefore, there may still be much technological advances that one cannot really follow in-depth.

But there are some general trends that we can speculate on. For example, perhaps optical imaging methods in the life sciences and medicine will surely continue to grow, mostly because healthcare is

---

[*] In this book's context, by "radiation" we simply mean electromagnetic waves.

on everyone's minds. These would include (but certainly not limited to) optical coherence tomography (OCT), Raman imaging, and label-free detection (i.e., detection without the use of "markers" such as fluorescent dyes). Perhaps even super-resolution imaging methods could develop further and enter into this market. Communication – in particular, the use of mobile devices – is also big. I believe that smart imaging (also called "embedded vision") and mobile phone photography will continue, driven particularly by social media. Therefore, advances in optics related to these devices will likely continue, including mobile phone imaging attachments and possibly 3D displays. The gaming industry is also phenomenal, and we may see further developments in so-called "RGBD" (Red, Green, Blue, and depth) sensing. There are plenty more optical technologies that are evolving, and we are most definitely not limited to the above.

Yet despite what has been mentioned, note that those technologies are what one can read in the news. They are what everyone else is doing. Therefore, they aren't the future. They are today. In fact, they are already yesterday. The future is unknown, and you, the technopreneur, determine it. Since nobody has a crystal ball, it seems reasonable that if we carry the responsibility of developing a technology roadmap for ourselves or our company, putting everyone else's technology on a roadmap and attempting to extrapolate may not be the best approach. The question really is: what do *we* want to do? In this spirit, it makes sense to create a roadmap that emphasizes more on capability rather than technology (Fig. 1.4).

The capabilities in such a roadmap could be design capabilities, or they may be manufacturing capabilities, or both. Manufacturing capabilities are not restricted to new advanced hardware. They also include developing new ways to use such hardware, because if you can purchase the hardware, then so can others. But if you can develop novel ways to use the hardware, then it is *you* who has made the future.

The rationale for a capability roadmap is that human resource is by far the greater resource than technology resource. If we put great minds together, great things can happen. If you're an engineer of any specialization, you may be reading technology and trade magazines. Every now and then, you realize that people are moving ahead – very quickly. If one were a software developer for instance, how does one keep up? Learning new codes every quarter of the year may not be very realistic. But if one develops mastery on perhaps a couple of codes, with focus on the fundamentals, then I believe that one will succeed. Similarly, as optical designers and engineers, focusing on new technologies isn't the key to developing ourselves. Rather, keeping perhaps just a quarter of an eye on technological advances, and focusing more on improving our fundamental understanding of optics will help us succeed in understanding and developing new technologies.

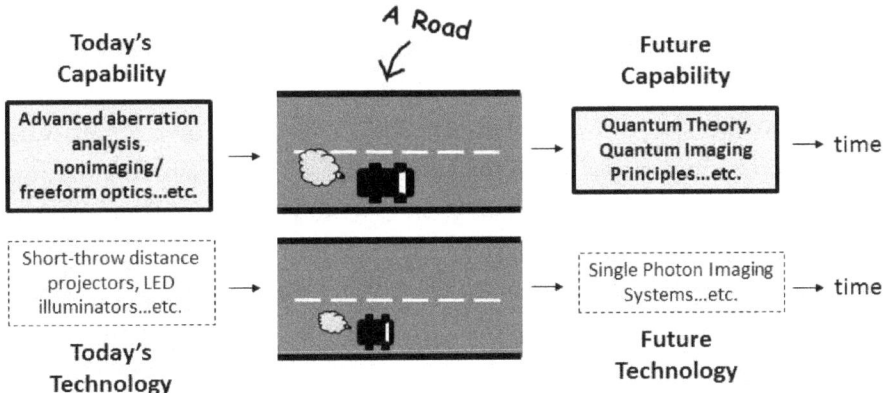

FIGURE 1.4 Example of a "capability-technology" roadmap, with emphasis on capability development over time.

1.3.2 Who will be the quantum optical engineers?

There is one other skillset that seems worth considering now and for the future, and it is quantum optical engineering, which could include quantum imaging applications, and also quantum nonlinear optics. Consider, for example, the concept of "slow-light", whose fundamental processes are described by the principles of quantum nonlinear optics. As a consequence of those processes, the group velocity of light in a medium can be made to reduce, because the group refractive index of the medium is increased [35, 36]. One potential application of this is in the miniaturization of spectrometers [37]. As Boyd [37] states, "Anyone can miniaturize a spectrometer, but the trick is to miniaturize it without losing spectral resolution". Let us see how this works.

The spectral resolution of a dispersing element is effectively proportional to the path length difference between interfering beams. Let us call this path length difference "L" a sort of "characteristic dimension" for any spectrometer. For a prism-based spectrometer, L would be the base length of the prism. For a grating-based spectrometer, L would be the grating width. For an interferometer, L is the optical path length difference between interfering beams. Now, if a spectrometer's physical dimensions are reduced for miniaturization, then all component dimensions of the spectrometer are also reduced, including the characteristic dimension L, which means that the spectral resolution is also reduced. However, it turns out that if a slow-light medium (i.e., a medium in which quantum nonlinear optical phenomena may be effected) is placed into one of the paths within a Fourier transform spectrometer (say, in a Mach-Zehnder interferometer configuration), the spectral resolution would be proportional to both L and the group refractive index of the slow-light medium (Fig. 1.5).

Following Boyd's analysis [37], one examines the instantaneous rate of change of the introduced phase ϕ from the slow-light medium in Fig. 1.5 with respect to the light beam's angular frequency ω:

$$\frac{d\phi}{d\omega} = \frac{d}{d\omega}\left(\frac{\omega n L}{c}\right) = \frac{L}{c}\left(n + \omega\frac{dn}{\omega}\right) = \frac{L n_g}{c}. \qquad (1.1)$$

Here, c is the speed of light in vacuum, n is the phase refractive index of the slow-light medium, and n_g is the group refractive index of the slow-light medium. The rate $d\phi/d\omega$ is a measure of spectral resolution because it is an indication of the sensitivity of phase changes in the interferogram with respect to light frequency. The more rapid it changes, the more sensitive is the interferometer at detecting small variations in light frequency.

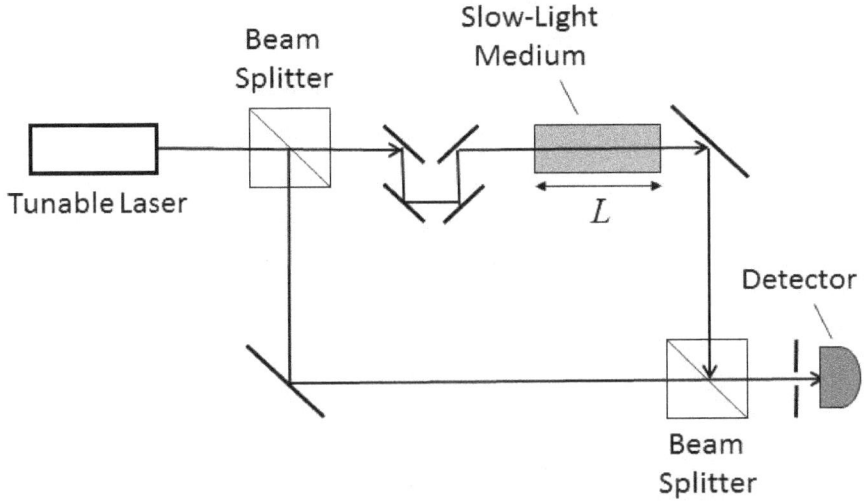

FIGURE 1.5 Mach-Zehnder interferometric spectrometer (re-sketched according to Refs. 35, and 36) with a slow-light medium.

Now, if this interferometer's dimensions are reduced, the group refractive index n_g can be made to increase using the slow-light medium in order to compensate for the reduction in L, thus, maintaining the spectral resolution. This is far from being an obvious mechanism for most "classically trained" optical engineers (including myself). Therefore, if the miniaturization of spectrometers into chip-sized devices becomes a reality in the market, who will be the "quantum optical engineers" for such devices? Moreover, quantum optical engineering would possibly include many of the "weird" features of quantum physics, such as quantum entanglement, and therefore requires a whole different way of thinking, problem solving, and intuition [38].

Perhaps classically trained optical engineers who have kept up with the fundamentals of nonlinear and quantum nonlinear optics (including quantum optics principles) may possess sufficient knowledge to become quantum optical engineers, but it would surely take much continuing self-study (and perhaps, even attending university or professional development courses at trade shows and conferences). Hence, the more ready-skilled quantum optical engineers of the future would possibly start with optical physicists who, like it once was during the telecom bubble, may flee from their PhD programs and run into "the Valley", or spin off some start-ups on their own. In any event, I believe this is another example of the importance of emphasizing the fundamentals rather than the technology.

1.3.3 Why geometrical optics is well and alive and probably forever

Once, sometime between the years 2002 – 2003, a college professor said to me, "I don't understand why people are saying that, because of trends like nanotechnology and nanophotonics, classical lens design is going away. But you still need to form an image!" At least, that was approximately what he had said. That was almost 15 years ago, and he was right. Today, we still need people who know how to form images onto image sensors, and how to balance aberrations or even to eliminate them. Additionally, lens design software will very likely never replace lens designers, for the software is not where the real thinking takes place. Optical design involves more than optimization, and optimization is not just about damped least squares and genetic algorithms. It involves people making intuitive decisions. It involves people connecting the dots across departments and functions. Smart cameras need smart optical engineers to design them and to integrate them with other technologies. It won't be done by robots, because robots won't usually know how to consider alternative perspectives to solve a problem unless those alternatives aren't already programmed into a chip. Hence, optical designers, optical engineers, and optical technicians will always be relevant [e.g., see also Ref. 39].

At the core of all imaging devices are the principles of geometrical optics, even if Fourier transform theory and Fourier optics may be applied for the design of such devices. Even if extended depth of field technologies such as wavefront coding [40] and light field imaging [41] applies Fourier optics principles, the concept of depth of field may be understood in a simple fashion through the application of geometrical optics principles. Moreover, in wavefront coding, phase masks are introduced into the path of an imaging system in order to modify the system's point spread function. Therefore, lenses or mirrors for imaging are still used – and therefore, geometrical optics is still applied. If, for example, quantum optical engineers are developing a system, and if they need a lens, they would need the assistance of lens designers. If they need to integrate optics with other systems, they would need the assistance of today's "classical" optical engineers.

One usually seeks the simplest tools to perform tasks. In imaging, geometrical optics is by far the simplest, and the most intuitive. Geometrical optics is therefore, quite generally the backbone of optical engineering. Throughout this book, we will continue to see this theme exemplified in a variety of modern industrial imaging applications. So, let us forge ahead.

References

1. P. W. Milonni, "Answer to Question #21 ["Snell's law in quantum mechanics," Steve Blau and Brad Halfpap, Am. J. Phys. **63**(7), 583 (1995)]," *Am. J. Phys.* **64**, p. 842 (1996).
2. B. A. Sherwood, "Answer to Question #21 ["Snell's law in quantum mechanics," Steve Blau and Brad Halfpap, Am. J. Phys. **63**(7), 583 (1995)]," *Am. J. Phys.* **64**, p. 840 (1996).
3. R. P. Feynman, *QED: the strange theory of light and matter*, (Princeton University Press, 1985), pp. 36 – 54.
4. E. Hecht, *Optics*, 4th ed., (Addison Wesley, 2002), pp. 95 – 141.
5. Victor F. Weisskopf, "How Light Interacts with Matter," *Scientific American* **219** (1968), pp. 60 – 71.

6. H. Fearn, D. F. V. James, and P. W. Milloni, "Microscopic approach to reflection, transmission, and the Ewald-Oseen extinction theorem," *Am. J. Phys.* **64**(8) (1996), pp. 986 – 995.

7. V. C. Ballenger and T. A. Weber, "The Ewald-Oseen extinction theorem and extinction lengths," *Am. J. Phys.* **67**(7) (1999), pp. 599 – 605.

8. M. Born and E. Wolf, *Principles of Optics*, 6th ed., (Cambridge University Press, 1980), pp. 100 – 108.

9. M. O. Scully and M. S. Zubairy, *Quantum Optics*, (Cambridge University Press, 1997), pp. 1 – 2, and pp. 20 – 39.

10. E. Wolf, "Einstein's Researches on the Nature of Light," *Optics News*, (winter, 1979), pp. 24 – 39.

11. See Ref. 4, pp. 1 – 9.

12. W. T. Welford, *Aberrations of Optical Systems*, (IOP Publishing, 1986), p. 10, 107

13. See Ref. 4, pp. 105 – 106.

14. J. W. Goodman, *Introduction to Fourier Optics*, 2nd ed., (McGraw-Hill, 1996), p. 126.

15. S. G. Lipson, H. Lipson, and D. S. Tannhauser, *Optical Physics*, 3rd ed., (Cambridge University Press, 1995), pp. 10 – 11.

16. P. M. Duffieux, *The Fourier Transform and its Applications to Optics*, 2nd ed., (Wiley, 1983).

17. O. H. Schade, "Electro-optical characteristics of television systems," *RCA Review*, (1948).

18. H. H. Hopkins, *Wave Theory of Aberrations*, (Oxford University Press, 1950).

19. D. Gabor, "A new microscopic principle," *Nature* **161** (1948), pp. 777 – 778.

20. See Ref. 9, pp. 20 – 22.

21. P. W. Milonni and J. H. Eberly, *Lasers*, (Wiley, 1988), pp. 243 – 260.

22. B. E. A. Saleh and M. C. Teich, *Fundamentals of Photonics*, (Wiley, 1991), pp. 645 – 648.

23. R. W. Boyd, K. W. C. Chan, A. Jha, M. Malik, C. O'Sullivan, H. Shin, P. Zerom, "Quantum Imaging: Enhanced Image Formation Using Quantum States of Light," in *Quantum Information and Computation VII*, E. J. Donkor, A. R. Pirich and H. E. Brandt (editors), *Proc. SPIE* **7342** (2009).

24. R. W. Boyd and P. J. Reynolds (editors), Special Issue: Quantum Imaging in *Quantum Information Processing* **11**(4), (Springer, 2012), pp. 887 – 1011.

25. R. W. Boyd, "Quantum Imaging: New Methods and Applications," presented at SPIE (August, 2005), available for download at: http://www.optics.rochester.edu/workgroups/boyd/archive/presentations/ Boyd_SPIE_05_2.pdf.

26. C. Roychoudhuri, A. F. Kracklauer, and K. Creath (editors), *The Nature of Light: What is a photon?* (CRC Press, 2008).

27. C. Roychoudhuri and K. Creath (editors), The Nature of Light: What is a Photon? *Proc. SPIE* **5866** (2005).

28. C. Roychoudhuri, "The nature of light: what are photons?" *SPIE Newsroom*, (26th December, 2006), DOI: 10.1117/2.1200611.0480.

29. R. Bennett, T. M. Barlow, and A. Beige, "A physically motivated quantization of the electromagnetic field," *Eur. J. Phys.* **37** 014001 (2016).

30. W. E. Lamb Jr., "Anti-photon," *Appl. Phys. B* **60**(2), pp. 77 – 84 (1995).

31. P. C. D. Hobbs, *Building Electro-Optical Systems: Making It All Work*, (Wiley, 2000).

32. S. J. Boege, J. A. Hoshizaki, M. F. Oldham, and L. Ilkova, "Fluorescence Detector with Automatic Changing Filters," United States Patent No. **7,295,316 B2** (Nov. 13, 2007).

33. M. Gladwell, *Outliers: The Story of Success*, (Back Bay Books, 2008), pp. 38 – 76.

34. B. Guenter, N. Joshi, R. Stoakley, A. Keefe, K. Geary, R. Freeman, J. Hundley, P. Patterson, D. Hammon, G. Herrera, E. Sherman, A. Nowak, R. Schubert, P. Brewer, L. Yang, R. Mott, and G. McKnight, "Highly curved image sensors: a practical approach for improved optical performance," *Opt. Express* **25**, 13010-13023 (2017).

35. Z. Shi, R. W. Boyd, D. J. Gauthier, and C. C. Dudley, "Enhancing the spectral sensitivity of interferometers using slow-light media," *Opt. Lett.* **32**(8), pp. 915 – 917, (2007).

36. M. E. Holmes, M. Mirhosseini, and R. Boyd, "Slow Light: Moving Out of the Lab," *Photonics Spectra*, July 2014, pp. 42 – 44.

37. R. W. Boyd, "Quantum Nonlinear Optics: Nonlinear Optics Meets the Quantum World," presented at SPIE Photonics West 2016, DOI: 10.1117/2.3201602.17.

38. C. R. Stroud, Jr., "Quantum Optical Engineering," in *A Jewel in the Crown: Essays in Honor of the 75th Anniversary of the Institute of Optics, University of Rochester*, edited by Carlos R. Stroud Jr., Meliora Press of the University of Rochester, (2004), pp. 392 – 394.

39. P. Daukantas, "The Optics Workforce: Looking to the Future," *Optics & Photonics News*, July/August 2017, pp. 26 – 33.

40. E. R. Dowski and W. T. Cathey, "Extended depth of field through wave-front coding," App. Opt. 34(11), pp. 1859 – 1866 (1995).

41. R. Ng, M. Levoy, M. Bredif, G. Duval, M. Horowitz, and P. Hanrahan, "Light Field Photography with a Hand-held Plenoptic Camera," *Stanford Tech Report* CTSR 2005-02.

2. Optics and Imaging in Theory

*"I'd give up this flawed theory
if it weren't so beautiful."*

Fortunately, we have had a rather successful theory in optics. In this section, we briefly review some of the more essential fundamentals of imaging that are relevant to understand and appreciate the selection of modern imaging systems covered in Sec. 3. In those cases where Zemax OS examples are provided, readers are assumed to have some familiarity with the Zemax OS program (version 16) in both the sequential and non-sequential modes. Some excellent introductory references are, for example, Smith [1], and Geary [2].

2.1 Geometrical optics: The backbone of optical engineering

Few would doubt that the first thing one would normally do when visualizing the propagation of light through an imaging system is to imagine "rays" that are either bending at the interfaces between air and glass, or reflecting off mirror surfaces.* In fact, optical designers and engineers know that such rays are actually lines that are

---

* Although curved mirrors form images too, we'll focus on lenses throughout most of this book.

orthogonal to wavefronts (Figs. 2.1a and 2.1b). This is known as the ***Theorem of Malus and Dupin*** [e.g., 3, 4]. Hence, the foundation for geometrical optics is actually a wave theory of light, and so another way to think about geometrical optics is "physical optics, minus diffraction and interference". A source for light rays is, in physical terms, a source of oscillating dipoles generating transverse electromagnetic waves. The wiggles of an oscillating dipole resemble the wiggles of a finger in a pool of water. This physical picture of light rays yields a rich theory of imaging, and it provides much of our understanding of imaging aberrations.

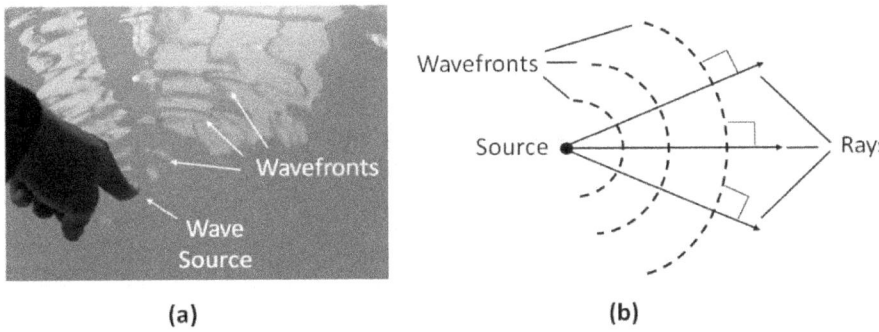

(a)   (b)

FIGURE 2.1 Wavefronts and rays. (a) A wiggling finger in a pool of water creates ripples of water waves, whose crests are the so-called "wavefronts". (b) Rays are lines orthogonal to wavefronts.

2.1.1 What a lens is, and the "imaging principle"

As the sketches in Fig. 2.2 point out, an imaging system is most useful if the image were both sharp and bright. Hence, we shall have the principle that, for any source of rays emitted from a point at the object, an image of that point is formed whenever rays intersect *after* passage through an imaging system. Let us call this ***The Imaging Principle***.*

---

* When rays intersect at precisely a single point after passage through a lens, it means that they all took the same time to get there. This is a consequence of Fermat's principle, and is a basic tenet of ideal imaging.

There was a time when images were formed using 'holes'

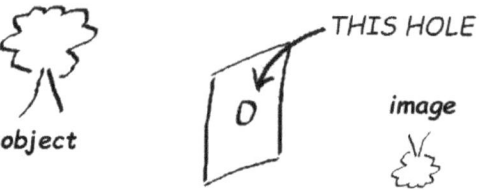

Big holes resulted in **bright** but **blur** images

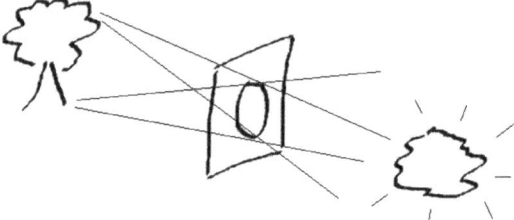

Small holes resulted in **sharp** but **dim** images

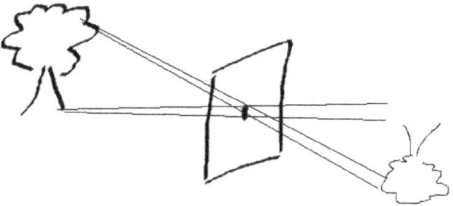

A Lens is a BIG hole made out of material which can bend rays of light so that both **bright** AND **sharp** images may be formed

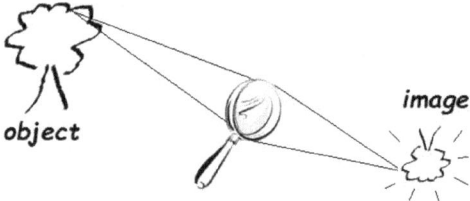

FIGURE 2.2 Illustration of what a lens is.

Intersecting rays can arise from a set of converging rays (Fig. 2.3) or a set of diverging rays (Fig. 2.4), in which case the diverging rays would appear to originate from a virtual point where they intersect, yielding a virtual point image. In either case, the image point is said to be a ***conjugate*** of its object point. Clearly, since every point across the surface of an object is a source for rays, there will appear across the image plane a collection of conjugates of all the object points, and this collection of image points forms the complete image of the object (Fig. 2.5). A' is a conjugate of A, B' is a conjugate of B…etc. Incidentally, the lens systems shown thus far are rotationally symmetric about the optic axis. Such rotationally symmetric imaging systems (which are by far the most common systems) shall be the focus of this book.

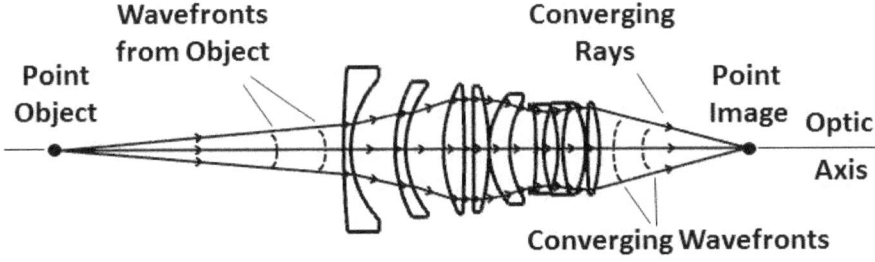

FIGUIRE 2.3 Converging rays.

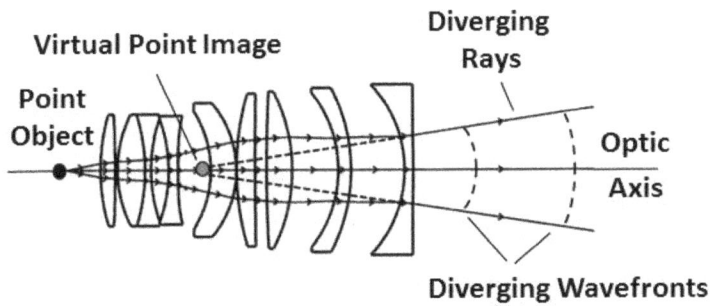

FIGURE 2.4 Diverging rays.

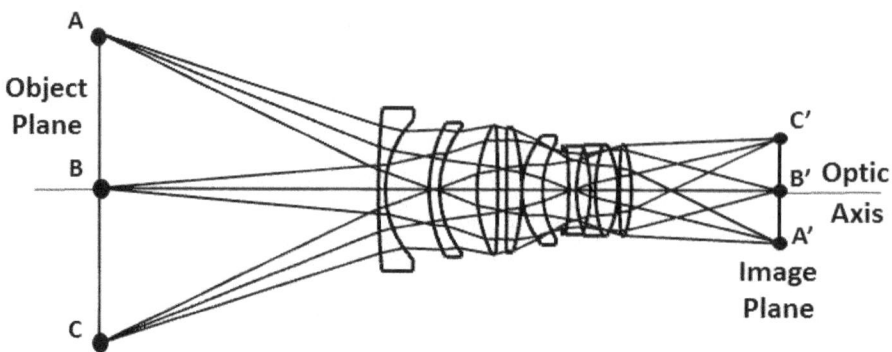

FIGURE 2.5 A complete image at the image plane.

If, in Fig. 2.5, we shift a little farther away from the image plane, the rays now diverge from points C', B', and A', which now appear to be sources of rays (Fig. 2.6). In fact, they *are* sources of rays. Therefore, placing a second lens system somewhere behind this new source of rays shall form an image of points A', B', and C' across a new image plane behind the second lens system, as shown in Fig. 2.7 (so it follows that any real or virtual image formed by an imaging system serves as the object for another imaging system). But wait! The front lens element of the second lens system is too small to allow the outermost off-axis rays to pass through. The solution is to use a *field lens*, placed at or near the first image, to bend the ***chief ray*** and aim it into the ***pupil*** of the second lens system (Fig. 2.8).

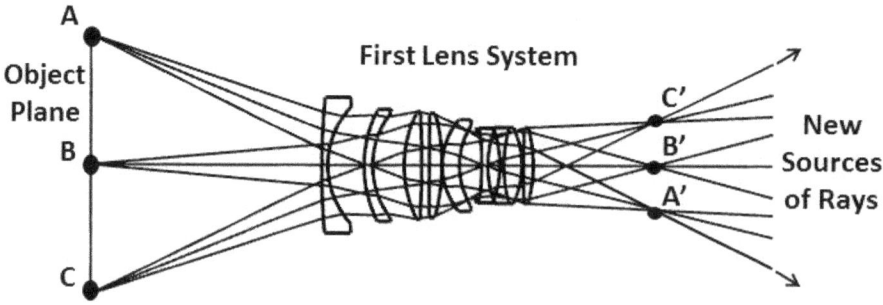

FIGURE 2.6 New sources of rays from the first image plane.

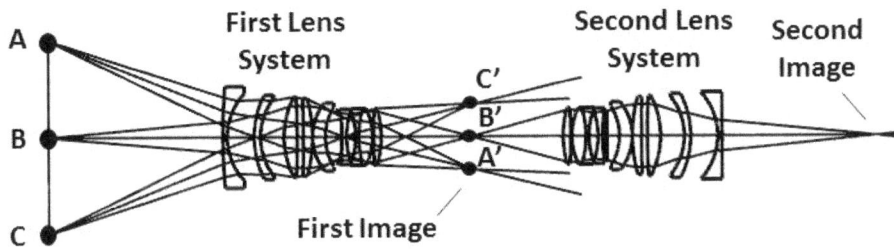

FIGURE 2.7 A secondary lens system forming a second image.

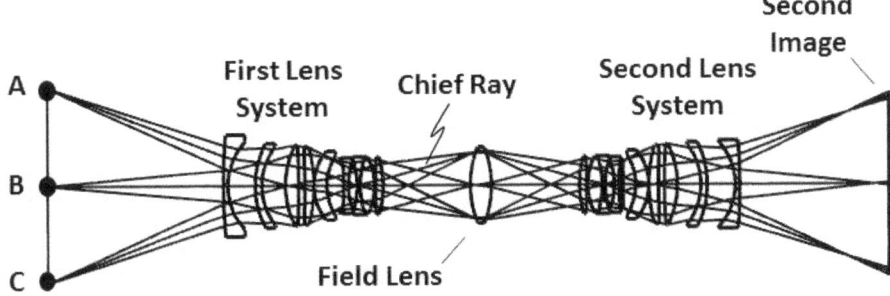

FIGURE 2.8 A field lens refracts the chief ray (and all rays surrounding this chief ray) back into the second lens system.

Lens systems are generally re-optimized upon introducing a field lens, but our simple example is illustrative (we will review pupils in Sec. 2.1.3). So much for the more qualitative and physically appealing characteristics of imaging theory. We now start to review the more quantitative features of optical imaging principles. Returning to the lens system in Fig. 2.3, in the limit that a point object source is at infinity, wavefronts effectively appear as planes, whose rays are therefore parallel to the optic axis. Such *plane waves* converge to a focus at the *rear infinite conjugate focal point* of the lens system (Fig. 2.9a). Similarly, tracing rays from an infinitely distant point object source behind the lens yields rays that converge to the *front infinite conjugate focal point* of the lens system (Fig. 2.9b). In the limit that the ***numerical aperture*** (NA) of the converging rays is small [i.e., in the limit that the small angle approximation is valid, such as when $\sin \theta' \approx \theta'$ in radian units], all focusing rays "back-intersect" at planes known as the ***principal planes*** of the lens system. Because it is rather difficult to illustrate this intersection of rays at small NA, a reasonable approximation is illustrated at large NA in Figs. 2.10a and 2.10b.

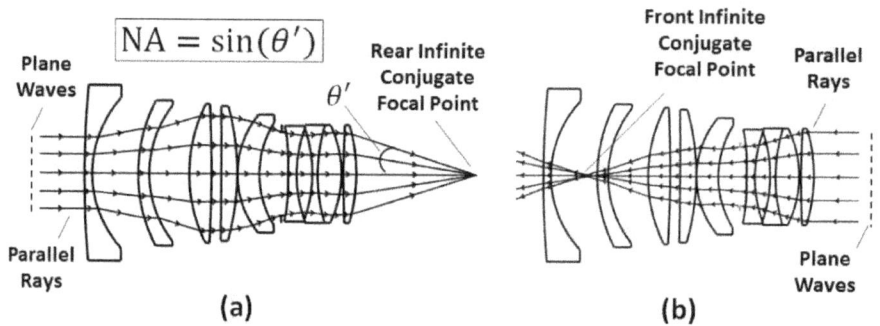

FIGURE 2.9   Infinite conjugate focal points. (a) Rear. (b) Front.

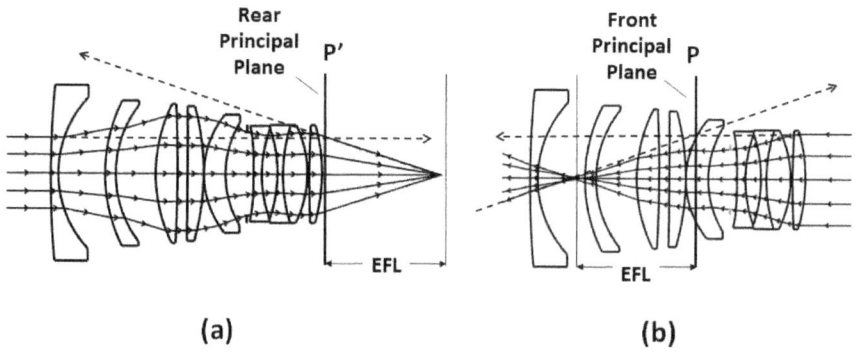

FIGURE 2.10 Definition of lens EFL. (a) From the rear principal plane. (b) From the front principal plane. They are equal if the spaces to the left and right of the lens system have equal refractive index.

For any imaging system, the condition of small angle approximation (i.e., small NA) is called the ***paraxial approximation*** of rays through the imaging system. Under the paraxial approximation, the distance between the rear principal plane (P') and the rear infinite conjugate focal plane is called the ***effective focal length*** (EFL) of the lens system. If both the front and rear spaces of the lens system are of the same medium (usually air, at a refractive index of approximately unity, which shall be the assumption used throughout this book, unless stated otherwise), the distance between the front principal plane and the front infinite conjugate focal plane is of the same magnitude as the EFL of the lens system. Under these conditions, the EFL of an imaging system fully characterizes the paraxial imaging properties (often called ***first order*** properties) of the system, such as image location and magnification. The term "first order" refers to the idea that, in a Taylor polynomial series expansion of, say, the sine of a ray angle, only the first order term of that expansion is retained on account of the small angle approximation. This means that the principal planes of an imaging system can fully represent the entire imaging system, at least in terms of the first order properties. Hence, let us replace the lens system in Fig. 2.10 by its principal planes (Fig. 2.11).

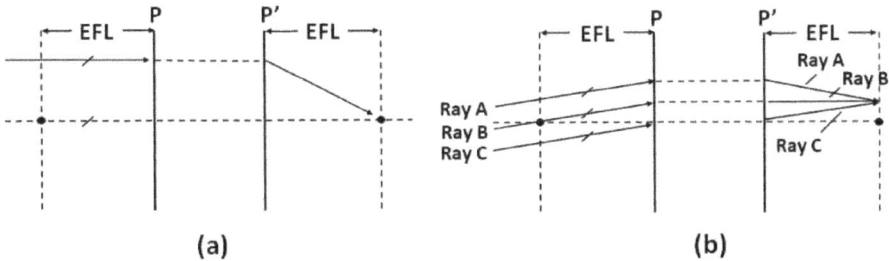

(a)   (b)

FIGUIRE 2.11 Representation of an imaging system by its principal planes. (a) Rays parallel to the optic axis converge to the rear focal point. (b) Parallel off-axis rays intersect at an off-axis point on the rear focal plane (see continuing Fig. 2.11c on the next page).

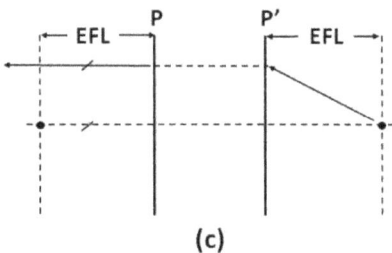

(c)

FIGURE 2.11 *Continued.* (c) Reversing any ray traces its path back.

From Fig. 2.11, let us note the following three characteristics, which we shall call the "paraxial lens laws" (PLL):

1. All rays parallel to the optic axis intersect at the infinite conjugate focal points (Fig. 2.11a).
2. In the absence of image distortion, all rays parallel to each other at an angle to the optic axis intersect at an off-axis point on the infinite conjugate focal planes of the imaging system (Fig. 2.11b).
3. Reversing the rays traces their paths precisely back from where they came (Fig. 2.11c).

From the three PLLs, and applying the *imaging principle* from before, we can graphically determine the location of an image, and its magnification (Fig. 2.12).

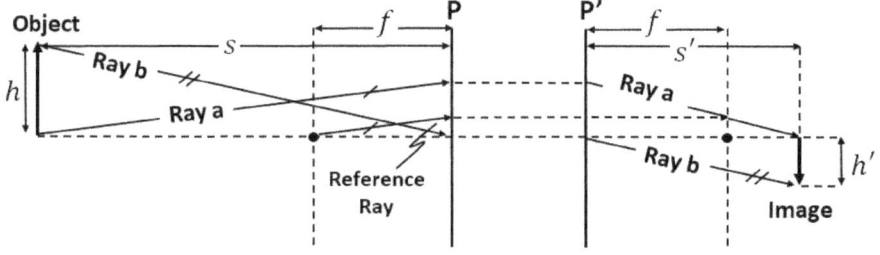

FIGURE 2.12 Location of an image, and determining its magnification.

From the geometry of rays in Fig. 2.12, one derives the well-known expression*

$$\frac{1}{f} = \frac{1}{s} + \frac{1}{s'}.$$ (2.1)

Here, we have denoted the system EFL by f. One also notes that the *paraxial image magnification* m_p (conventionally defined as the ratio of the image height to the object height) may be expressed as

$$m_p = \frac{h'}{h} = \frac{s'}{s}.$$ (2.2)

In modern optical engineering, one uses optical design programs (such as Zemax OS) to trace any ray, whether parabasal (i.e., rays that satisfy the small angle approximation and therefore obey paraxial optics rules), or real (i.e., rays that do not satisfy the small angle approximation). Therefore, all first order quantities such as principal plane locations and EFL may be obtained from optical design programs. However, knowing the PLLs is useful and provides intuition. Other analytical methods for computing first order quantities (such as determining and applying the so-called "ABCD matrix") are not covered in this book, and the reader should consult those textbooks that provide a thorough introduction to optics, such as the excellent classic by Hecht [5], or by Pedrotti and Pedrotti [6].

2.1.2 The paraxial thin lens model (PTLM)

Suppose we replace the lens system in Fig. 2.10 (which has 9 lens elements) with a bi-convex single element lens of the same EFL (Fig. 2.13a). Although this single element lens (aka "singlet") has reduced image quality (due to the reduction in the number of lens elements

---

* We are applying the usual sign convention that f > 0 represents positive powered lenses, and f < 0 for negative powered lenses. Hence, s > 0 occurs for real images, and s' < 0 occurs for virtual images.

32

used for balancing image aberrations), it possesses precisely the same first order imaging properties as the multi-element lens in Fig. 2.10. Upon examining the singlet, one will note that its principal planes are much closer together than the multi-element lens system of Fig. 2.10. In fact, reducing the center thickness of this lens further (and reducing the refractive index accordingly or increasing the lens's radii of curvatures) reduces the separation between the principal planes. In the limit that the lens has infinitesimally zero thickness, the principal planes coincide. Further, under the paraxial approximation, at zero lens thickness, one obtains the so-called *paraxial thin lens model* (PTLM) of the lens (Fig. 2.13b), which is familiar to most students of high school physics. The PTLM is by far the simplest and most intuitive model of a lens. Everyone sketches thin lenses on the white board when explaining optics and imaging to colleagues. In Zemax OS, a PTLM is simply called a "Paraxial Lens Model" (PLM), and so we shall use the abbreviations PTLM and PLM synonymously in those cases where we use PLMs in Zemax OS for some imaging system analysis.

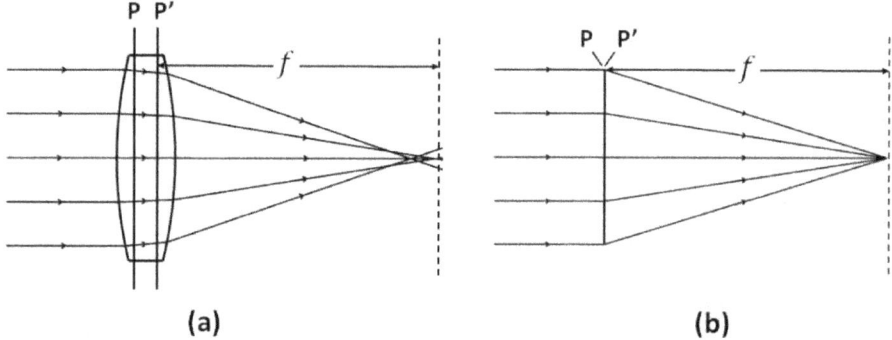

FIGURE 2.13 Single element lens (singlet) replacement for the lens in Fig. 2.10 (not at the same scale as Fig. 2.10). (a) Real singlet lens. (b) Its paraxial thin lens model (PTLM).

Pause for insight: → In arriving at the PTLM, I stated that one would "reduce the singlet's refractive index accordingly" whilst reducing the lens's center thickness. Just in case that wasn't clear, take a look at the well-known *lens maker's formula*, which expresses the reciprocal EFL of a singlet as a function of its two radii of curvatures,[*] R_1, and R_2, center thickness t, and refractive index n:

$$\frac{1}{f} = (n - 1) \left[\frac{1}{R_1} - \frac{1}{R_2} + \frac{t(n-1)}{nR_1R_2} \right]. \tag{2.3}$$

This formula may be re-expressed as the sum of two terms:

$$\frac{1}{f} = (n - 1) \left[\frac{1}{R_1} - \frac{1}{R_2} \right] + \left[\frac{t(n-1)^2}{nR_1R_2} \right]. \tag{2.4}$$

If, on the right-hand side of Eq. (2.4), the center thickness is zero for the second term, then we are left with the first term, and the singlet's refractive optical power (which may be thought of as the reciprocal of the lens's EFL) would be *increased* (a little) by the absence of that second term[†]. In order to maintain the lens's EFL (or equivalently, its refractive optical power), one may either reduce the refractive index, or increase the radii of curvatures.

2.1.3 Aperture stops, pupils, and f-number

Returning to the lens system in Fig. 2.10a, the aperture stop is the smallest physical hole in an imaging system, and it determines the largest angle θ' at which the axial rays converge onto the image plane (Figs. 2.14a and 2.14b). This is generally the so-called "iris" in a camera lens. Setting it at its maximum yields the system's brightest image (Fig. 2.14a). Reducing it yields a dimmer image (Fig. 2.14b).

---

[*] We are using the sign convention that the radius of curvature is positive when the center of curvature is to the right of the surface, and negative when it is to the left. Thus, for the lens in Fig. 2.13a, $R_1 > 0$ and $R_2 < 0$.

[†] Remember that $R_2 < 0$, so that makes the second term negative.

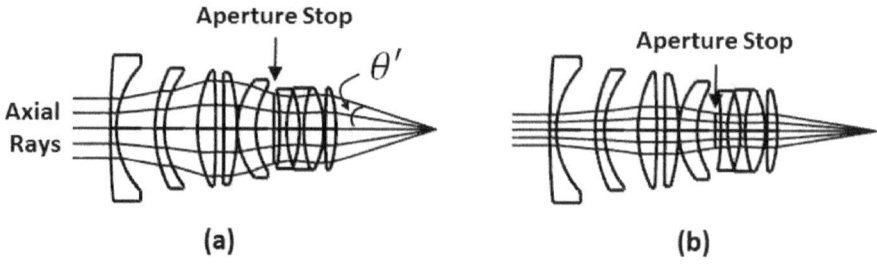

FIGURE 2.14 Aperture Stop. (a) At maximum diameter (brightest image). (b) At reduced diameter (dimmer image).

When one looks through the lens system in Fig. 2.14 from the left side, one sees a "virtual aperture" floating in space (but collinear to the optic axis if you are also collinear to the optic axis). This is an image of the aperture stop formed by the combined lens elements to the left of the aperture stop, and it is called the ***entrance pupil*** of the imaging system. Conversely, when one looks through the lens system from the right side, one sees the image of the aperture stop formed by the combined lens elements to the right of the aperture stop, and this is the ***exit pupil*** of the imaging system. It is the exit pupil's semi-diameter that subtends the angle θ' from the focal plane. As we shall see in Sec. 2.3, the brightness of an image (i.e., the image irradiance) is proportional to the square of the numerical aperture (NA) of the rays in image space, which is defined in air as

$$NA = \sin \theta'. \tag{2.5}$$

A quantity that is more familiar to photographers is the system's f-number, though they use a different definition for it. In optical design, the quantity that most suits radiometric calculations for an imaging system is the ***working f-number*** (WFN), defined as

$$WFN = [2(NA)]^{-1}. \tag{2.6}$$

NA is given by Eq. (2.5). In contrast to this, photographers ordinarily use an approximation, which is the infinite conjugate paraxial image space f-number (ISFN$_\infty$), defined as

$$\text{ISFN}_\infty = \left(2\tan\theta_p{'}\right)^{-1}. \qquad (2.7)$$

In contrast to θ' in Fig. 2.14a, $\theta_p{'}$ in Eq. (2.7) is the angle subtended by the **paraxial** exit pupil's semi-diameter from the rear infinite conjugate focal point (determined by paraxial ray tracing). Equivalently, under the paraxial approximation, Eq. (2.7) may be expressed as the ratio of the lens EFL to the paraxial entrance pupil diameter D_{EN}:

$$\text{ISFN}_\infty = \text{f-stop} = \frac{\text{EFL}}{D_{EN}}. \qquad (2.8)$$

This is the definition that is most familiar to photographers, and they call it the "f-stop" or "focal ratio". Since image brightness is proportional to the square of the NA [given by Eq. (2.5)], it is only approximately proportional to the square of $\tan\theta_p{'}$. But, since it is approximately proportional to the square of $\tan\theta_p{'}$ it is also approximately proportional to the reciprocal of the square of the f-stop. Camera lenses in photography provide iris diameter settings in steps that change (approximately) the image brightness by factors of 2. Since image brightness is approximately proportional to the reciprocal of the square of the f-stop, the f-stop number sequences for a photographer's lens change in steps of factors of the square root of 2. For example, suppose the maximum f-stop setting is f/1. It then follows that the next dimmer image is achieved at $f/(1 \times \sqrt{2}) \approx$ f/1.4. The next dimmer image is achieved at $f/(1 \times \sqrt{2} \times \sqrt{2}) = $ f/2, and so on. So, we have the relationship that the f-stop settings on a photographer's lens are given by

$$\text{f-stop}_n = \left(\sqrt{2}\right)^{n-1}, \quad n = 1, 2, 3 \dots \qquad (2.9)$$

Pause for thought: → Entrance and exit pupils play significant roles in optical design and in the characteristics of imaging systems. In particular:

1. Image aberrations are often formulated in terms of how rays emerge from the exit pupil (see Sec. 2.1.5)
2. They impact image brightness and the illumination across the image (e.g., see Secs. 2.1.4 and 2.3.7).
3. In visual systems and imaging attachments (e.g., mobile phones), they require pupil matching conditions (e.g., see Sec. 3.3.2).
4. Their locations can be important for depth sensing (e.g., see Sec. 3.2.7).

2.1.4 Relative illumination (Part One): Vignetting

Although the aperture stop limits the diameter of axial rays through the imaging system, something else limits the off-axis rays. The (sometimes deliberate) clipping of off-axis rays by all other mechanical parts in the system is known as ***vignetting***. In most "fast" imaging systems (i.e., systems with low f-number, such as between, say, f/1 to f/2), vignetting plays a significant role in the ***relative illumination*** of the image (Fig. 2.15a). The relative illumination at any point across the image plane is the ratio of the off-axis image irradiance to the axial image irradiance (i.e., the irradiance at the center of the image). For practical optical design, relative illumination is computed by tracing a number of "rim" rays (i.e., rays that sample the circumference of the exit pupil) through the imaging system [7]. At high f-number, other factors dominate in their impact on relative illumination (Fig. 2.15b). These factors include the so-called "cos^4th effect" (not a law, as we will see in Sec. 2.3.7), image distortion, differential distortion, and aberrations of the entrance and exit pupils. We will examine these in further detail in Sec. 2.3.7.

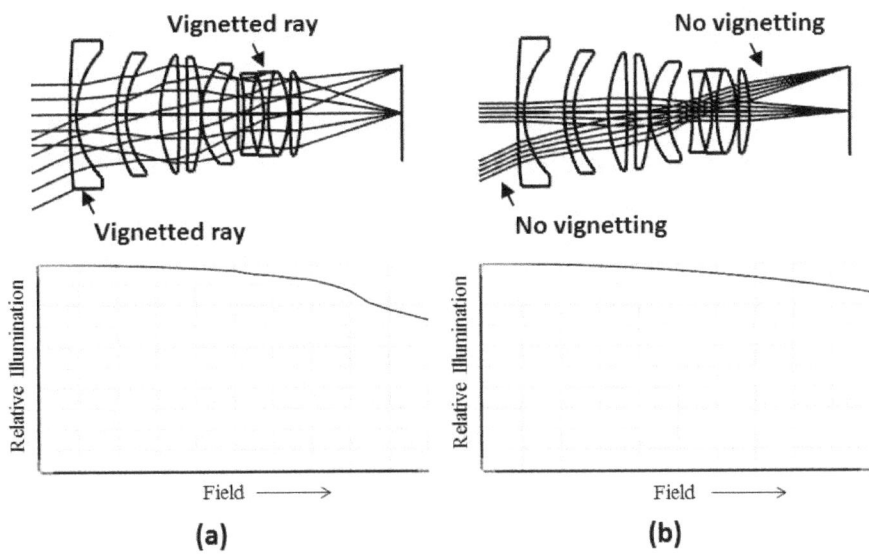

FIGURE 2.15 (a) Vignetted rays at low f-number (above) and resulting relative illumination (below). (b) No vignetting at high f-number (above), and the resulting relative illumination (below).

2.1.5 Aberrations and "apparitions" of imaging systems

By the Theorem of Malus and Dupin, a spherical wavefront yields converging or diverging rays that intersect precisely at a single point, yielding perfect imagery (Fig. 2.16a). Imperfect imagery therefore results from a deformed (aka "aberrated") wavefront, whose rays do not intersect at a single point (Fig. 2.16b).[*] Consequently, for any object height h (aka "field height"), there will result an ***optical path difference*** (OPD) between a ray from a perfect spherical wavefront and a ray from an aberrated wavefront emerging from a height ρ at the exit pupil (Fig. 2.16c).

---

[*] Equivalently, rays that originate from a single object point and do not intersect at a single image point are not all taking the same amount of time to get to the ideal image point, which results in a spread of rays at the image.

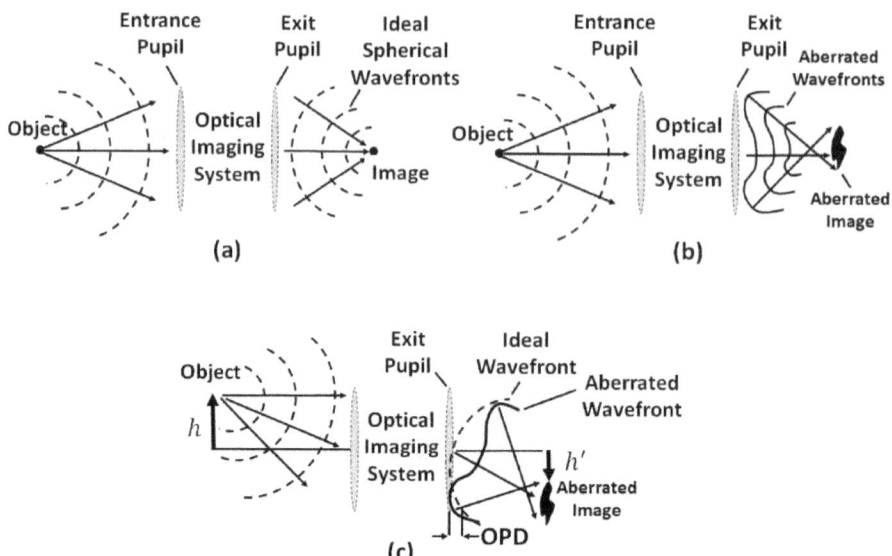

FIGURE 2.16 (a) Ideal imaging. (b) Non-ideal imaging. (c) OPD.

Pause for insight: → Suppose the center of an image formed by an optical system were defocused. What is the mathematical expression for the OPD as a function of ρ for axial rays that are converging to a focus position that is closer to the exit pupil from the expected focus position? From Fig. 2.17, if we take the sag height difference to be the OPD, then we have:

$$OPD(\rho) = z_2 - z_1$$

$$= (R_2 - R_1) + R_1 \sqrt{1 - \left(\frac{\rho}{R_1}\right)^2} - R_2 \sqrt{1 - \left(\frac{\rho}{R_2}\right)^2} \qquad (2.10)$$

$$\approx (R_2 - R_1) + R_1 \left(1 - \frac{\rho^2}{2R_1^2}\right) - R_2 \left(1 - \frac{\rho^2}{2R_2^2}\right) \qquad (2.11)$$

$$= \frac{\rho^2}{2} \left(\frac{1}{R_2} - \frac{1}{R_1}\right). \qquad (2.12)$$

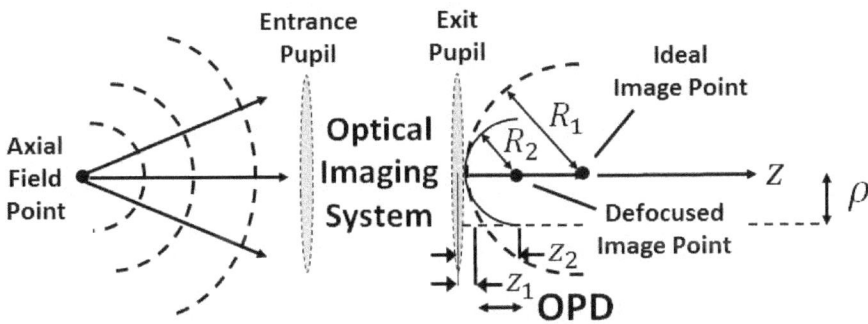

FIGURE 2.17 Axial rays converging to a focus closer to the exit pupil.

So, Eq. (2.12) tells us that, to an approximation, axial defocus results from an OPD function that is quadratic with exit pupil height ρ. But surely, the OPD for any arbitrary aberrated wavefront would be a more complex function of exit pupil radial height ρ and field height h. In fact, it can be shown [e.g., 8 – 10] that the OPD may be expanded as a series in normalized heights h ($0 \le h \le 1$) and ρ ($0 \le \rho \le 1$):

$$\text{OPD}(\rho) = w_d\rho^2 + w_s\rho^4 + w_c h\rho^2\rho_y + (w_a + w_f)h^2\rho_x^2$$

$$\downarrow \qquad \downarrow \qquad \downarrow \qquad \downarrow$$

Defocus Spherical Coma   Astigmatism

$$+(3w_a + w_f)h^2\rho_y^2 + w_D\rho_y h^3 + HO... \qquad (2.13)$$

$$\downarrow \qquad \qquad \downarrow \qquad \qquad \downarrow$$

Field Curvature Distortion   Higher Order Terms...

These are the most well-known monochromatic imaging aberrations[*]. Each of the coefficients[†] in Eq. (2.13) may be identified by its effect on the OPD between an aberrated wavefront and a perfect spherical wavefront, and each effect on the OPD may be understood in terms of the product of the aberration coefficient on h and ρ in Eq. (2.13). For

---

[*] We will consider chromatic aberration in Sec. 3.2.4.

[†] We are not using standard notation for the aberration coefficients in Eq. (2.13).

instance, suppose all of the coefficients were positive for some given imaging system. Under this condition, we have the following:

1. **Defocus:** We already know from Eq. (2.12) that an OPD function that is given by the product of a positive coefficient with ρ^2 implies that a wavefront emerging from the exit pupil is converging to a closer focus position than its intended position. Hence, the term "defocus" is appropriate for the first aberration term in Eq. (2.13).

2. **Spherical Aberration** may be understood by observing that an OPD that varies as ρ^4 would yield rays at the outer radial positions of the exit pupil to converge more than those rays that emerge near the optic axis.

3. The effect from **coma** is understood by observing that an OPD that varies with the product of $h\rho^2$ results in a linear variation of focus for off-axis rays across the image plane. Additionally, due to the presence of ρ_y, the defocused spot for coma grows linearly in size in the y dimension, yielding a "coma tail".

4. **Astigmatism** may be understood by observing that, in its presence, a wavefront in the ρ_x dimension converges less than in the ρ_y dimension. Since this varies with field height, the image plane is curved more strongly in the x than in the y dimension.

5. **Field Curvature:** In the absence of astigmatism, the 4th and 5th terms in Eq. (2.13) are each left with equal contributions from the field curvature coefficient, so the image plane is still curved, but it is now curved equally in both the x and y dimensions.

6. **Distortion** is understood by noting that an OPD varying as $\rho_y h^3$ yields a wavefront that tilts and focuses onto ever higher image positions as the field of view increases (i.e., as h increases). But since the "transverse ray error" at the image plane is proportional to the first derivative of the OPD wavefront error, the image height either increases (if there is pincushion distortion) or reduces (if there is barrel distortion) as the square of the image height.

Of course, the wavefront aberration function of Eq. (2.13) is not the only way to understand the impact of the aberration coefficients. Lens designers routinely analyze other useful graphs such as ray intercept plots [e.g., 11], spot diagrams, and modulation transfer function (MTF)[*].

Image aberrations aren't the only defects in an optical imaging system. "Apparitions" may also occur. By this I mean "ghost images" and stray light that result from partial reflections off lens surfaces in the optical imaging system (Figs. 2.18a and 2.18b). For most systems, ghost images and stray light are minimized by taking the following measures:

1. Apply anti-reflection (AR) coatings on all polished lens element surfaces.
2. Blacken all edges of lens and mirror components.
3. Blacken everything else.
4. Apply baffles onto internal walls of the lens housing.

In some applications, even the above four measures may not be sufficient to minimize the occurrence of stray light. A good example is in fluorescence detection systems, which involves the use of a bright light source to illuminate and excite fluorescent dyes that are tagged to DNA molecules (Fig. 2.19). In fluorescence detection systems, the magnitude of the emitted fluorescence is very low relative to the emittance or intensity of the excitation light source. Hence, ghost reflections that originate from the excitation may become significant. One way to manage this is to specify an appropriate level of reflective coating for the emission filters. Another is to design good baffles or clever light traps [12].

---

[*] MTF is a measure of the contrast (aka "modulation") of the image. Strictly, it is the ratio of the image's contrast to the object's contrast when the object has a sinusoidal irradiance pattern.

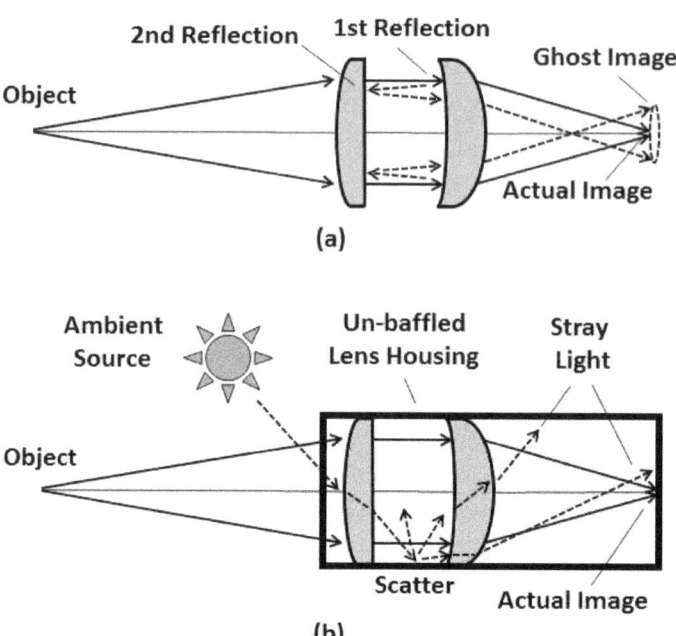

FIGURE 2.18 (a) Ghost image of the object. (b) Stray ambient light.

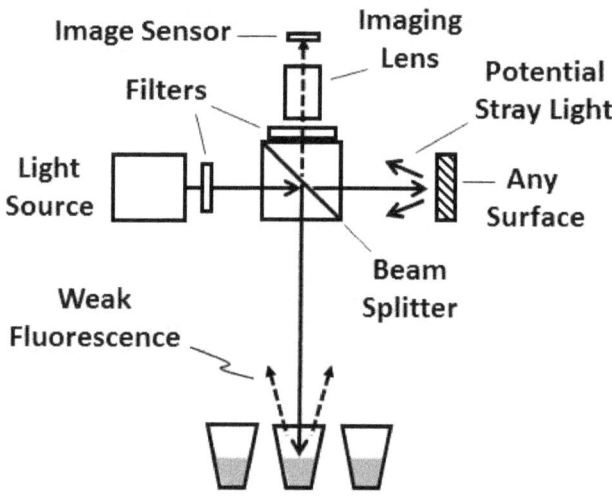

FIGURE 2.19 Fluorescence detection system with potential stray light.

Pause for insight: → We ordinarily believe that black paint minimizes stray reflections, but we have all also experienced seeing specular reflections (aka "glare") from black surfaces. Paint is comprised of dyes (not-so-closely-packed macro-molecules) immersed in some transparent medium (closely packed molecules). Black paint must therefore be comprised of light-absorbing not-so-closely-packed macro-molecules immersed in some transparent medium. Hence, at the interface between air and the transparent medium, Fresnel reflections occur. So, black paint yields partial specular reflection.[*] Additionally, diffuse reflections may occur from black "matte" surfaces. At the opposite end of "blackness", consider the fine scattering of light by smooth surfaces. Even the surfaces of polished optical glasses and first surface mirrors scatter light. The appearance of "silver", for instance, shouldn't even be visible if a mirror doesn't scatter. Generally, scatter of any magnitude occurs whenever rays within the optical system strike any surface, even if that surface is not within the path of image forming rays. The typical lens design "environment" in sequential ray tracing ordinarily displays only those rays that pass through the aperture stop and mechanical spacers in the imaging path (Fig. 2.20a). But in reality, the object illuminates the entire first physical opening of the imaging system, and rays that do not make it through strike other parts and surfaces in the optics housing. Those other parts and surfaces are therefore potential sources of stray light. So, place a large "dummy surface" in front of the first physical opening of your system and set it as the system's entrance pupil (in Zemax OS, you set the dummy surface as the aperture stop, and set "Aperture Type" to either "Entrance Pupil Diameter" or "Float by Stop Size"). Then, take the brightest source in the scene [aka "region of interest" (ROI)] and trace rays sequentially through this new entrance pupil, and watch where

---

[*] There are perhaps other black paint options, such as *Avian DS Black* (www.aviantechnologies.com). See also Ref. 67, pp. 132 – 133, and S. M. Pompea and R. P. Breault, "Black Surfaces for Optical Systems," in *OSA Handbook of Optics Vol. 2*, edited by M. Bass, E. W. Van Stryland, D. R. Williams, and W. L. Wolfe, (McGraw-Hill, 1995), Chap. 37.

the rays get stopped or clipped (Fig. 2.20b), including lens element edges. The most suspicious areas should then become sources (if tracing from source to detector) or detectors (if tracing from detector to sources) in non-sequential ray tracing.

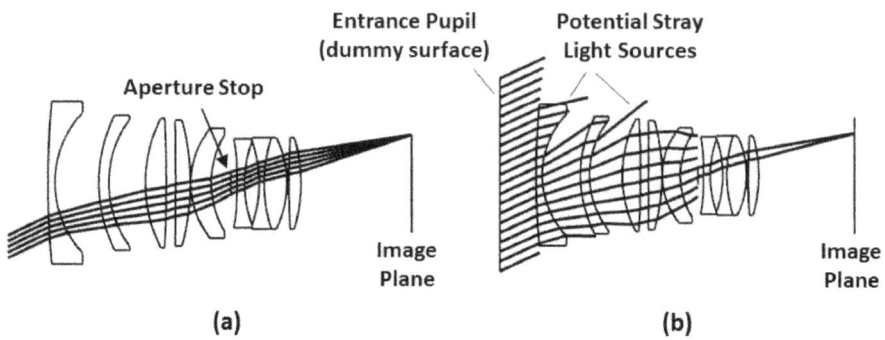

FIGURE 2.20 How to start a stray light analysis. (a) Typical lens design environment does not tell the whole story. (b) Identifying potential sources of stray light in sequential mode by setting a front dummy surface as a large entrance pupil.

2.1.6 Depth of field and focus

Figs. 2.21a and 2.21b show a well-corrected lens with 1 mm EFL, and working f-numbers at f/4.2 and f/8.4 respectively. Their image MTF curves are shown below the layouts. The object is 10 mm from the first surface, and the image is at optimum focus. When the object distance is at 100 mm, without refocusing, the MTF at f/4.2 has dropped significantly (Fig. 2.22a) compared to the MTF at f/8.4 (Fig. 2.22b). So, the *depth of field* for this lens at f/8.4 is greater than that at f/4.2. How much greater is rather arbitrary, depending on what spatial frequency the MTF is read at, and on what level of MTF is considered acceptable image quality. For example, at 50 cycles per mm, the depth of field in the forward direction for this lens at f/4.2 is perhaps as large as +90 mm (i.e., 100 mm – 10 mm = +90 mm) if one assumes an acceptable image blur of 20% modulation, while the depth of field for the same lens at f/8.4 seems significantly greater. But at

10 cycles per mm, the MTF is virtually identical between both f-number conditions, and in this case, the depth of field at f/4.2 is the same as at f/8.4.

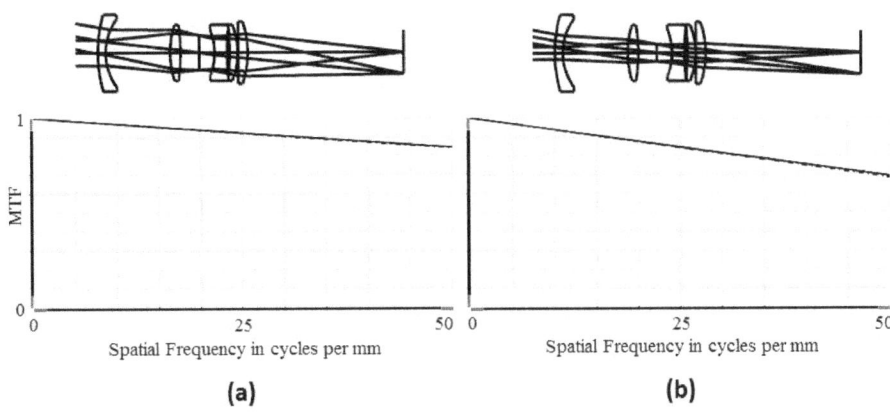

FIGURE 2.21 Lens at different working f-numbers and their corresponding image MTF plots (for an object at 10 mm distance). (a) f/4.2 (b) f/8.4.

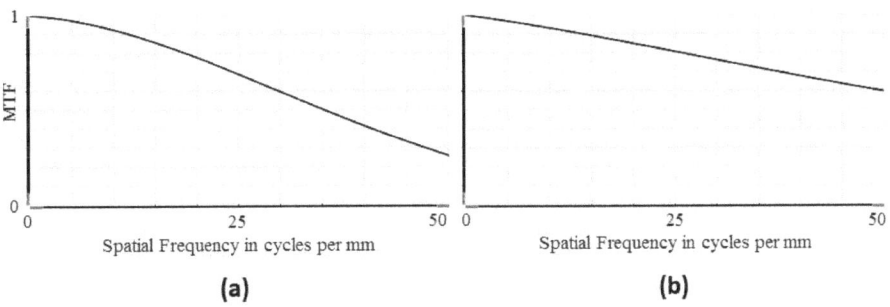

FIGURE 2.22 MTF at object distance of 100 mm. (a) f/4.2. (b) f/8.4.

So much for depth of field in the forward direction (i.e., when the object is farther from its normal operating distance or *working distance* of 10 mm). How about when the object is closer to the lens? Suppose we now bring the object to a distance of about 5 mm from the first surface of the lens, then the MTF plots become as shown in Figs. 2.23a and 2.23b.

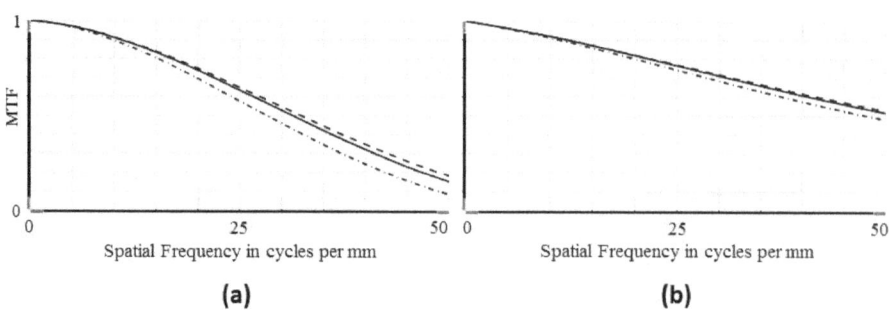

FIGURE 2.23 MTF at object distance of 5 mm. (a) f/4.2. (b) f/8.4.

Note that the MTF curves in Figs. 2.23a and 2.23b drop faster than in Figs. 2.22a and 2.22b. One may therefore say that, at an aperture of f/4.2, this lens's *far point* is 100 mm, and its *near point* is 5 mm, based on an acceptable modulation of 20% at 50 cycles per mm. Equivalently, based on the mentioned acceptable image blur criterion, the lens's acceptable field range is +90/-5 mm about its working distance of 10 mm. The full depth of field of this lens is therefore 95 mm, but the variation of the acceptable object range is not +/- half of 95 mm, due to the asymmetry between the lens's far point (100 mm) and near point (5 mm). So, it seems that image blur is asymmetric about the best focus position. However, this result is only incidental to the example provided, and it is a natural consequence of geometrical optics. Let us pause to see why.

Pause for insight: → The cause for the asymmetry in focal range between the far and near points of object distances is the nonlinear variation of image focal distance with object distance, based on the laws of geometrical optics. Applying Eq. (2.1), we may express the image focal distance s' as a function of object distance s (measured with respect to the first order principal planes) as

$$s' = \frac{f}{1 - \dfrac{f}{s}} \ . \tag{2.14}$$

Differentiating s' in Eq. (2.14) with respect to object distance yields

$$\frac{ds'}{ds} = -\left(\frac{f}{s-f}\right)^2.$$

(2.15)

The slope is negative because the farther the object is from the lens, the closer the image focal distance is to the lens. But more importantly, we note from Eq. (2.15) that the instantaneous rate of change of image distance with respect to object distance is unity only when $s = 2f$. Thus, image focus (and hence, image blur) is symmetric about the best focus position only when the object distance (as measured from the front principal plane) is twice the EFL of the lens. For any lens, at object distances of $s > 2f$, the instantaneous rate of change of the image focal distance reduces as s increases. This means that, at object distances greater than twice the lens's EFL (which is the case for the lens example in Figs. 2.21 – 2.23), the image would defocus at a lower rate as the object gets farther from the lens. This is the reason for the asymmetry between the far point and near point object distances for this lens.

What about the lens's ***depth of focus***? I would determine depth of focus for the current lens example by setting the image distance from the lens as a variable in optical design software, and then optimize the image's MTF at the perturbed object distances (e.g., 100 mm and 5 mm for the lens in Fig. 2.2.1), taking note of the change in optimum image plane position between the far and near object positions from the lens. The total change in image focus position is the depth of focus, based on an agreed criterion for acceptable MTF at some agreed image or object spatial frequency. This gives us depth of focus for a corresponding depth of field. In some cases, only the image plane's position may be in question (i.e., the image plane's position may be uncertain to some degree, perhaps due to production tolerances), and so one may be interested to know what is the acceptable range of image plane distances. In this case, I would also usually perform the depth of focus analysis by examining changes in MTF as a function of the shift in image plane position.

But there is also a simple way to "guesstimate" depth of focus and depth of field, given some agreed acceptable geometric image blur spot radius R' (whose definition is also rather arbitrary, depending on one's application). From the geometry of rays in Fig. 2.24, we have:

$$ds' = \frac{R'}{\tan \theta'}. \qquad (2.16)$$

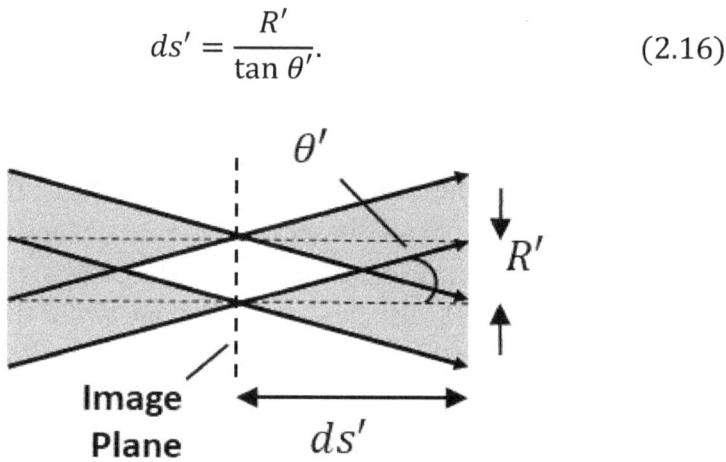

FIGURE 2.24 Geometric depth of focus for two point-images.

If we take the approximation that $\tan \theta' \approx \sin \theta'$, then we have

$$ds' \approx \frac{2R'}{2\sin \theta'} = 2R'(\text{WFN}). \qquad (2.17)$$

The WFN here is given by Eq. (2.6). Hence, for an acceptable blur radius R', one may say that the depth of focus (DOFc) is approximately $+/- ds'$, or

$$\text{DOFc} \approx \pm 2R'(\text{WFN}). \qquad (2.18)$$

This means that the full depth of focus range is $\approx 4R'(\text{WFN})$. We can convert DOFc into an estimate of the depth of field (DOFd) by recalling from Fig. 2.12 and Eq. (2.2) that the paraxial image magnification is given by

$$m_p = \frac{h'}{h} = \frac{s'}{s}.$$ (2.19)

Substituting s' from Eq. (2.14) into Eq. (2.19) yields

$$m_p = \frac{f}{s - f}.$$ (2.20)

Cross-referencing Eq. (2.20) with Eq. (2.15), we may write

$$\left|\frac{ds'}{ds}\right| = m_p{}^2.$$ (2.21)

Therefore,

$$\text{DOFd} \approx \pm|ds| = \pm\frac{|ds'|}{m_p{}^2} = \frac{\pm 2R'(\text{WFN})}{m_p{}^2}.$$ (2.22)

Since it could seem more practical to specify a blur radius R at the object, and since we can say that $R'/R = m_p$, we may write

$$\text{DOFd} \approx \frac{\pm 2R(\text{WFN})}{m_p}.$$ (2.23)

Strictly speaking, the validity of Eqs. (2.18) and (2.23) is restricted to small variations in image and object displacements respectively, and they are also restricted to the imaging of point-like sources (such as perhaps in microscopy). In most other cases of imaging, objects being imaged are of finite extent (i.e., they are **extended sources**), and of some range of spatial frequencies. At the image plane, it is the **convolution** between the blur spot [aka **point spread function** (PSF)] and the ideal image distribution that yields the full actual image. By the *convolution theorem*, the PSF is related to the imaging system's MTF. In fact, the normalized modulus of the Fourier transform of the PSF is the MTF. Therefore, for extended objects, a more accurate representation of image blur (and therefore, image contrast or modulation) is provided by the MTF curve. However, it also seems unreasonable to generalize. Neither the MTF curve nor the blur spot radius can be said to be applicable to all cases. My advice: Return to the fundamentals, and rationalize what is the most suitable analytic tool. Let us pause to consider an example.

Pause for insight: → The image quality for a lens whose aberrations are zero or negligible would be limited only by diffraction effects. Such a lens would image a point source into a so-called "Airy disk" diffraction pattern at the image plane (see Sec. 2.2 for a review of diffraction and the Airy disk). It may happen that the geometric point image blur radius R' due to some amount of image defocus is smaller than the radius of the diffraction Airy disk. The radius r' of the Airy disk in an imaging system is given by

$$r' \approx 1.22\lambda(\text{WFN}). \tag{2.24}$$

Geary [13] provides an expression for *diffractive depth of focus* (DDOFc) as

$$\text{DDOFc} \approx \pm 2(\text{WFN})^2(\lambda). \tag{2.25}$$

Substituting Eq. (2.24) into Eq. (2.25) and re-arranging terms we obtain

$$\text{DDOFc} \approx \frac{\pm 2r'(\text{WFN})}{1.22} \approx \pm 1.64r'(\text{WFN}). \tag{2.26}$$

Comparison between Eqs. (2.26) and (2.18) shows certain similarities between the two expressions. Since the numeric factor of 2 in Eq. (2.26) is close to the factor of 1.64 in Eq. (2.26), one may say that, for any given lens at some fixed f-number, diffractive depth of focus is determined by the Airy disk radius r', while geometric depth of focus [i.e., given by Eq. (2.18)] is determined by the geometric blur spot radius R'. Hence, it seems reasonable that, at some fixed f-number, when R' is less than r', the more appropriate estimator of depth of focus is given by Eq. (2.26).

With all that's said and done concerning depth of field, if focus adjustment is provided for a lens, then the lens usually performs rather well for imaging objects that are at any distance, regardless of whether or not it had been designed for imaging at one specific object distance, or if it had been optimized over a range of object distances (see Sec. 3.2.6). In the former case, the lens would perform optimally at some specified object distance, but focusing the lens would also make it function sufficiently well at all other object distances. In the latter case, focus adjustability would make the lens function pretty well over a range of object distances, but at none of those object distances would this lens function as well as the lens that was optimized at one object distance. In either case, if the lens also comes with an adjustable iris, then there will be negligible differences in image quality between an infinity focus lens and a finite focus lens when the iris is set to a large f-number, because depth of field increases with f-number [Eq. (2.23)]. Therefore, even if a catalog or "off-the-shelf" lens has been advertised for imaging objects at a finite working distance, as long as there is focus adjustability, it can usually be used sufficiently well for imaging objects at infinity. And, if an off-the-shelf lens has been advertised for imaging objects at infinity, as long as there is focus adjustability, it can usually function sufficiently well for imaging objects at a finite distance away. This is especially true for miniature lenses, because:

1. Some aberrations scale with size [16] (smaller lenses yield lower aberration).
2. At constant f-number, smaller lenses naturally yield smaller EFLs, which naturally yields smaller magnification, and this leads to larger depth of field [e.g., see Eq. (2.23)]. We will revisit these ideas in Sec. 3.2.8.

In some cases, a camera may not provide focus adjustability for its lens. For such *fixed focus lenses*, perhaps one might have set the image plane at the factory such that it is at some fixed distance, say, at s_1' (Fig. 2.25) corresponding to an object at s_1. Consequently, when a point object is at infinity, there will result a blur radius R' at s_1'.

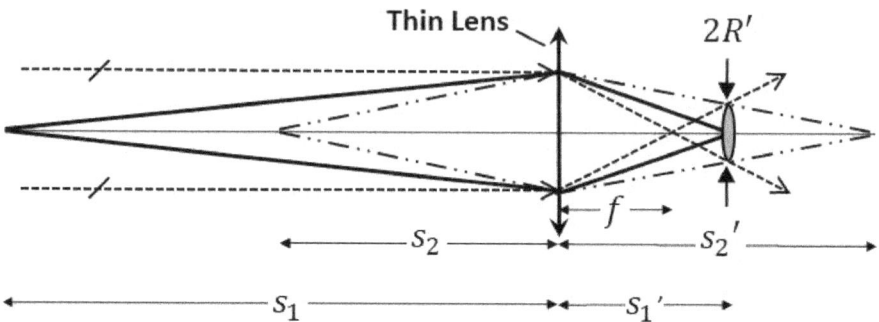

FIGURE 2.25 Blur radius R' resulting from setting the image plane at s_1', and viewing an object at infinity or at s_2 without refocusing.

From the geometry of rays in Fig. 2.25, it is straightforward to determine that the distance s_1 is

$$s_1 = f\left[1 + \frac{f}{2R'(\mathrm{ISFN}_\infty)}\right]. \tag{2.27}$$

Note that ISFN_∞ is given by Eq. (2.7). One also finds that the same blur radius R' at s_1' results when a point object is at $s_1/2 = s_2$. As a consequence of geometrical optics, this property holds for any object distance. Some suggest that a decision should be made for an acceptable R', which of course constrains s_1, and results in a far point at infinity, and a near point at $s_1/2$. Clearly, this restricts the best image focus position to s_1', making s_1 the optimal object distance for a chosen R'. Some call s_1 the lens's *hyperfocal distance* [14, 15], and make the approximation that $f/[2R'(\mathrm{ISFN}_\infty)] \gg 1$, so the "1" is ignored in Eq. (2.27).

In any event, if one were choosing an off-the-shelf fixed focus lens for imaging at a specific primary object distance, it is most sensible to adjust the image plane of the chosen lens at the factory to focus at that intended object distance. The depth of field depends on an agreed acceptable image contrast at a spatial frequency of the primary *region of interest* (ROI), so one should determine it experimentally. For those who use optical design software, it is of course best to obtain (if possible) the lens file, and then one can study and determine the depth of field theoretically. For those using Zemax OS, some vendors provide so-called "Zemax Black Box" files, which contain the actual lens data, but protects the prescription from being visible.

Food for thought: → A lens that is optimized for an object at infinity is essentially a "far sighted lens". A lens that is optimized to focus at some finite object distance between some near point and infinity is perhaps a "mid-sighted" or "finite-sighted" lens. And a lens that is optimized for close-up "macro" imaging is a "near-sighted" lens. Thus, just like the human eye, any lens has a far point and a near point, and the range between these two points is the depth of field. For a fixed focus lens that is set to focus at objects at infinity, it is the lens's near point that sets the limit at which the lens yields a clear image, at some spatial frequency. This near point distance is not necessarily the so-called "hyperfocal distance", whose definition is rather restrictive (and theoretically only applies to point objects). Instead, one should define a lens's near point for its intended target's spatial frequency. Next, a lens's far and near points depend on an agreed criterion for acceptable image quality, which depends on the spatial frequency of the intended target. If, for example, a lens is used primarily to image hippos in the wild, then one can invent a new criterion, such as "the farthest distance at which a lens can resolve two grown hippos". Such a criterion may be called, for example, the lens's "hippo-focal distance". For gorillas, we could have "gorilla-focal distance". And so on.

2.2 Physical optics: The backbone of geometrical optics

In the context of applying the laws of physical optics to imaging, one delves into the subject known as Fourier optics, whose principles account for the finer details of the wave character of imaging. That is, in Fourier optics, imaging is not just seen to be the result of the refraction of wavefronts, but is *effected* by the nature of diffraction and interference. In fact, diffraction and interference become necessary and fundamental ingredients for imaging. In Fourier optics, features on an object are like diffraction gratings. Light incident on an object is diffracted by the features on the object into spatial frequency components that are synthesized by a lens to form the final image. However, due to the finite extent of a lens, not all frequency components are collected and synthesized, resulting in an image of reduced quality. Lens aberrations reduce this quality even further. These ideas are derived from the fact that, in any imaging system, the product of the Fourier transform of the ideal image and the Fourier transform of the image of a point source (i.e., the system's impulse response) yields the Fourier transform of the final image distribution. Thus, an imaging system's impulse response filters spatial frequency components from the ideal image. Equivalently, by the convolution theorem, the final image distribution is given by the convolution of the ideal image distribution predicted by geometrical optics with the system's impulse response function – a very useful result that we will apply in Sec. 2.3.4 to describe the radiometry of images of any size, formed by optical systems possessing virtually any level of diffraction and aberrations. To that end, let us review some of the essential principles of Fourier optics.

2.2.1 Diffraction and interference in imaging: A brief review of Fourier optics principles

Within the realm of Fourier optics, our picture of light passing through holes evolves from simple wavefronts and rays (Fig. 2.26a), to wavelets with amplitude and phase arising from points distributed across the area of a hole* (Fig. 2.26b).

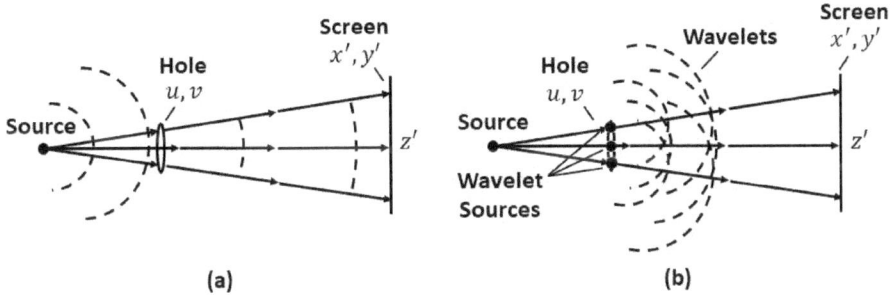

FIGURE 2.26 (a) Wavefronts passing through holes. (b) Wavefronts diffracted by holes introduce wavelets propagating towards a screen.

At a distant screen, the summing of wavelet amplitudes† from an infinite number of infinitesimally spaced wavelet sources bounded by the area of the hole yields the scalar diffraction amplitude distribution [18] given by

$$A(x',y') \propto \iint_{-\infty}^{+\infty} A(u,v)e^{i\frac{k}{2z'}(u^2+v^2)}e^{-i\frac{k}{z'}(ux'+vy')}dudv. \quad (2.28)$$

$$\downarrow \qquad\qquad \downarrow$$

(Fresnel factor)   (Fraunhofer factor)

In Eq. (2.28), the variable $k = 2\pi/\lambda$. In the absence of the Fresnel factor, the integral in Eq. (2.28) is recognized to be the Fourier transform of the amplitude distribution across the u, v plane bounded by the area of the hole. Since a single thin lens is a big aperture (Sec. 2.1.1), for a distant source, wavelets across the surface of a lens's exit

---

* Equivalently, a physically appealing model is to think of scattered wavelets from particles at the edge of the hole. The superposition of the scattered wavelets with the un-scattered wave yields the diffraction pattern [17].

† We are, of course, assuming (in here and throughout the book) that only those waves and wavelets whose electric field amplitudes are polarized in the same direction are summing.

pupil propagate towards a screen at the lens's focal plane, where we might expect to see an amplitude distribution given by Eq. (2.28). But within the paraxial approximation, a lens introduces a **phase term** that exactly cancels the phase in the Fresnel factor, leaving a distribution at the focal plane given by a scaled Fourier transform of the amplitude distribution across the lens's exit pupil [19]. For a lens with a circular aperture (and hence, a circular exit pupil), this distribution is a "Bessel function of the first kind", whose squared modulus is the well-known "Airy disk". The disk's radial size is often defined by the radial position at which the Bessel function first reaches zero [i.e., Eq. (2.24)]. Thus, at the focal plane, the image of a distant point source formed by an aberration-free lens is not a point image. Rather, it is the Airy blur spot, which limits the lens's image quality to the effects of diffraction. It is a **diffraction-limited** lens, and its **point spread function** (PSF) is the Airy disk diffraction blur spot. It turns out that this doesn't just apply to PSFs formed at the focal plane from imaging point sources at infinity. To a reasonable approximation, it also applies when the point source is at a finite distance from the lens. In this case, a PSF formed at the image plane is given by the squared modulus of the Fourier transform of the amplitude distribution at the exit pupil [19].

Pause for insight: → What's a "phase term"? Consider a plane wave of amplitude A_o, and wavelength λ, traveling in the z' direction at speed c:

$$A(z,t) = A_o \cos(kz' - \omega t). \qquad (2.29)$$

Here, $k = 2\pi/\lambda$, $\omega = 2\pi f$, $f = c/\lambda$, c is the speed of light in air (which we approximate here as a vacuum), and t is time. As time progresses, this plane wave propagates in the $+z'$ direction. The argument $(kz' - \omega t)$ contains two **phase terms**, kz' and ωt. If a glass slab with thickness d and refractive index n is placed in its path, this plane wave travels at a speed c/n inside the glass slab, which means that it would take an amount of time $t' = (nd)/c$ for a crest (or trough) to travel through the glass slab. During this amount of time, a plane wave that isn't traveling through this glass slab would have traveled a distance $D = ct' = nd$. Therefore, the wave that's outside the glass would be ahead of the wave inside the glass by an amount of distance $D - d = nd - d = d(n - 1)$. Equivalently, one can say that, relative to the wave outside the glass, the wave that's inside the glass has been shifted by a distance $d(n - 1)$ backwards (i.e., in the $-z'$ direction). To shift the plane wave in Eq. (2.29) in the $-z'$ direction by a spatial amount equal to $d(n - 1)$, one needs to add a positive phase term of the correct magnitude in the argument of the cosine. That correct phase term is given by $kd(n - 1)$, where $k = 2\pi/\lambda$. Hence, the presence of the glass slab in the path of this plane wave modifies Eq. (2.29) into

$$A(z, t) = A_o \cos[kz' - \omega t + kd(n - 1)]. \qquad (2.30)$$

So, we have the general rule that any clear homogeneous optical medium of refractive index n and thickness d that's in the path of a wave introduces a phase term $kd(n - 1)$ into the argument of that wave's equation. Returning to Eq. (2.28), recall that within the paraxial approximation, a lens (say, a thin bi-convex lens) introduces a phase term that cancels the Fresnel factor. That phase term is

$$\phi(u, v) = kd(u, v)(n - 1). \qquad (2.31)$$

Here, $d(u, v)$ is the thickness of the lens across its plane. It can be shown [18] that, within the paraxial approximation,

$$d(u, v) \approx d_o - \frac{(u^2 + v^2)}{2}\left(\frac{1}{R_1} - \frac{1}{R_2}\right). \qquad (2.32)$$

Here, d_o is the center thickness of the lens. When Eq. (2.32) is substituted into Eq. (2.31), followed by substituting Eq. (2.31) into Eq. (2.28), the center thickness d_o is of no practical interest as far as the irradiance distribution $|A(x', y')|^2$ is concerned. But the second term in Eq. (2.32) yields

$$\phi(u, v) = -\frac{k(u^2 + v^2)}{2}\left[(n-1)\left(\frac{1}{R_1} - \frac{1}{R_2}\right)\right]$$

$$= -\frac{k(u^2 + v^2)}{2f}. \tag{2.33}$$

Note that, at the lens's focal plane, $z' = f$ in Eq. (2.28), so the phase in Eq. (2.33) exactly cancels the phase in the argument of the Fresnel factor in Eq. (2.28).

Let us now return to Eq. (2.30). Due to the presence of the added phase term $kd(n-1)$ in Eq. (2.30), constructive optical interference between a plane wave outside a glass slab with a plane wave that had traveled through the glass slab (assuming that they overlap without any tilt angle between them) would depend on the **phase difference** $kd(n-1)$ between the two waves. Whenever the phase difference is $kd(n-1) = 2m\pi$, where $m = 0, 1, 2, 3...$, there would be constructive interference. This is the origin of the concept of the optical path difference (OPD), which we discussed in Sec. 2.1.5 on aberrations. That is, one may say that the phase difference $kd(n-1)$ is the product of the wave propagation constant k with the OPD between two waves (one in a medium with index n, the other outside), where OPD $= d(n-1)$. Another way to see this is to note that the wave that traveled outside the glass slab traveled an **optical path length** (OPL) of d, while the wave inside the glass slab *effectively* traveled a distance of nd. This "effective" distance for the wave inside the glass slab is the OPL for that wave. The OPD between the two waves is therefore $nd - d = d(n-1)$, which, when multiplied by k, yields the phase difference. This is why physical optics is the backbone of geometrical optics. Since wavefronts possess phase information, so do rays. This is the origin of the wavefront aberration function in geometrical optics given by Eq. (2.13), which is an OPD between an ideal wavefront and an aberrated wavefront. Equivalently, it is the OPD between an ideal ray and an aberrated ray.

Now, if the wavefront aberration function of Eq. (2.13) is an OPD, then, in addition to introducing the phase term $\phi(u, v)$ that cancels the Fresnel factor in Eq. (2.28), a lens would also introduce an aberration phase error given by kOPD(u, v) into the argument of the Fraunhofer factor in Eq. (2.28). In fact, it does [20]. This means that a lens's PSF is a consequence of the "competition" between the aberration phase error term kOPD(u, v) and the Fraunhofer phase term $(k/z')[(ux') + (vy')]$. Whichever term dominates would yield a lens that's either diffraction-limited or "aberration-limited", or somewhere between. For an aberration-limited lens, the PSF would be defined by the combined effect of any or all of the aberrations given by Eq. (2.13). A PSF that's primarily affected by coma would look like a comet. A PSF that's primarily affected by astigmatism would look like a thin line in one dimension at one focal position, and would change into a thin line in the orthogonal dimension at some other focal position. And so on.

At this point, we have come to appreciate that, due to the effects of diffraction, or aberrations, or both, the image of a point object source is a PSF. But what does the image of a complete object look like in the presence of diffraction and aberrations? Surely, it would not have sharp edges. That is, since we know that a distribution of point images across the image plane produces a complete image with sharp edges, it should follow that a collection of PSFs would produce an image with blurred edges. In fact, this is indeed the case. It turns out that if the form of an imaging system's PSF were approximately constant across an area of the image (it's generally not), it is found that the relative irradiance distribution $E'(x', y')$ of the image is the convolution of the PSF with the ideal image pattern $F(x'', y'')$ given by geometrical optics:

$$E'(x', y') \propto \int_{-\infty}^{\infty} \int_{-\infty}^{\infty} \text{PSF}(x' - x'', y' - y'')F(x'', y'')dx''dy''.$$
$$(2.34)$$

Eq. (2.34) is one of the most useful analytic expressions in Fourier optics. It tells us that an imaging system's PSF serves to "smooth out" the sharp step function $F(x'', y'')$. Of course, the PSF of an aberration-limited lens varies across the entire image area. But even so, the PSF is usually approximately constant over small areas of the image, and in those areas (aka "isoplanatic patches"), Eq. (2.34) would be valid.[*] We could then divide the entire image area into isoplanatic patches, and in each patch, we may use an appropriate PSF for the convolution. The final image distribution would be the sum of all of the convolutions performed in each isoplanatic patch. In this way, Eq. (2.34) actually provides the connection between physical optics and geometrical optics in imaging systems. But that's not all. Eq. (2.34) also provides the connection among geometrical optics, physical optics, and radiometry. That is, note that geometrical optics helps us determine image location and size, while physical optics determines the image's shape due to diffraction and aberrations. But neither tells us anything about the brightness magnitude of an image. It is the subject of radiometry that provides the means for computing the magnitude of image brightness, but radiometry ordinarily does not account for diffraction and aberrations. It turns out that the complete picture requires combining radiometry with geometrical and physical optics, and this is achieved through the application of Eq. (2.34) to the mathematical machinery of radiometry, which we will do in Sec. 2.3.4.

2.2.2 Physical optical imaging systems: Some examples

It is instructive to acquire an appreciation for the practical application of physical and Fourier optics to imaging systems. We shall attempt this with two specific examples. First, we briefly review the basic operating principle of a diffractive lens, where image formation is achieved completely through means involving diffraction and

---

[*] A constant PSF over an area in the image is "space-invariant" (or, also "shift-invariant").

interference.[*] Second, we briefly review the technique known as apodization, where one alters an imaging system's spatial frequencies to effect or even improve the quality of an image.

2.2.2.1 Diffractive optics

Refraction is not the only way to form an image. Diffraction can do it too. To understand the principle of a diffractive lens, we start by thinking of it as a transmissive diffraction grating with circular grooves. Between successive grooves, annular spaces transmit rays with $m\lambda$ optical path differences ($m = 1, 2...$), thereby providing constructive interference at discrete axial distances from the lens. Suppose one of these axial distances is f (see Fig. 2.27).

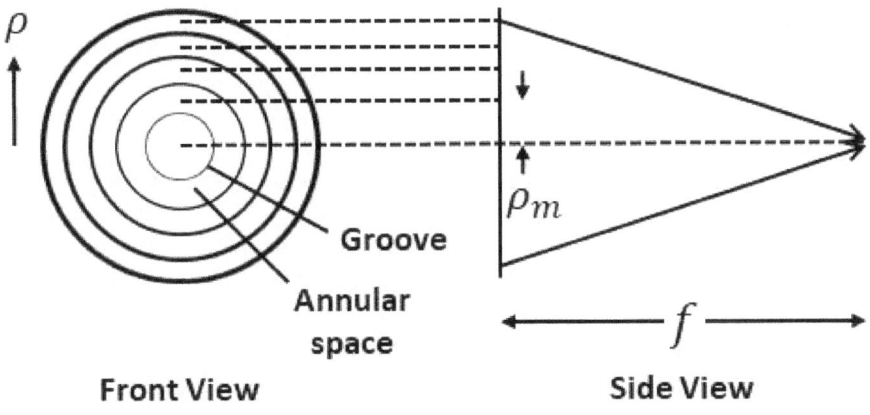

FIGURE 2.27 Annular spaces between grooves diffract light to f.

For incident plane waves, constructive interference at f is achieved if radial heights ρ_m at the m'th annular space satisfy

$$\rho_m{}^2 = 2m\lambda f. \tag{2.35}$$

---

[*] Note that holography is also an image forming process that completely utilizes diffraction and interference.

Note that f is the intended focal length of the diffractive lens [21]. But we know that gratings diffract light into multiple orders, with most of the incident light passing straight through into the 0th order. In order to direct the 0th order light into any m'th order of choice, we need to *blaze* the light there. So, let's pause to understand blazing.

Pause for insight: → What is *blazing*? To simplify, let us only consider diffraction and blazing in one dimension. Consider a plane wave incident on a slit of width w (Fig. 2.28).

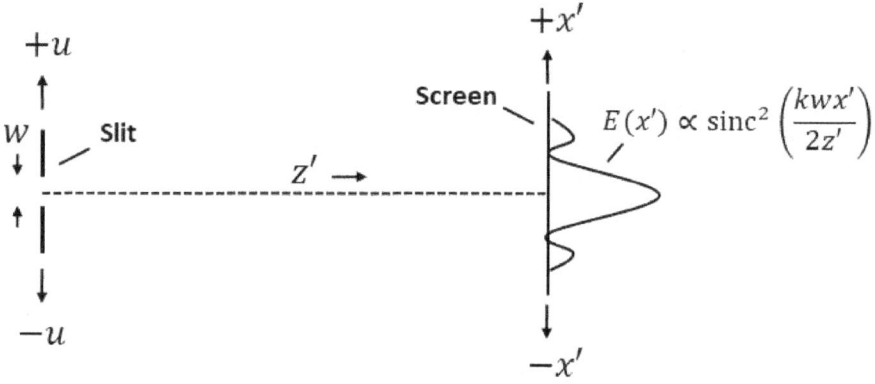

FIGURE 2.28 Single slit diffraction at the far-field.

Passage of a plane wave through the slit in Fig. 2.28 results in a far-field irradiance distribution at a distant screen given by

$$E(x') = |A(x')|^2 \propto \text{sinc}^2\left(\frac{kwx'}{2z'}\right). \qquad (2.36)$$

As usual, we have $k = 2\pi/\lambda$. Clearly, the peak irradiance is found at $x' = 0$. Placing a tiny prism component with refractive index n over the slit (Fig. 2.29) introduces a one-dimensional varying phase term $\phi(u)$ given by

$$\phi(u) = -ku(\tan\theta)(n-1). \qquad (2.37)$$

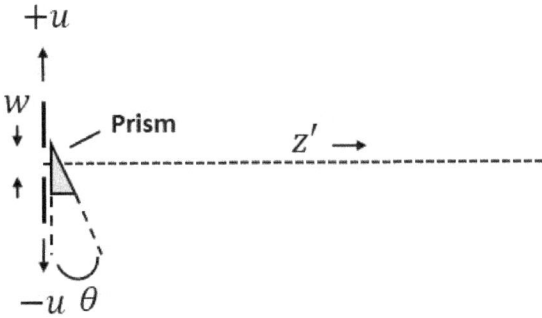

FIGURE 2.29 A tiny prism of index n in front of a slit.

Applying Eq. (2.28) in one dimension (but without the Fresnel factor), the far-field amplitude distribution as a result of the introduced phase is

$$A(x') \propto \int_{-w/2}^{+w/2} e^{-iu\frac{k}{z'}[x'+z'(\tan\theta)(n-1)]} \, du$$

$$\propto -\frac{e^{-iw\frac{k}{2z'}[x'+z'(\tan\theta)(n-1)]} - e^{iw\frac{k}{2z'}[x'+z'(\tan\theta)(n-1)]}}{i\frac{k}{z'}[x'+z'(\tan\theta)(n-1)]}$$

$$\propto (-w)\frac{\sin\left\{\frac{kw}{2z'}[x'+z'(\tan\theta)(n-1)]\right\}}{\frac{kw}{2z'}[x'+z'(\tan\theta)(n-1)]}$$

$$\propto (-w)\mathrm{sinc}\left\{\frac{kw}{2z'}[x'+z'(\tan\theta)(n-1)]\right\}. \qquad (2.38)$$

The far-field irradiance distribution is therefore

$$E(x') = |A(x')|^2 \propto \mathrm{sinc}^2\left\{\frac{kw}{2z'}[x'+z'(\tan\theta)(n-1)]\right\}. \qquad (2.39)$$

Hence, from Eq. (2.39), the new peak irradiance is found at $x' = -z'(\tan \theta)(n - 1)$. For small angles, $\tan \theta \approx \theta$ so $x' \approx -z'\theta(n - 1)$. We recognize $\theta(n - 1)$ as the angle of deviation for a prism with a small apex angle. Thus, the peak irradiance has been deviated towards the position shown in Fig. 2.30.

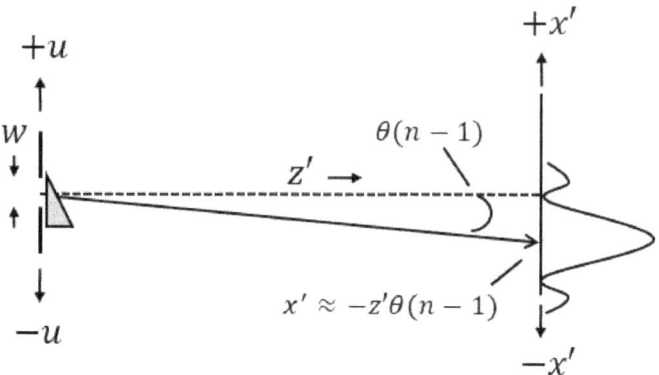

FIGURE 2.30 Blazed diffraction peak by a tiny prism component.

Interestingly, the deviation $\theta(n - 1)$ has a characteristic reminiscent of "refraction" by a prism. Hence, at least for small angles, transmissive diffractive blazing by a tiny prism optical component is *mathematically* equivalent to prism refraction, which simplifies how we imagine blazing in diffractive optics.

Returning to the diffractive lens, blazing must occur at each radial zone, at the radial heights given by Eq. (2.35), so that 0th order light energy is re-directed to an axial position at a focal distance f from the lens. Therefore, at those radial heights, we need tiny prisms. To direct light towards the focal point, each prism's blazing direction must in theory be a function of the radial height ρ_m. Applying Eq. (2.33), each prism must therefore introduce a phase term $\phi(\rho_m)$:

$$\phi(\rho_m) = -k\left(\frac{\rho_m^2}{2f}\right)(n - 1). \tag{2.40}$$

The focal length f in Eq. (2.40) is given by

$$\frac{1}{f} = \frac{(n-1)}{R}. \tag{2.41}$$

We recognize that the phase in Eq. (2.40) arises from a plano-convex lens surface of radius R, whose sag height is $\rho_m{}^2/2R$ [e.g., see Eqs. (2.12) and (2.32)]. In order to have a diffractive optical surface (i.e., a nice flat surface with diffractive grooves rather than a thick curved lens surface), a decision must be made on an appropriate fixed sag height at each zone, which would serve as the zone's so-called **etch depth** d. Applying Eqs. (2.35) and (2.41), for any diffracted order m, the etch depth is therefore

$$d_m = \frac{\rho_m{}^2}{2R} = \frac{2mf\lambda}{2R} = \frac{mf\lambda}{f(n-1)} = \frac{m\lambda}{(n-1)}. \tag{2.42}$$

In theory, any m'th order may be used, though things get a little more complicated for $m > 1$ [22]. So, we might start with a choice of $m = 1$, yielding $d = \lambda/(n-1)$. The resulting diffractive optical element (DOE) would appear as depicted in Fig. 2.31. We then walk to a factory and attempt to request for the fabrication of a DOE with a physical sag height profile $z(\rho)$ given by

$$z(\rho) = \frac{m\lambda}{(n-1)} - \frac{\rho^2}{2R}$$

$$= \frac{m\lambda}{(n-1)} - \frac{\lambda\rho^2}{2f\lambda(n-1)}$$

$$= \frac{\lambda}{(n-1)}\left[m - \frac{\rho^2}{2f\lambda}\right]. \tag{2.43}$$

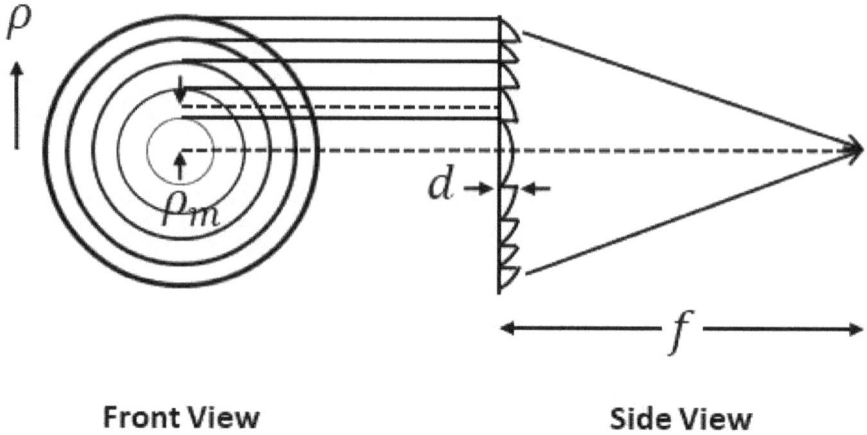

Front View   **Side View**

FIGURE 2.31 Kinoform DOE.

A DOE with the profile illustrated in Fig. 2.31 is called a ***kinoform***. In practice, it might be difficult to fabricate kinoforms, especially at large radial zones. Therefore, much of diffractive optics deals with finding the most practical way to design and produce such parts, other than applying them to correct the aberrations of refractive lenses[*], and dealing with stray light from other diffracted orders. Still, the kinoform DOE profile is a good place to start in understanding how to design DOEs. If you are using Zemax OS, there are a number of excellent articles available online at the Zemax Knowledge Base site [e.g., 24 – 26].

2.2.2.2 The approximately diffraction-free lens: Apodization

Recall that the amplitude PSF of an imaging system is given by the Fourier transform of the amplitude distribution across the exit pupil. Therefore, if the amplitude distribution at the exit pupil were a Gaussian function, then the amplitude distribution of the PSF would be a Gaussian[†]. This being the case, a Gaussian PSF would have little

---

[*] Due to the nature of diffraction, a DOE's dispersion is opposite that of refractive lenses.

[†] Recall from Fourier theory that the Fourier transform of a Gaussian function is a Gaussian function.

to no diffraction "ringing" side lobes. And that's not all. Gaussian PSFs can also produce increased resolution beyond the limits imposed by the Airy disk at certain spatial frequencies. This seems quite interesting. How does one have a Gaussian amplitude distribution at the exit pupil? One way is to place a variable density filter at the aperture stop. This is called **apodization** [e.g., 23] of the aperture. If this filter possesses variable transmittance from its center to the edge given by a Gaussian distribution, then so would the exit pupil. Let us get some appreciation for this fascinating characteristic by taking a quick pause to think about the theory, and then try it out in Zemax OS.

Pause for insight: → Suppose we have some diffraction-limited lens imaging at a working f-number of about f/5.6 operating at a central wavelength of 587.56 nm. Being diffraction-limited, the PSF would be an Airy disk, and its radius [applying Eq. (2.24)] would be:

$$r' \approx 1.22\lambda(\text{WFN}) \approx 1.22(0.58756)(5.6) \approx 4 \text{ μm.} \quad (2.44)$$

The MTF at spatial frequency v for this lens may be expressed in closed-form [27]:

$$\text{MTF}_a(v) \approx \frac{2}{\pi}\left[\text{acos}\left(\frac{vr'}{1.22}\right) - \frac{vr'}{1.22}\sqrt{1 - \left(\frac{vr'}{1.22}\right)^2}\right]. \quad (2.45)$$

The subscript "a" denotes that it is the MTF for the Airy disk. If the PSF were a Gaussian with a 4 μm 1/e^2 blur radius, this Gaussian PSF may be expressed as

$$\text{PSF}_g(r) = e^{-2\left(\frac{r^2}{r'^2}\right)}. \quad (2.46)$$

The optical transfer function (OTF) for this Gaussian PSF would be its Fourier transform:

$$\text{OTF}_g(v) = \int_{-\infty}^{+\infty} \text{PSF}_g(r)e^{-i2\pi vr}dr = \int_{-\infty}^{+\infty} e^{-2\left(\frac{r^2}{r'^2}\right)}e^{-i2\pi vr}dr$$

$$= \int_{-\infty}^{+\infty} e^{-\left[2\left(\frac{r^2}{r'^2}\right)+i2\pi vr\right]}dr = r'\sqrt{\frac{\pi}{2}}e^{-\frac{(\pi r'v)^2}{2}}. \qquad (2.47)$$

Note that the Fourier transform in Eq. (2.47) is evaluated by applying the relation [28]

$$\int_{-\infty}^{+\infty} e^{-(ax^2+bx+c)}dx = \sqrt{\frac{\pi}{a}}e^{\frac{b^2-4ac}{4a}}. \qquad (2.48)$$

The MTF is the normalized modulus of the OTF, yielding

$$\text{MTF}_g(v) = e^{-\frac{(\pi r'v)^2}{2}}. \qquad (2.49)$$

Note in Eq. (2.49) that the 1/e radius of the Gaussian MTF is $\sqrt{2}/(\pi r')$, signifying that smaller Gaussian PSF widths yield larger Gaussian MTF widths, which is expected in analogy to the MTF for Airy disk PSFs. A comparison of Eqs. (2.45) and (2.49) for $r' = 4\mu m$ is shown in Fig. 2.32. Note the slight increase in MTF for the Gaussian MTF at lower spatial frequencies. We are not restricted to letting the Gaussian PSF radial size to be the same as that of the Airy disk. In fact, doing so effectively results in a slightly wider Gaussian PSF (because the radius of the Airy disk is defined as the point where the Airy diffraction distribution first reaches zero). As we shall see in the following Zemax OS example, we can tune the apodization at the lens's entrance pupil such that higher MTF results over a useful range of spatial frequencies.

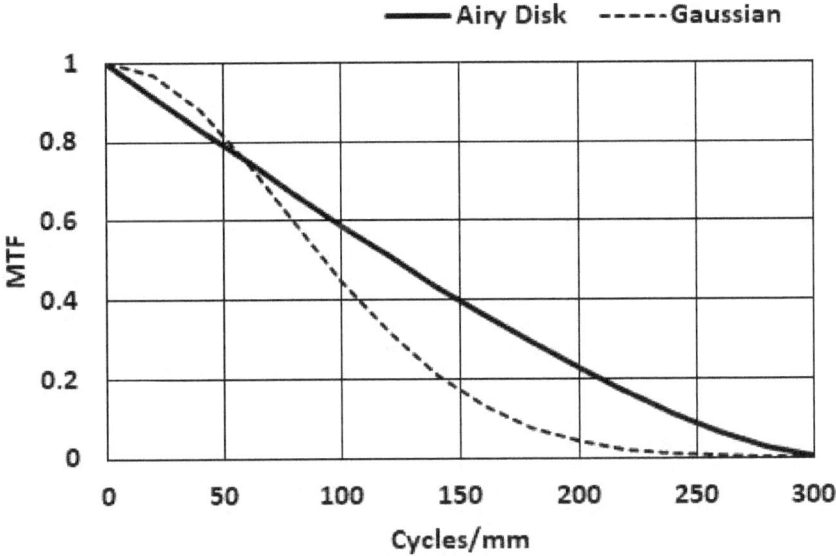

FIGURE 2.32 MTF from an Airy disk PSF [Eq. (2.45)], and a Gaussian PSF [Eq. (2.49)], for $r' = 4$ µm and working f-number of f/5.6.

2.2.2.3 Zemax OS example: Apodization and resolution enhancement

Consider a diffraction-limited lens system with 1 mm EFL at wavelength 587.56 nm, WFN = f/5.6, and object at 10 mm from the first lens's surface (Fig. 2.33). The prescription for this lens[*] is provided in Table 2.1. We are going to apodize this lens with a Gaussian apodizing distribution, and then compare the lens's resulting MTF for the condition of the apodized pupil with the condition of an un-apodized pupil, at a variety of Gaussian widths.

---

[*] Derived from US patent 4,203,653 (I. Mori, 1980. Re-optimized with a cemented doublet and Schott glasses).

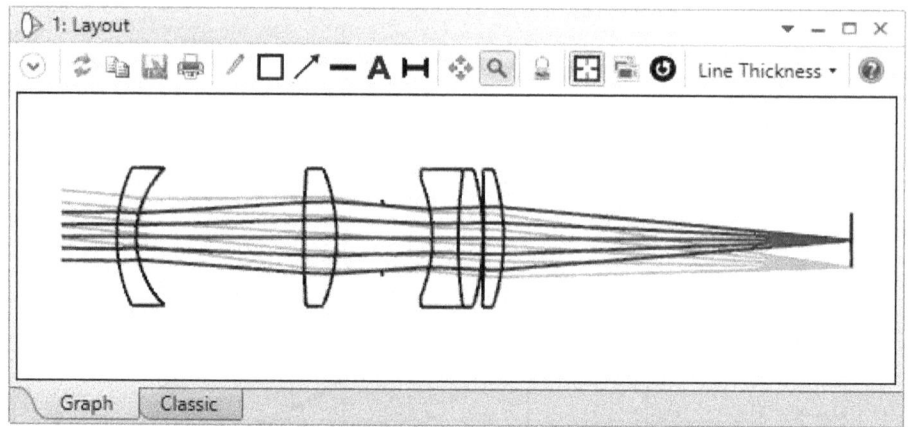

FIGURE 2.33 Zemax OS print-screen of the lens system layout for the apodization example. This lens has EFL = 1 mm, and working f-number f/5.6. Its prescription is given in Table 2.1.

Table 2.1 Lens prescription for apodization example (units in mm).

Surf	Radius	Thickness	Glass	Diameter
OBJECT	Infinity	9.8		2
1	Infinity	0.2		0.335
2	0.5837831	0.071	N-LAK9	0.5
3	0.3553177	0.608		0.5
4	3.40934	0.120	N-BAF52	0.5
5	-0.6676893	0.165		0.5
STOP	Infinity	0.181		0.236
7	-0.5593936	0.091	N-SF11	0.42
8	1.794006	0.089	N-LAF2	0.5
9	-0.9474117	0.002		0.5
10	-10.51463	0.072	N-SSK5	0.5
11	-0.8711321	1.257		0.5
IMAGE	Infinity			0.198

In Zemax OS, Gaussian apodization is applied to the *entrance pupil* of the imaging system. To apodize, one clicks on the "System Explorer" menu and selects "Gaussian" for the "Apodization Type" under the "Aperture" pull-down menu. If you're trying to search for this feature, there is a neat "Feature Finder" you can use (click on the "Help" icon on the main menu bar and you'll see it). Once "Gaussian" is selected for the Apodization Type, one enters an "Apodization Factor" to generate a Gaussian apodizing distribution at the entrance pupil. What's this? The apodization Factor G in Zemax OS is defined by the apodizing amplitude distribution

$$A(\rho_N) = e^{-G\rho_N^2}. \tag{2.50}$$

The quantity ρ_N is a normalized radial coordinate at the entrance pupil such that $0 < \rho_N < 1$. Now, if we want to obtain the result shown in Fig. 2.32, we can determine an appropriate value for G by trial and error, or we can estimate G analytically, which is quite instructive. So, let's do that.

Analytic approach: $\rightarrow$ Let us first refer to the amplitude distribution of a PSF by the term "amplitude spread function" (ASF). Since a Gaussian ASF at the image plane is the Fourier transform [defined by the Fraunhofer diffraction integral of Eq. (2.28)] of the amplitude distribution at the exit pupil, let us first take the Fourier transform of a Gaussian amplitude distribution of radial width w_{ex} and radial coordinate ρ at the exit pupil of the lens system in Fig. 2.33. Applying the diffraction integral from Eq. (2.28) without the Fresnel term, we have

$$\text{ASF}_g(r) = \int_{-\infty}^{+\infty} e^{-\left(\frac{\rho}{w_{ex}}\right)^2} e^{-\frac{i2\pi r\rho}{\lambda z\prime}} d\rho = w_{ex}\sqrt{\pi}e^{-\left(\frac{\pi w_{ex} r}{\lambda z\prime}\right)^2}. \tag{2.51}$$

We have once again made use of the integral relation from Eq. (2.48) to evaluate the integral in Eq. (2.51). If we normalize Eq. (2.51), it would represent the square root of the Gaussian PSF given by Eq. (2.46). Hence, cross-referencing Eq. (2.51) with the square root of Eq. (2.46) we have

$$e^{-\left(\frac{\pi w_{ex} r}{\lambda z'}\right)^2} = e^{-\left(\frac{r}{r'}\right)^2}. \tag{2.52}$$

This means that the radial width w_{ex} of the Gaussian amplitude distribution at the exit pupil is related to the Gaussian ASF's radial width r' by

$$w_{ex} = \frac{\lambda z'}{\pi r'}. \tag{2.53}$$

Normalizing the radial coordinate ρ at the exit pupil means that $\rho_n = \rho/\rho_{exmx}$, where ρ_{exmx} is the maximum semi-diameter of the exit pupil. This means that

$$e^{-\left(\frac{\rho}{w_{ex}}\right)^2} = e^{-\left(\frac{\rho_n \rho_{exmx}}{w_{ex}}\right)^2} = e^{-G\rho_n^2}. \tag{2.54}$$

Cross-referencing all three exponents in Eq. (2.54), we note that

$$G = (\rho_{exmx}/w_{ex})^2. \tag{2.55}$$

Substituting Eq. (2.53) into (2.55), we have

$$G = \left(\frac{\pi r' \rho_{exmx}}{\lambda z'}\right)^2. \tag{2.56}$$

Note that the result in Eq. (2.56) applies to apodization at the exit pupil, and z' is the distance from the exit pupil to the image plane. But note also that, by simple scaling, $(w_{en}/w_{ex}) = (\rho_{enmx}/\rho_{exmx})$ where w_{en} is the radial width of the Gaussian apodization amplitude distribution at the entrance pupil, and ρ_{enmx} is the maximum semi-diameter of the entrance pupil. This makes G in Eq. (2.55) applicable to both the exit and entrance pupils. Hence, G is given by Eq. (2.56). In our lens example, $\lambda = 0.58756 \ \mu m$ and $r' \approx 4 \ \mu m$. In Zemax OS, clicking on the "Reports" button at the main menu and then on "System Data" provides z' and ρ_{exmx}, which are about 1721.918 μm and 153.8744 µm respectively. Putting all these into Eq. (2.56), we obtain $G \approx 3.6$. Fig. 2.34 shows Zemax OS plots for the resulting on-axis Gaussian MTF plots for $G = 3.6$ and $G = 1$, superimposed with the on-axis diffraction-limited MTF plot for a non-apodized entrance pupil.

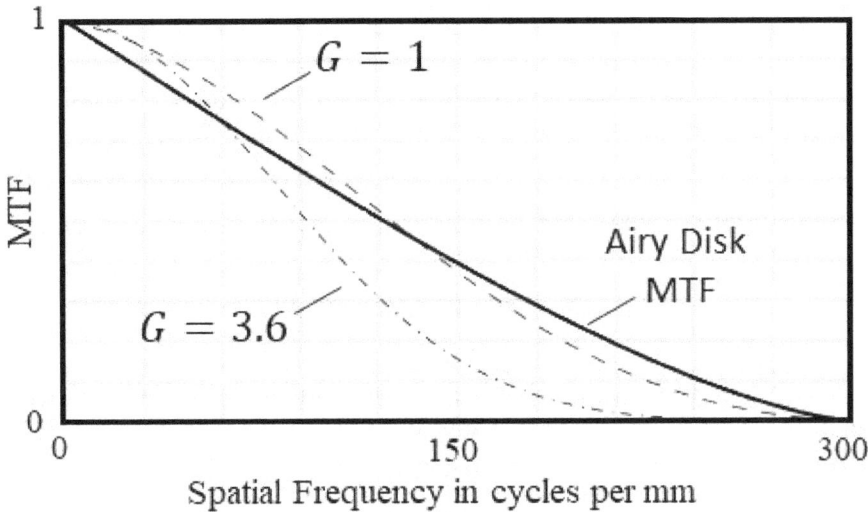

FIGURE 2.34 MTFs at $G = 1$ and $G = 3.6$ for the Gaussian apodized pupil of the lens in Fig. 2.33, superimposed with the MTF for this lens without apodization (Airy Disk MTF).

Note the increased MTF at spatial frequencies $0 - 50$ cycles/mm for $G = 3.6$ (compare this with our analytic result in Fig.2.32), and at $0 - 120$ cycles/mm for $G = 1$, relative to the MTF for a lens without apodization of the pupil. It is also instructive to look at Zemax OS's plots for the PSFs for these MTF curves (Fig. 2.35), as well as the ray densities through the lens system layouts (Fig. 2.36).

There is another way to achieve a Gaussian PSF without apodization: Pass a TEM_{00} laser beam through the lens system, and the PSF at the focal plane will be approximately a Gaussian distribution! This is effectively what laser scan lenses do.

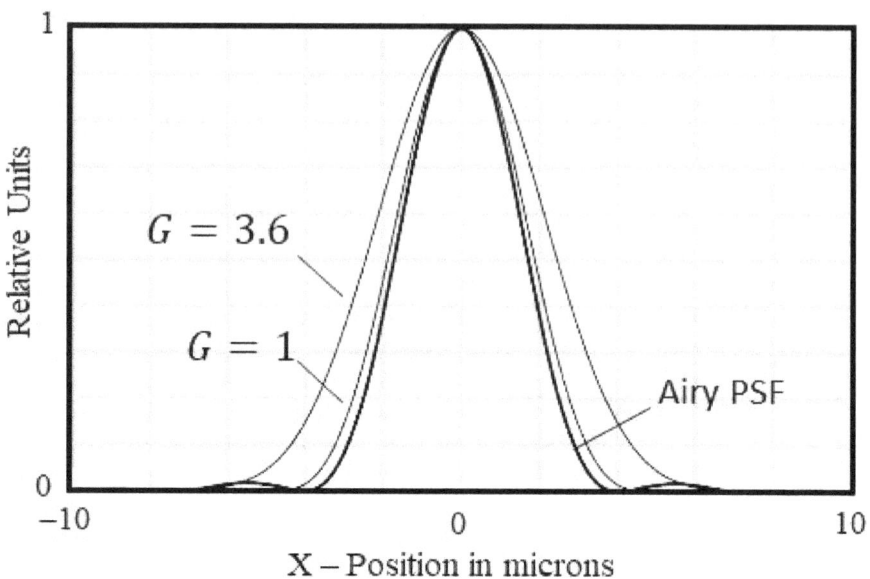

FIGURE 2.35 PSFs for the Gaussian apodized pupil of the lens system in Fig. 2.33, superimposed with this lens's PSF without apodization.

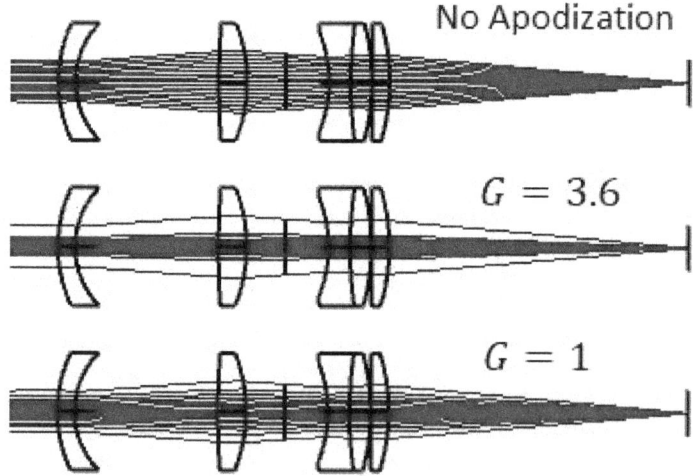

FIGURE 2.36 Density of apodized rays through the lens system.

2.2.2.4 A short note about coherence

In all of Sec. 2.2, we have made some assumptions about the coherence properties of the source, the object, and the image. For example, in the Gaussian apodization exercise of Sec. 2.2.2.3, we were computing PSFs at a monochromatic wavelength of 588 nm, yet we analyzed the MTF of the system rather than its *coherent transfer function* (CTF). We had a Gaussian distribution at the entrance pupil, seemingly analogous to the condition of a TEM_{00} beam entering it, yet we did not perform the analysis in terms of the propagation of a laser Gaussian beam. Why is that? The rationale is as follows:

1. I assumed that 587.56 nm was simply the central wavelength of the light source, whose spectral bandwidth is finite. Hence, the source is quasi-monochromatic. In this case, its light will not propagate like any transverse modes of a laser beam. The source is also large (e.g., ambient quasi-monochromatic light from all over a room is lighting up the object being imaged), so there is virtually no correlation between any two points across the object plane. Hence, the object is a source of spatially incoherent light.

2. Although the object is spatially incoherent, a point object source will illuminate the entrance pupil with a high degree of spatial coherence. The exit pupil is a conjugate of the entrance pupil, so it possesses a high degree of spatial coherence for point images [29]. Hence, the amplitude distribution of the PSF (i.e., the ASF) at the image plane must be given by a Fraunhofer amplitude diffraction integral [i.e., Eq. (2.28)] over the exit pupil. However, the image is incoherent (it's a conjugate of the incoherent object), so its distribution is given by the convolution of the incoherent impulse response (i.e., the PSF) with the geometric image distribution. By the convolution theorem, the image's transfer function is therefore the OTF, not the CTF.

But conditions would be different for a scene that's illuminated by, for example, a laser whose beam is made to spread over an area. In this case, all points across the illuminated scene would be highly correlated, resulting in *laser speckle*. Coherence is a very big topic and is beyond the scope of this book. Interested readers may refer to some rather excellent texts on the subject [e.g., 29, 30].

2.3 Radiometry: The backbone of illumination and nonimaging optics

As mentioned near the end of Sec. 2.2.1, the ray tracing principles of geometrical optics provide the means to determine the location and size of an image, while the convolution integral from physical optics [i.e., Eq. (2.34)] tells us how to compute the relative light distribution across the image's area in the presence of diffraction and aberrations. But these formulations say nothing about how much optical power is contained in an image, nor how much power per unit area there is at some small region of the image, which would be helpful if we wanted to know, for example, what is the effect of the image's brightness on an area of film or image sensor. In other words, we need a means to compute the absolute magnitude of optical flux and flux density (aka "irradiance") of an image. This is provided by the subject of radiometry, whose key principles and concepts lay the foundations for the fields known as illumination engineering and nonimaging optics.

2.3.1 Radiance, etendue, intensity, and all that: A brief review

In our review of radiometry, we shall assume that the optical radiation is incoherent, though it should be noted that considerations for coherence in radiometry have been studied quite extensively [e.g., 32 – 47]. For an excellent overview of the impact of coherence on the foundations of radiometry, have a look at Ref. 38, where E. Wolf provides a short introduction to the theory of partial coherence, and he shares some rather fascinating reasons for why there remains some limited understanding of the connection between electromagnetic wave theory and radiometric quantities (e.g., what is radiance in terms of the Poynting vector?). These are all neglected for simplicity in our review of radiometry. We shall also assume that the imaging system is non-diffractive and aberration-free (but we will consider them in Sec. 2.3.4).

Our review begins with the quantity called *radiance*, which is perhaps the most useful quantity in analytic radiometry. By "analytic" I mean the formulary of radiometry, such as deriving closed-form equations that enable calculating the amount of *flux* (optical power, e.g., in units of Watts) from a light source, the flux in an image, and the image's *irradiance* (flux per unit area). It is useful to think of radiance as a "density function" for radiation, in analogy to a mass density function when computing the total mass for a solid object. For example, to compute the total mass for an object whose mass varies over its volume, one has to know what is the object's differential mass per unit volume dm/dV as a function of position everywhere in the space occupied by the object. The object's differential mass per unit volume is therefore the local mass density ρ or "mass density function" $\rho(x, y, z)$ of the object over space coordinates. Knowing this, the object's total mass is the scalar integral of the object's mass density function over the space bounded by the object's volume:

$$Total\ Mass = \int \int \int \rho(x, y, z) dx dy dz. \qquad (2.57)$$

In a manner similar to mass density, a source's radiance L is a "radiation density function" that enables the computation of radiometric quantities, and it is defined as

$$L = \frac{d^2\Phi}{dA\cos\theta d\Omega}. \qquad (2.58)$$

In Eq. (2.58), $d^2\Phi$ is an element of optical flux radiated by a source of elemental area dA, in the direction θ, within an elemental solid angle $d\Omega$ (Fig. 2.37).

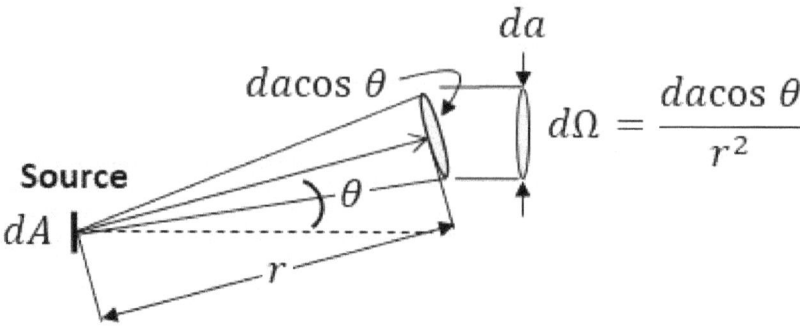

FIGURE 2.37 Geometry for the definition of radiance in Eq. (2.58).

The projected area $dA\cos\theta$ accounts for the effective reduction in source size in the direction θ. Hence, radiance is essentially the flux per unit projected area per unit solid angle in a beam of light. If the source radiance is constant with θ, the total flux Φ_{tot} emitted by the source is

$$\Phi_{tot} = \int\int L dA\cos\theta d\Omega = L \int\int dA\cos\theta d\Omega. \qquad (2.59)$$

It is known that, by the **radiance theorem** [e.g., 48], for rays propagating in a lossless medium in any direction through an optical system, the radiance is invariant[*], which means that the total double integral in Eq. (2.59) is also invariant (since, by energy conservation, the flux Φ_{tot} is invariant). Eq. (2.59) may then be expressed as

$$\Phi_{tot} = L \int\int dA\cos\theta d\Omega = L\varepsilon_{tot}. \qquad (2.60)$$

Here, ε_{tot} is the total **etendue** of the propagating radiation. For a source of elemental area dA, a lens captures an amount of optical

---

[*] Radiance is invariant upon refraction through any lens element, provided that the medium before and after refraction is the same. Otherwise, we have $L_1/n_1 = L_2/n_2$, where n is the refractive index [e.g., 49].

flux $d\Phi$ (Fig. 2.38) which may be expressed as the product of the source's radiance and the differential etendue $d\varepsilon$ of the captured beam:

$$d\Phi = L\left(dA\int_0^\theta \cos\theta d\Omega\right) = L(d\varepsilon). \qquad (2.61)$$

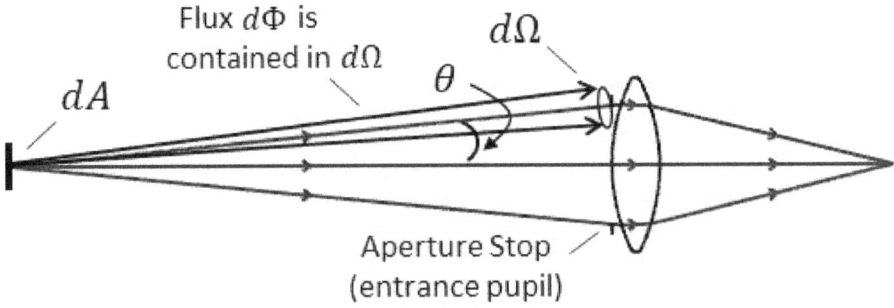

FIGURE 2.38 A lens captures flux from a small source.

The utility of Eq. (2.61) to optical systems is better appreciated when its integral is evaluated. To do this, it is useful to note that, for a symmetrical optical system,

$$d\Omega = 2\pi\sin\theta d\theta. \qquad (2.62)$$

Substituting Eq. (2.62) into Eq. (2.61), we have

$$d\Phi = 2\pi LdA\int_0^\theta \cos\theta\sin\theta d\theta = 2\pi LdA\int_0^\theta \sin\theta d(\sin\theta)$$

$$= 2\pi LdA\left(\left.\frac{\sin^2\theta}{2}\right|_0^\theta\right) = \pi LdA\sin^2\theta. \qquad (2.63)$$

For small source areas, $dA \approx A$ so the etendue is $\approx \pi A\sin^2\theta$, and one may say that the product of A and $\sin^2\theta$ is invariant throughout an optical system (Fig. 2.39).

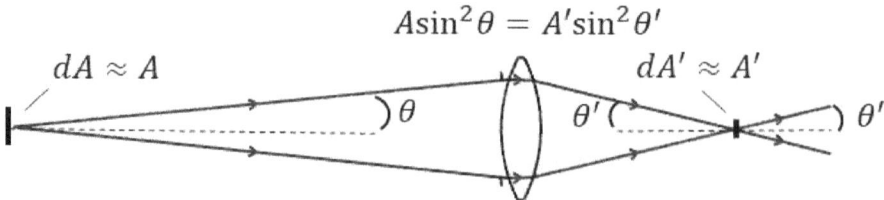

FIGURE 2.39 Invariance of "area × angle" through an optical system.

Alternatively, along the optic axis, for small pupil areas, the product of a lens's pupil area and the angle subtended by the source is invariant (Fig. 2.40). To see why, return to Eq. (2.61), which, without the integral, may be expressed as

$$d^2\Phi = LdAd\Omega = LdA\frac{da_{lens}}{r^2} = L\frac{dA}{r^2}da_{lens}. \qquad (2.64)$$

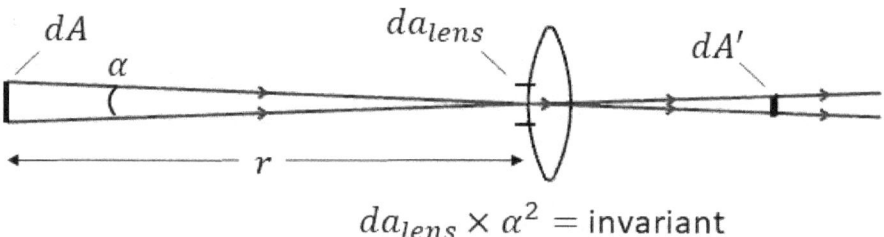

FIGURE 2.40 Invariance of pupil area × source angle.

For a large source, one may divide its area into subareas such that the total etendue is approximately $\pi \sum_i dA_i \sin^2\theta_i \cos^2\alpha_i$. The angle α accounts for the obliquity of the central ray. Thus, this sum would be approximately invariant (Fig. 2.41). If a source's radiance is a function of the direction θ, then one may subdivide the solid angles from each sub-areas into "sub-solid-angles" where the radiance within each sub-solid-angle is of a magnitude given by a radiance function (Fig. 2.42).[*]

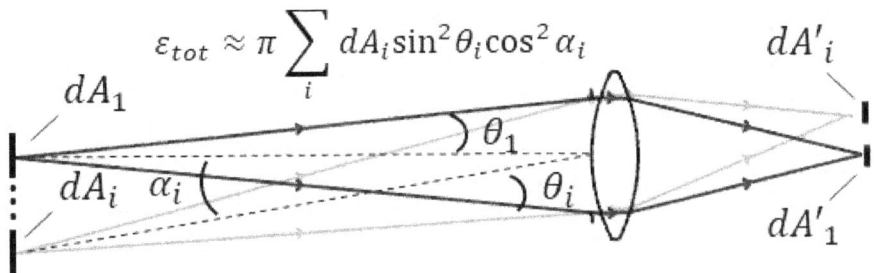

$$\varepsilon_{tot} \approx \pi \sum_i dA_i \sin^2 \theta_i \cos^2 \alpha_i$$

FIGURE 2.41 Invariance of a sum of "area × angles".

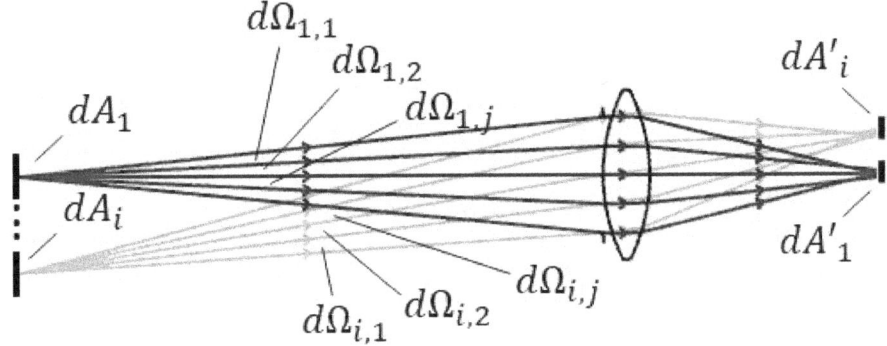

FIGURE 2.42 Rays with different radiances within sub-solid-angles.

In Fig. 2.42, note that, due to a directionally varying source radiance, the subdivision of fluxes into "sub-fluxes" within respective sub-solid-angles renders the concept of the total etendue in Eq. (2.60) to be not meaningful. Each sub-solid-angle is associated with its own "sub-etendue". Therefore, there is no such thing as total etendue for a source whose radiance varies directionally. To understand the concept of "sub-etendue", first note that the underlying physical quantity that's

---

[*] We are assuming (and shall continue to assume in this book unless stated otherwise) that the source's radiation distribution is symmetric about the optic axis.

always invariant is flux. So, we return to Eq. (2.59) and note that, for a source whose radiance varies with θ, we may write the total flux as the integral of the product of a directionally varying radiance $L(\theta)$ with an element of etendue $d^2\varepsilon$:

$$\Phi_{tot} = \int\int L(\theta)dA\cos\theta d\Omega = \int\int L(\theta)d^2\varepsilon. \qquad (2.65)$$

Accordingly, the "sub-flux" in a small beam of light subtending a solid angle $d\Omega$ in the direction θ is

$$d^2\Phi = L(\theta)d^2\varepsilon. \qquad (2.66)$$

The elemental etendue $d^2\varepsilon$ is therefore a "sub-etendue" for this small beam of light. Some refer to the right-hand side of Eq. (2.66) as the *physical etendue* [50] of a beam, but as can be seen, it is simply the flux of the beam. It is also important to note that, by the radiance theorem, $L(\theta)$ is invariant. Therefore, within each sub-solid-angle, the sub-etendue is invariant.

Evidently, things are always a little more complicated when radiance is a function of θ. A source[*] whose radiance is independent of θ is called a *Lambertian source*. Diffusive scattering or "matte" surfaces are often reasonable approximations to Lambertian sources. A good example is a sheet of ordinary writing paper. The radiation pattern of light emitting diodes (LEDs) are often not Lambertian, but those that emit light over a wide angle can sometimes be approximately Lambertian. A quick way to check is to look at a plot of their *radiant intensity* (flux per unit solid angle) or, in photometric units, *luminous intensity* (luminous flux or "lumens" per unit solid angle) in their datasheet. Let us pause to see how this may be done.

Pause for insight: → Suppose, as in Fig. 2.38, we have a small LED with area dA at some distance from a lens that is capturing its

---

[*] By "source" we mean either a radiating source or a non-radiating surface that's diffusely scattering light.

light over an angle θ. If this LED's radiation pattern is not Lambertian, then the total flux captured by the lens would be given by Eq. (2.65). Therefore, if we wish to apply Eq. (2.65) to estimate the magnitude of captured flux, we need to know the LED's radiance function $L(\theta)$. But, one will often not find $L(\theta)$ in a manufacturer's datasheet. Rather, LED datasheets often provide the "radiant intensity" (if in radiometric units) or "luminous intensity" (if in photometric units). To understand what radiant intensity is, one may first return to Eq. (2.65) and express it in differential form:

$$d^2\Phi = L(\theta)dA\cos\theta d\Omega. \qquad (2.67)$$

Carrying $d\Omega$ to the left side of Eq. (2.67), we have

$$d^2\Phi/d\Omega = d(d\Phi/d\Omega) = dI = L(\theta)dA\cos\theta. \qquad (2.68)$$

The quantity I in Eq. (2.68) is the radiant intensity (flux per unit solid angle) of the source. If the source is small relative to the distance to the lens, then $dA \approx A$, so

$$I(\theta) \approx L(\theta)A\cos\theta. \qquad (2.69)$$

This is the approximation that LED manufacturers make in accordance with the LED measurement standard called "CIE127-2007" [51, 52]. Note that the radiant intensity measurement provided by a manufacturer includes the combined variation from the product $L(\theta)\cos\theta$. This data can then be applied to Eq. (2.65) whereby performing the integrals over a finite angle subtended by the lens enables an estimate of the flux captured by the lens. Now, if the LED were Lambertian, then Eq. (2.69) becomes

$$I(\theta) \approx LA\cos\theta. \qquad (2.70)$$

So, a Lambertian source's radiant intensity varies only with the cosine of the angle. One can then check this against the LED's radiant intensity in the datasheet. For a quick analytical test, one can simply note that $\cos(60^0) = 0.5$. Therefore, if the LED's data shows that the intensity at 60 degrees is approximately 50% of its peak at 0 degrees, then it's approximately Lambertian. Of course, one should decide at what level of approximation one can allow for one's application.

When a LED's radiant intensity is Lambertian, analytic calculations become very simple. For instance, manufacturers often provide the peak radiant intensity in absolute units (or in absolute photometric units). For a Lambertian LED emitter, this peak refers to the quantity LA in Eq. (2.70). So, knowing the LED's emitter area A, one may estimate its radiance L. Now, it is known that the total *exitance* (i.e., the total emitted flux per unit area) from a Lambertian source is πL [e.g., 53], which means that the total emitted flux by a Lambertian LED is πLA. Applying Eq. (2.63), the flux captured by a lens is therefore $\approx \pi LA\sin^2\theta$.

2.3.2 Radiance theorem for real and "fake" but useful lenses

By the radiance theorem, radiance is invariant through a lossless optical system. However, we recall from Sec. 2.1.2 that one often performs basic optical imaging system analysis with the paraxial thin lens model (PTLM). Moreover, it is often illustrative to trace rays at large angles through PTLMs even though a PTLM's optical properties are derived from rays making small angles relative to the optic axis. In fact, in Zemax OS, a PTLM admits rays at arbitrarily large incident angles, and it traces them in a manner that satisfies the first order simple lens equation [i.e., Eq. (2.1)]. This is very useful when one sets up an initial first order optical system layout to model imaging systems of any complexity, which may contain any number of paraxial thin lens elements. In some cases, it is even quite useful to set up PTLMs for an initial basic illumination system analysis. However, the PTLM is a rather fictitious lens in the sense that rays that refract through it do not obey Fermat's principle (because, unlike real diffraction-limited lenses, rays that pass through PTLMs focus onto ideal image points without having to take the same amount of time to travel from the object point to the image point). Consequently, rays traveling through PTLMs generally do not obey the radiance theorem even though their flux is conserved. Let us pause to understand why.

Pause for insight: → Consider a PTLM that's imaging a small source with area dA as shown in Fig. 2.43.

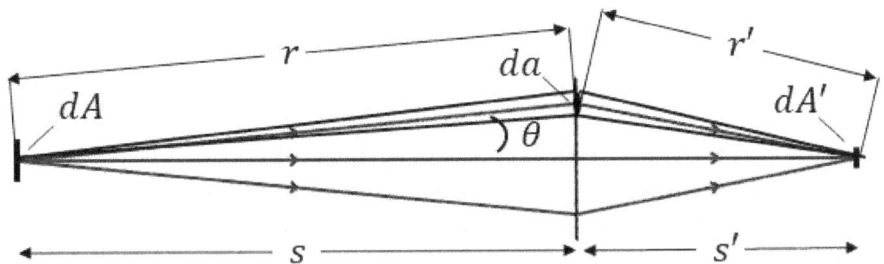

FIGURE 2.43 A Paraxial Thin Lens Model (PTLM) forms an image of a small source.

The PTLM in Fig. 2.43 refracts rays according to Eq. (2.1). Therefore, image magnification is defined by Eq. (2.2), which also means that

$$m_p{}^2 = \frac{dA'}{dA} = \frac{s'^2}{s^2}.$$ (2.71)

Now, note that the element of flux $d^2\Phi$ in an oblique ray bundle striking the top of the PTLM is

$$d^2\Phi = \frac{LdA\cos^2\theta da}{r^2}.$$ (2.72)

Upon refraction, this flux is conserved (equivalently, one notes that the number of rays within the refracted solid angle is constant), so it must be that

$$\frac{LdA\cos^2\theta da}{r^2} = \frac{L'dA'\cos^2\theta'da}{r'^2}.$$ (2.73)

Since $\cos\theta = s/r$ and $\cos\theta' = s'/r'$, we have

$$\frac{LdA\cos^4\theta}{s^2} = \frac{L'dA'\cos^4\theta'}{s'^2}.$$ (2.74)

Finally, applying Eq. (2.71) to Eq. (2.74) yields

$$L\cos^4\theta = L'\cos^4\theta'.$$ (2.75)

Eq. (2.75) reveals that radiance is generally not conserved for PTLMs, except for small ray angles, whereas for real lenses, the radiance theorem proves that $L = L'$ along a ray at any angle upon refraction. On the other hand, flux is clearly conserved through a PTLM (which is also why the Lagrange invariant – a paraxial quantity – is an invariant quantity for PTLMs). For convenience, let us call Eq. (2.75) the *radiance theorem for paraxial thin lens models* (RTPTLM).

Why go through the trouble of deriving a "law" for a fictitious lens? Because PTLMs are often used for analyzing many optical systems. Sometimes this includes using them for setting up "first order" layouts for illumination systems. Accordingly, it is important to keep track of when our models break down, and Eq. (2.75) will help keep us in check.

2.3.3 Invariance of image brightness with distance

At a fixed numerical aperture, within the conditions imposed by geometrical optics (i.e., in the absence of diffraction and aberrations), the brightness of the image of an extended Lambertian source is invariant with source distance, and it is not only due to the invariance of radiance. To see why, consider a real lens (i.e., not a PTLM) producing the image of a flat extended Lambertian source as shown in Fig. 2.44.

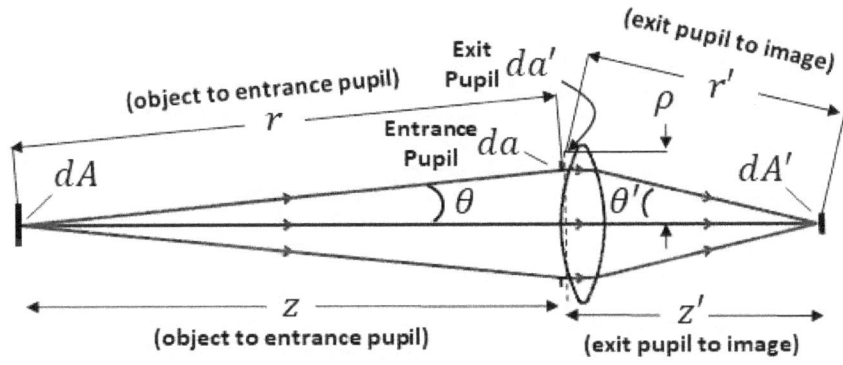

FIGURE 2.44 Lens imaging an extended Lambertian source.

The flux $d^2\Phi$ from an elemental area dA of the source entering an elemental area da at the entrance pupil is

$$d^2\Phi = \frac{LdAda\cos^2\theta}{r^2}.$$ (2.76)

By conservation of energy, this same flux emerges from the exit pupil:

$$d^2\Phi = \frac{LdA'da'\cos^2\theta'}{r'^2}.$$ (2.77)

The axial image irradiance E' is $d\Phi/dA'$, so we have

$$dE' = \frac{Lda'\cos^2\theta'}{r'^2} = \frac{Lda'\cos^4\theta'}{z'^2}.$$ (2.78)

Integrating over the exit pupil area, we have

$$E' = \frac{L}{z'^2}\int \cos^4\theta' da' = \frac{L}{z'^2}\int_0^{2\pi}\int_0^{\rho}\cos^4\theta' \rho d\rho d\phi$$

$$= \frac{2\pi L}{z'^2}\int_0^{\rho}\cos^4\theta' \rho d\rho.$$ (2.79)

Now, note that $\rho = z'\tan\theta' = z'(\sin\theta'/\cos\theta')$, which means that $d\rho = z'(1 + \tan^2\theta')d\theta'$. Substituting these into Eq. (2.79), we have

$$E' = \frac{2\pi L}{z'^2}\int_0^{\theta'}\cos^4\theta' z'^2\left(\frac{\sin\theta'}{\cos\theta'} + \frac{\sin^3\theta'}{\cos^3\theta'}\right)d\theta'$$

$$= 2\pi L\int_0^{\theta'}\cos^4\theta'\left(\frac{\sin\theta'\cos^2\theta'}{\cos^3\theta'} + \frac{\sin^3\theta'}{\cos^3\theta'}\right)d\theta'$$

$$= 2\pi L\int_0^{\theta'}\cos\theta'(\sin\theta'\cos^2\theta' + \sin^3\theta')\,d\theta'$$

$$= 2\pi L\int_0^{\theta'}\sin\theta'\cos\theta'(\cos^2\theta' + \sin^2\theta')\,d\theta'$$

$$= 2\pi L\int_0^{\theta'}\sin\theta'\cos\theta'\,d\theta' = 2\pi L\int_0^{\theta'}\sin\theta'd(\sin\theta'). \quad (2.80)$$

Finally, applying the process of integration from Eq. (2.63) to Eq. (2.80), we have

$$E' = \pi L\sin^2\theta'. \quad (2.81)$$

Eq. (2.81) is the axial image irradiance at any aperture diameter for a real lens. It is also instructive to derive an expression for the axial image irradiance E'_p for a PTLM. To do this, we substitute the radiance L' from Eq. (2.75) for L in Eq. (2.78), yielding

$$dE'_p = \frac{L\cos^4\theta\,da'\cos^4\theta'}{\cos^4\theta'\quad z'^2} = \frac{Lda'\cos^4\theta}{z'^2}. \quad (2.82)$$

Substituting Eq. (2.71) into Eq. (2.82), and noting that the exit and entrance pupils of a PTLM share the same plane (therefore, $da' = da$), we have

$$dE'_p = \frac{1}{m_p{}^2} \frac{Ldacos^4\theta}{z^2}.$$ (2.83)

The differential axial image irradiance for a PTLM is therefore seen to be dependent only on the geometry of rays passing through the entrance pupil. Since the factor involving $(Ldacos^4\theta)/z^2$ in Eq. (2.83) has the same form as in Eq. (2.78), integrating the right-hand side of Eq. (2.83) over the entrance pupil yields

$$E'_p = \frac{\pi L \sin^2\theta}{m_p{}^2}.$$ (2.84)

Alternatively, since, for a PTLM, $m_p{}^2 = \tan^2\theta/\tan^2\theta'$, substituting this into Eq. (2.84) yields

$$E'_p = \pi L \cos^2\theta \tan^2\theta'.$$ (2.85)

In the limit of low numerical aperture, Eqs. (2.85) and (2.81) are approximately equal to one another – a result that is useful for illumination analysis using PTLMs.

Back to real lenses. In Eq. (2.81), note that $\sin^2\theta'$ is the square of the image's axial numerical aperture. Since radiance is invariant, the axial image irradiance would be invariant with source distance if the lens's numerical aperture in image space is held constant. This would indeed be the case for an imaging system that uses a liquid lens, whose focal length is adjustable for any object distance. Hence, it would also be the case for the human eye if its pupil's diameter is somehow held constant. Since image irradiance is the flux per unit area at the image, an image's irradiance directly influences its brightness at an image sensor, such as the human eye's retina. Therefore, the brightness of the image of an extended Lambertian source is invariant with source distance under the condition of constant image numerical aperture. But how could this be? Surely it is more uncomfortable for the eye to look at a 60-Watt bulb at close range than when it is farther away. Yet

Eq. (2.81) tells us that this is not so. Actually, the reason for this contradiction is because Eq. (2.81) is only valid for conditions in which the size of the image remains extended (i.e., when the image size is significantly larger than the size of the imaging system's PSF) at all object distances. Surely, as an object gets farther away from a lens, its image size would reduce until it gradually becomes comparable to or less than the size of the imaging system's PSF, and then one would be left wondering what happens at this stage, because Eq. (2.81) does not account for this. It turns out that, when the geometric image size tends towards the size of the PSF, the irradiance of the image begins to reduce. And as the geometric image size tends to zero, the image irradiance also tends to zero. This property of imaging is not obvious by any means, unless one accounts for the diffractive wave character of imaging, which is done by combining the principles of Fourier optics – in particular, Eq. (2.34) – with the radiometric formula for image irradiance [i.e., Eq. (2.81)]. This we shall do next.

2.3.4 Impact of diffraction and aberrations on image irradiance

To account for diffraction and aberrations in the radiometry of an image, we shall take an approach similar to what I had done in an earlier published study [54]. First, we make the definition that the function $F(x'', y'')$ in Eq. (2.34) takes on the value of unity within the entire surface area that it contains. With this definition, the act of performing the convolution operation in Eq. (2.34) is to compute the volume under the PSF. If the function $F(x'', y'')$ spans an infinite space, then $E'(0,0)$ yields the total volume of the PSF at the center of the image. If, however, the function $F(x'', y'')$ spans an area that limits the extent of the PSF, then $E'(0,0)$ yields a fraction of the total on-axis PSF volume. This all means that the absolute magnitude of the image irradiance $E'(x', y')$ must be an appropriately scaled convolution between the PSF and $F(x'', y'')$. This is done by normalizing the convolution by the volume of the PSF, and multiplying the result by Eq. (2.81):

$$E'_d(x',y') = \frac{E'}{V}\int_{-\infty}^{\infty}\int_{-\infty}^{\infty} PSF(x'-x'',y'-y'')\,F(x'',y'')dx''dy''.$$

(2.86)

Here, the subscript "d" denotes that E'_d is the image irradiance that includes effects from diffraction and aberrations, while E' is given by Eq. (2.81), and V is the volume of the PSF. That is,

$$V = \int_{-\infty}^{\infty}\int_{-\infty}^{\infty} PSF(x',y')\,dx'dy'. \qquad (2.87)$$

Eq. (2.86) has pedagogical appeal in that it is a compact formula that contains the principal ingredients of geometrical optics [i.e., the function $F(x'',y'')$], physical optics (i.e., the PSF), Fourier optics (i.e., the convolution operation), and radiometry (i.e., the irradiance E'). Does this equation truly compute the radiometric properties of an image? Let us see. What is the total flux in the image given by Eq. (2.86)? To answer this question, we integrate Eq. (2.86) over all image space:

$$\Phi_{tot} = \int_{-\infty}^{\infty}\int_{-\infty}^{\infty} E'_d(x',y')\,dx'dy'$$

$$= \frac{E'}{V}\int_{-\infty}^{\infty}\int_{-\infty}^{\infty}\int_{-\infty}^{\infty}\int_{-\infty}^{\infty} PSF(x'-x'',y'-y'')\,F(x'',y'')dx''dy''\,dx'dy'.$$

(2.88)

The integral of the PSF over the entire span of $dx'dy'$ is its volume, so we have

$$\Phi_{tot} = E' \int_{-\infty}^{\infty} \int_{-\infty}^{\infty} F(x'', y'') \, dx'' dy''. \tag{2.89}$$

The integral of $F(x'', y'')$ over all of $dx'' dy''$ is the surface area A' of the image, so

$$\Phi_{tot} = E' \int_{-\infty}^{\infty} \int_{-\infty}^{\infty} F(x'', y'') \, dx'' dy'' = E'A'. \tag{2.90}$$

Eq. (2.90) makes sense. The total flux in the image is the product of its irradiance with its total surface area. Hence, Eq. (2.86) is consistent with radiometry and, indeed, the conservation of energy.

Let us also consider what the axial image irradiance of an ideal image is. The PSF for an ideal image is a single point, which may be described mathematically by a Dirac delta function $\delta(x'', y'')$ in spatial coordinates:

$$\delta(x'', y'') = \begin{cases} \infty, \text{ for } x'' = 0, y'' = 0 \\ 0, \qquad\qquad \text{othewise} \end{cases}, \tag{2.91}$$

$$\int_{-\infty}^{\infty} \delta(x'', y'') dx'' dy'' = \int_{-\infty}^{\infty} \delta(-x'', -y'') dx'' dy'' = 1. \tag{2.92}$$

Accordingly, the volume of a delta function PSF is unity. Applying Eqs. (2.91) and (2.92) to Eq. (2.86) for the axial image irradiance of an ideal image, we have

$$E'_d(0,0) = \frac{E'}{V} \int_{-\infty}^{\infty} \int_{-\infty}^{\infty} \delta(-x'', -y'') \, F(x'', y'') dx'' dy'' = E'. \tag{2.93}$$

Once, again, this is consistent with radiometry, as we'd expect the ideal image to possess the axial irradiance given by Eq. (2.81).

We are ready to tackle the problem encountered near the end of the previous section concerning the apparent contradiction implied by Eq. (2.81), where it seemed as if image irradiance is always invariant with source distance. Our approach is to first recognize that for any imaging system, in the limit that the image size is significantly larger than the blur extent of its PSF, the geometric image distribution $F(x'', y'') = 1$ within and beyond the entire area spanning the blur radius of the PSF. This is equivalent to taking the limit that the area spanning x'' and y'' approaches infinity, so the axial image irradiance is

$$\lim_{\substack{\text{Area for} \\ x'', y'' \to \infty}} [E'_d(0,0)] = \frac{E'}{V} \int_{-\infty}^{\infty} \int_{-\infty}^{\infty} \text{PSF}(-x'', -y'') \, dx'' dy'' = E'.$$

$$(2.94)$$

The result in Eq. (2.94) confirms that the image of an extended source possesses the axial irradiance given by Eq. (2.81). Now, let us take the opposite limit, where the object is very distant such that the area spanning x'' and y'' approaches "close" to zero. In this case, we would have $F(x'', y'') = 1$ over an infinitesimally tiny area that is significantly less than the blur radius of the system's PSF. Under this condition, the value of the PSF would not vary much with x'' and y'', so it may be taken out of the convolution integral in Eq. (2.86), so we have

$$\lim_{\substack{\text{Area for} \\ x'', y'' \to 0}} [E'_d(0,0)] = \frac{E'}{V} \text{PSF}(0,0) \int_{-\infty}^{\infty} \int_{-\infty}^{\infty} F(x'', y'') \, dx'' dy''$$

$$= \frac{E'A'}{V} \text{PSF}(0,0). \qquad (2.95)$$

In Eq. (2.95), A' is the area of the geometric image, so the product $E'A'$ is equal to the total flux in the image. Equivalently, by energy conservation, this product is also equal to the total flux Φ_{EN} passing through the system's entrance pupil. So, we have

$$\lim_{\substack{\text{Area for} \\ x'',y''\to 0}} [E'_d(0,0)] = \frac{\Phi_{EN}}{V} \text{PSF}(0,0). \qquad (2.96)$$

Note in Eq. (2.96) that when the PSF is normalized to its peak [i.e., when $\text{PSF}(0,0) = 1$], the quantity V possesses units of area rather than volume, and in this case, Eq. (2.96) indeed possesses units of flux per unit area. The result in Eq. (2.96) tells us that a significantly distant extended source effectively appears like a point source, relative to the imaging system, and the axial irradiance of this source's image is essentially given by the amount of flux from the point source that passes through the system's entrance pupil. This makes sense. Clearly, the more distant is the point source, the less is the flux that gets into the entrance pupil, and so the less bright is the image. This is consistent with our experience in looking at stars, and it is also consistent with telescopes imaging stars. Widening a telescope's aperture increases the brightness of point-like stellar objects (such as stars), but not for extended objects (such as the Sun or the Moon).

An even more compelling practical example of the application of Eq. (2.86) to explain the variation of image irradiance with source distance (or equivalently, image size) in an imaging system is to derive an expression for the axial image irradiance that explicitly shows this variation. To do this, suppose the PSF for an imaging system is a Gaussian distribution (we have seen in Secs. 2.2.2.2 – 2.2.24 that this is indeed physically realizable) given by

$$\text{PSF}(x',y') = e^{-2\left(\frac{x'^2+y'^2}{r'^2}\right)}. \qquad (2.97)$$

In Eq. (2.97), we shall regard the $1/e^2$ radius r'^2 to represent the blur radius of the PSF. Due to circular symmetry, let us make a change of variables such that $R'^2 = x'^2 + y'^2$, and express Eq. (2.97) in circular coordinates:

$$\text{PSF}(R') = e^{-2\left(\frac{R'^2}{r'^2}\right)}. \qquad (2.98)$$

Substituting Eq. (2.99) into Eq. (2.86) and switching to circular coordinates we have

$$E'_d(R') = \frac{E'}{V} \int_0^{2\pi} \int_0^{\infty} e^{-2\left[\frac{(R' - R'')^2}{r'^2}\right]} F(R'', \phi'') R'' dR'' d\phi''. \quad (2.99)$$

The volume of the Gaussian PSF is determined by noting that

$$V = \int_0^{2\pi} \int_0^{\infty} e^{-2\left(\frac{R'^2}{r'^2}\right)} R' dR' d\phi' = \frac{\pi r'^2}{2}. \quad (2.100)$$

Substituting this result into Eq. (2.99) we have

$$E'_d(R') = E' \frac{2}{\pi r'^2} \int_0^{2\pi} \int_0^{\infty} e^{-2\left[\frac{(R' - R'')^2}{r'^2}\right]} F(R'', \phi'') R'' dR'' d\phi''.$$

$$(2.101)$$

If the source is a flat circular Lambertian source of radius R, then in the absence of distortion, its image would be a circle of radius $m_p R$, where m_p is the paraxial magnification of the imaging system. The axial image irradiance is therefore

$$E'_d(0) = E'_{d,A} = E' \frac{2}{\pi r'^2} \int_0^{2\pi} \int_0^{m_p R} e^{-2\frac{R''^2}{r'^2}} F(R'', \phi'') R'' dR'' d\phi''$$

$$= E' \left[1 - e^{-2\left(\frac{m_p^2 R^2}{r'^2}\right)} \right]. \quad (2.102)$$

In Eq. (2.102), I have used $E'_{d,A}$ to denote that we are computing the axial image irradiance in the presence of diffraction, or aberrations, or both. From this equation, note that a condition for extended source imaging may be regarded as the limit at which the product $m_p R$ would be significantly larger than the PSF blur spot radius r'. In this limit, the ratio $(m_p R)/r'$ in the exponent is large, so the axial image irradiance is $E'_{d,A} \approx E'$. When $m_p R = r'$, the axial image irradiance drops to a factor of $1 - e^{-2}$ of its peak. Finally, as the image size is reduced further, we have the limit that $E'_{d,A} \to 0$ as m_p approaches zero. So, let us try something interesting. Since m_p is the paraxial magnification, it is given by Eq. (2.2). So, let's substitute Eq. (2.2) into Eq. (2.102) for m_p, and normalize Eq. (2.102) by E':

$$\frac{E'_{d,A}(s)}{E'} = 1 - e^{-2\left(\frac{s'^2 R^2}{s^2 r'^2}\right)}. \tag{2.103}$$

Note that E' is invariant if one maintains a fixed numerical aperture (NA) in image space. If one uses an imaging system with variable effective focal length (EFL), such as a fluid lens, then both the image distance s' and image NA may be fixed, leaving the object distance s in Eq. (2.103) to be the only variable. So, let's suppose that $E'_{d,A}$ were the axial image irradiance of the Sun in the human eye, whose EFL is indeed variable. According to one model of the human eye [55], the blur spot diameter for an "average eye" with 4 mm pupil diameter is about 14 microns, therefore, $r' \approx 7 \times 10^{-6}$ m. According to this eye model, the physical distance between the eye lens and retina is about 17 mm. If we ignore the refractive index of the medium between the eye lens and retina, we can have $s' \approx 1.7 \times 10^{-2}$ m. The Sun's diameter is about 1.4×10^6 km [56], so we have $R \approx 7 \times 10^8$ m. Applying all of these numbers into Eq. (2.103), a plot of the variation of axial image irradiance of the Sun in an "average" human eye as a function of distance from the eye to the Sun is shown in Fig. 2.45.

The average distance between the Earth and the Sun is 1 astronomical unit (au), which is approximately 1.5×10^{11} m. In Fig. 2.45, I have plotted Eq. (2.103) starting from this distance. Note that the axial image irradiance remains invariant for a long distance, up to about 1×10^{12} m., where the irradiance begins to drop. At this distance, the image size is $(s'/s)R \approx 12$ microns, which is getting close to the PSF spot blur radius of 7 microns. This also means that the Sun's image brightness remains the same from where the Sun currently sits in space, right down to when it is placed directly in front of our eyes. Thus, for the eye model being considered, from a distance of zero meters up to about 1×10^{12} m., we have the condition of extended object imaging in the eye, and so Eq. (2.81) is sufficient to compute the image irradiance. Beyond this distance, aberrations dominate, and Eq. (2.81) no longer applies. Thus, Eq. (2.81) applies only to the imaging of extended objects, while Eq. (2.86) is more general and may be applied to image formation in the presence of diffraction and aberrations, under space-invariant conditions.

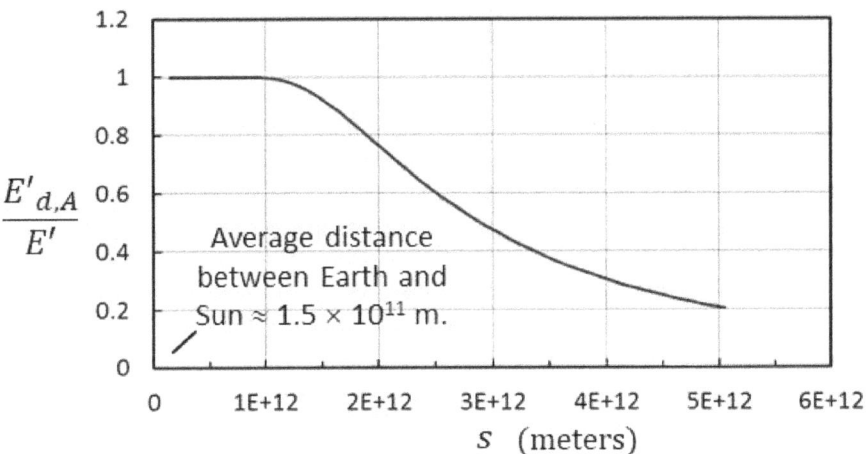

FIGURE 2.45 Plot of Eq. (2.103) for $r' = 7 \times 10^{-6}$ m., $s' = 1.7 \times 10^{-2}$ m., and $R = 7 \times 10^8$ m. Note the constant irradiance for $s < 1 \times 10^{12}$ m.

Pause for insight: → Did you notice something odd? The result in Eq. (2.93) tells us that the axial irradiance for an ideal image is given by Eq. (2.81), which seems reasonable, but according to the results from our exercise above concerning the image irradiance of the Sun in the eye, as well as the results shown in Eqs. (2.94) and (2.102), even aberrated images (i.e., images whose PSFs are not given by Dirac delta functions) possess axial irradiance given by Eq. (2.81).

Evidently, under space-invariant conditions, the axial irradiance for any image seems to be given by Eq. (2.81) as long as its image size is significantly greater than the spatial blur extent of its PSF. However, Eq. (2.81) has traditionally been derived [57, 58] by assuming that an imaging system is aplanatic (i.e., it is free from spherical aberration and coma). Hence, "tradition" has it that only an aberration-free lens would give rise to the axial image irradiance given by Eq. (2.81). Is Eq. (2.81) really consistent with aplanatism? Let us check this. We apply Eq. (2.63) and note that the differential irradiance at the image produced by an element of flux $d\phi$ from an elemental object area dA at the optic axis may be written as

$$dE' = \frac{d\phi}{dA'} = \frac{\pi L dA \sin^2\theta}{dA'}. \qquad (2.104)$$

The quantity dA'/dA in Eq. (2.104) is *related* to the magnification of the differential area elements, but it is not necessarily *the* magnification of the differential area elements, as we will see in Sec. 2.3.4.1. For now, let us recall that, by the Abbe sine condition, the magnification of an image is given by

$$m = \frac{h'}{h} = \frac{\sin\theta}{\sin\theta'}. \qquad (2.105)$$

Therefore, by this condition, it is implied that

$$m^2 = \frac{dA'}{dA} = \frac{\sin^2\theta}{\sin^2\theta'}. \qquad (2.106)$$

Substituting Eq. (2.106) into Eq. (2.104) yields Eq. (2.81). Hence, Eq. (2.81) seems to be consistent with the Abbe sine condition, and therefore, it would also seem to be consistent with aplanatism. Yet, as stated, Eqs. (2.94) and (2.102) show that an imaging system need not be aberration-free (and therefore, need not be aplanatic) to possess the axial image irradiance given by Eq. (2.81). The reason for this lies in conservation of energy and the effect of convolution in imaging. That is, note that the volume of a PSF represents its total flux (i.e., it is the total flux contained in the light that has passed through the entrance pupil from a point source at the object). Therefore, by conservation of energy, the volume of an ideal PSF (i.e., the Dirac delta function) must be equal to the volume of its aberrated form. This equality is assured by the volume normalization procedure performed in Eq. (2.86). For a Dirac delta function PSF, this normalization is a natural consequence of its definition. Now, note that, by performing the convolution operation in Eqs. (2.93) and (2.94) to determine the axial irradiance of an extended image, one is essentially computing the volume of the PSF. In Eq. (2.93), it is the volume of an ideal PSF. In Eq. (2.94), it is the volume of an aberrated PSF. In both cases, by conservation of energy, their volumes are the same. This is the reason why, under space-invariant conditions (a necessary condition for convolution in imaging), the axial irradiance of an extended image containing aberrations (or diffraction) is the same as the axial image irradiance for an ideal image. Even if an imaging system is not completely space-invariant, as long as it is approximately space-invariant over a small area of the image, the convolution approach to computing image irradiance in Eq. (2.86) would be reasonably accurate. The next section illustrates this fascinating characteristic of space-invariant imaging by way of a simple Zemax OS non-sequential illumination analysis of a real lens (i.e., not a PTLM) with aberrations.

2.3.4.1 Zemax OS example: Aberrations and image irradiance

Consider a simple bi-convex lens of Schott N-BK7 material, with 50 mm radii of curvature, 44 mm diameter, 12 mm center thickness, aperture stop diameter of 30 mm (located at the lens's front vertex), imaging a flat circular spatially incoherent Lambertian source of 50 mm diameter located 150 mm from the lens's front vertex, and emitting at a monochromatic wavelength of 587.56 nm, as shown in the Zemax OS print-screen layout in Fig. 2.46.

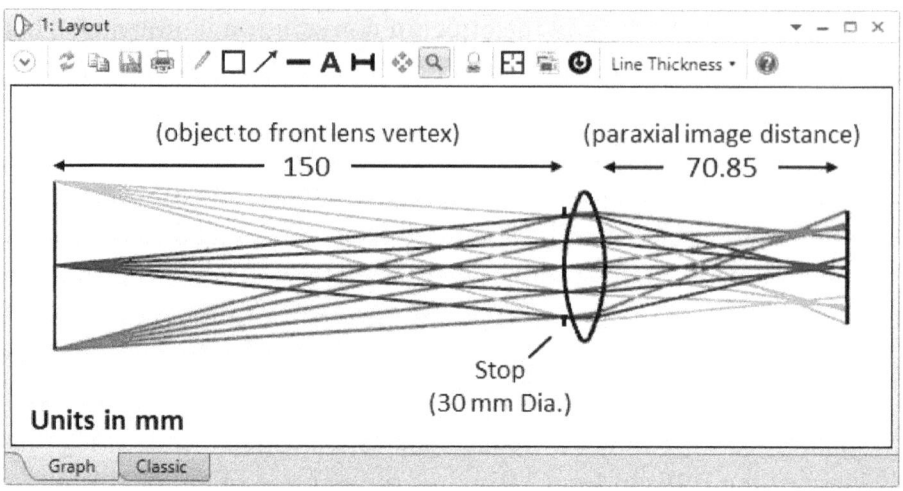

FIGURE 2.46 Zemax OS print-screen layout for a bi-convex lens imaging an extended source.

In Fig. 2.46, 70.85 mm is the distance from the lens's rear vertex to the paraxial image focus, for light at 587.56 nm wavelength (if you are setting up this system in an optical design program, use monochromatic wavelength of 587.56 nm and nothing else). Therefore, we are observing the image at a plane for which the marginal ray height is zero at a pupil zone of zero. At this plane, the image is highly out of focus (but it is *technically* not defocused in the sense that it is located at the paraxial image plane), due to about 202 waves of under-corrected third order spherical aberration, and 36 waves of Coma. Hence, this lens is clearly not aplanatic. The image blur is also quite evident in the ray trace of Fig. 2.46, where the focus of rays both axially and off-axis are "blobs" of light. Hence, the axial and off-axis PSFs have large blur radius, with a root mean square (RMS) axial spot radius of about 1.6 mm, and a full-field off-axis RMS spot radius of about 2.2 mm. A setup of this lens using Zemax OS's non-sequential mode us shown in Fig. 2.47. If we let the total flux of the source be 1-Watt (in Zemax OS non-sequential mode, enter "1" for the power in the source in the non-sequential component editor, and also enter "1" for the cosine exponent to make the source

Lambertian), and trace 20 million rays, one obtains the resulting cross-sectional image irradiance distribution (in Watts/cm²) as shown in Fig. 2.48.

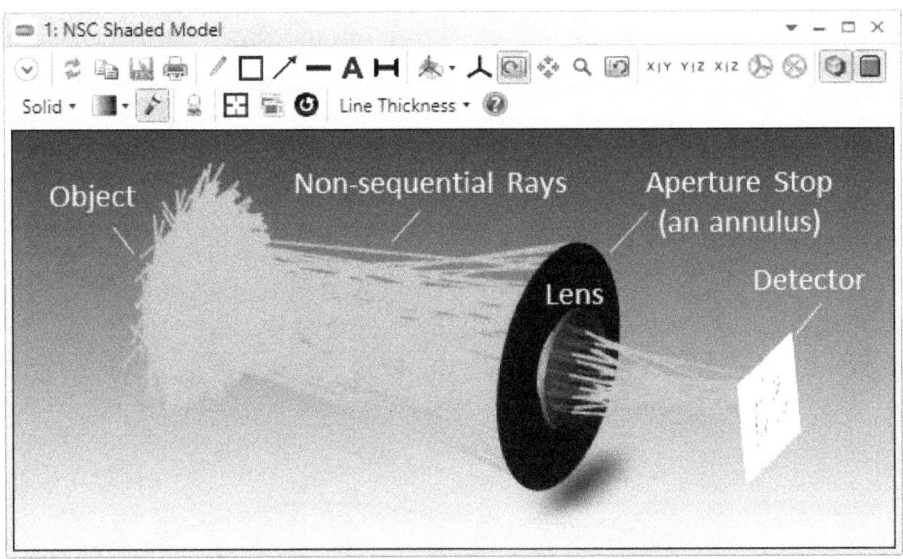

FIGURE 2.47 A 3D layout print-screen from Zemax OS of a non-sequential system model of the lens in Fig. 2.46.

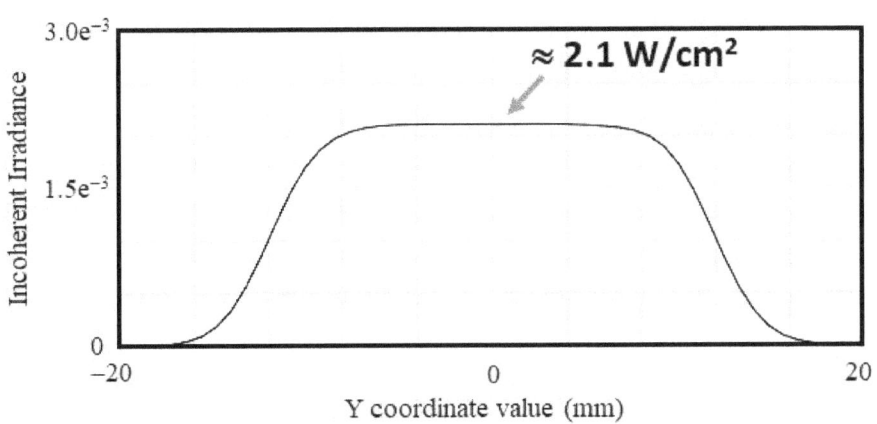

FIGURE 2.48 Image irradiance from the system in Fig. 2.47.

To obtain the plot display shown in Fig. 2.48 in Zemax OS non-sequential mode, I used an incoherent detector object with 40 mm square dimensions, 50 x 50 pixels, and a smoothing value of "5". The image's axial irradiance is observed to be approximately 2.1 Watts/cm^2. Now, suppose we were to compute this irradiance using Eq. (2.81). The object is a Lambertian source with area A, so its total flux is πLA. The object is also circular; hence, its radiance is given by $L = 1\ \text{Watt}/(\pi A) = 1/[\pi^2(2.5\ \text{cm})^2] = 0.01621\ \text{W/cm}^2\text{sr}^{-1}$. Next, in order to apply Eq. (2.81), we need to determine the marginal ray angle θ', which might seem rather ambiguous, due to the presence of significant spherical aberration. Or is it? Let us take a closer look at a trace of sequential rays from an off-axis object point close to the optic axis, in particular at a height of about 5.8 mm (Fig. 2.49).

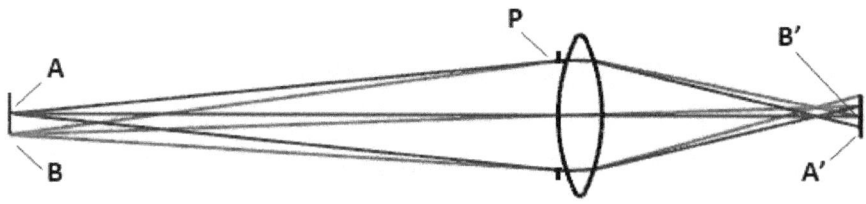

FIGURE 2.49 Sequential rays from an off-axis object point "B" trace through the lens in Fig. 2.46. Point B is 5.8 mm from Point A.

Due to spherical aberration, the axial ray traversing the path APA' in Fig. 2.49 misses the axial image point, so it does not contribute any flux to the axial image irradiance. However, this missing flux is contributed by the ray traversing the path BPB', where B' is the axial image position. Now, since ray AP and ray BP both intersect at P, which is the top of the entrance pupil, both rays emerge from the same differential area element at the exit pupil (assuming that the aberration of the pupils is small). If this area element is da', then ray PB' yields a differential irradiance at B' given by Eq. (2.78). But the angle θ' in Eq. (2.78) is precisely the angle we seek for Eq. (2.81) for computing the image's axial irradiance. In order to determine this angle from the layout in Fig. 2.49, one notes that the ray PB' is at a height of 14.69 mm at a plane at the back vertex of the lens. To obtain this height in Zemax OS's sequential mode, place a dummy surface at the lens's back vertex, followed by the use of the REAY operand in the merit function editor to trace a ray from a field height at -5.8 mm to the top of the dummy surface. Next, note that since the distance between the lens's back vertex and the image plane is 70.85 mm, we have $\theta' = \arctan(14.69/70.85) = 11.7137^0$. Applying Eq. (2.81) with $L = 0.016$ W/cm^{2}sr^{-1} from our earlier calculation, the axial image irradiance is given by $E' = \pi(0.01621)\sin^2(11.7137) = 2.099 \times 10^{-3}$ W/cm$^2 \approx 2.1 \times 10^{-3}$ W/cm^2. This is precisely the value obtained through non-sequential ray tracing (Fig. 2.48)! This result also implies that the lens is operating at approximately space-invariant conditions near the optic axis. Let us check this. Fig. 2.50 shows the image spot diagram for the axial point (left plot, Field 1) and for a -5.8 mm field height (right plot, Field 2). Note that the PSFs are nearly identical. This all proves that an imaging system need not be aplanatic for Eq. (2.81) to be valid. Rather, one only requires that its PSF is space-invariant, and that it is forming the image of an extended source.

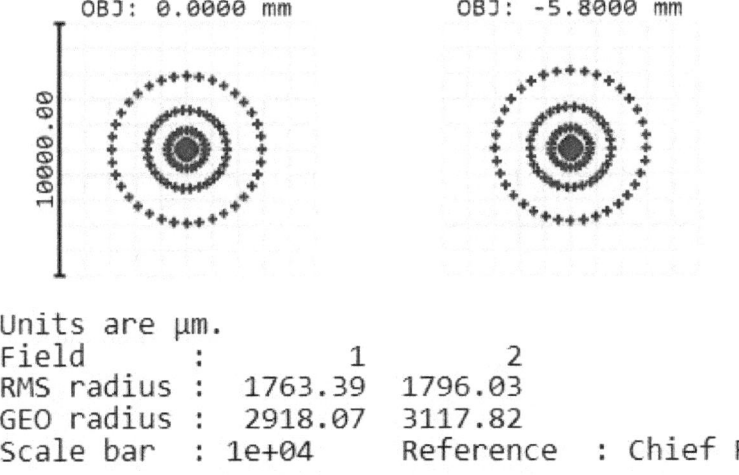

FIGURE 2.50 Image spot diagram for the lens in Fig. 2.49.

There is one final point to be made. If a lens need not be aplanatic for its axial image irradiance to be given by Eq. (2.81), then what exactly is the magnification given by Eqs. (2.105) and (2.106)? After all, both equations may be applied to derive Eq. (2.81). To answer this question, note that the quantity $\pi L \sin^2\theta$ in Eq. (2.104) possesses units of flux per unit area at the object, so it may be considered a sort of "local irradiance" dE at the object. Since, by Eq. (2.106), $m^2 = dA'/dA$, we may express Eq. (2.104) as

$$dE' = \frac{dE}{m^2}. \tag{2.107}$$

Thus, by Eq. (2.107), m^2 is seen to be a sort of "magnification of the irradiance" [59]. It is essentially an expression for how flux density propagates through an imaging system, whose fine details are perhaps related to the concept of "irradiance transport" [60 – 62]. Incidentally, by the same approach, note that for a PTLM, the differential image irradiance given by Eq. (2.84) may be recast as

$$dE'_p = \frac{dE}{m_p{}^2}. \tag{2.108}$$

In contrast to Eq. (2.107), the quantity m_p in Eq. (2.108) is the system's paraxial image magnification, given by Eq. (2.2). In general, the magnifications in Eqs. (2.107) and (2.108) are not the same, except under any of the following conditions:

1. The imaging system operates at low numerical aperture.
2. The imaging system has no aberrations (or is aplanatic).
3. The system is symmetric and at unit magnification.

Generally, practical imaging systems are not always aberration-free and do not all necessarily operate under any or all of the above conditions. However, we know from our earlier analyses that the quantity m in Eq. (2.107) applies even to systems possessing aberrations, as long as the systems are space-invariant, and provided that they are imaging extended objects. Therefore, for the purpose of radiometric calculations, Eq. (2.107) is generally applicable to all imaging systems.

2.3.5 Nonimaging characteristics of imaging lenses

Within the path of rays behind a focusing lens, a position that contains the highest ray density (number of rays per unit area) is a position of peak irradiance (POP). We have talked a lot about focusing and the irradiance of images, but generally, the POP for an imaging lens is not necessarily at the image plane, nor is it necessarily at a system's infinite conjugate focal plane. Let us examine this rather interesting lens property by way of a practical example using Zemax OS.

2.3.5.1 Zemax OS example: Axial nonimaging property of lenses

Let us consider, once again, the lens from Fig. 2.46 (or, equivalently, Fig. 2.47), which is imaging an extended incoherent flat circular Lambertian source at 587.56 nm. In Zemax OS's non-sequential mode, suppose we placed a rectangular detector of dimension 30 mm x 70.85 mm lying flat behind the lens as shown in Fig. 2.51.

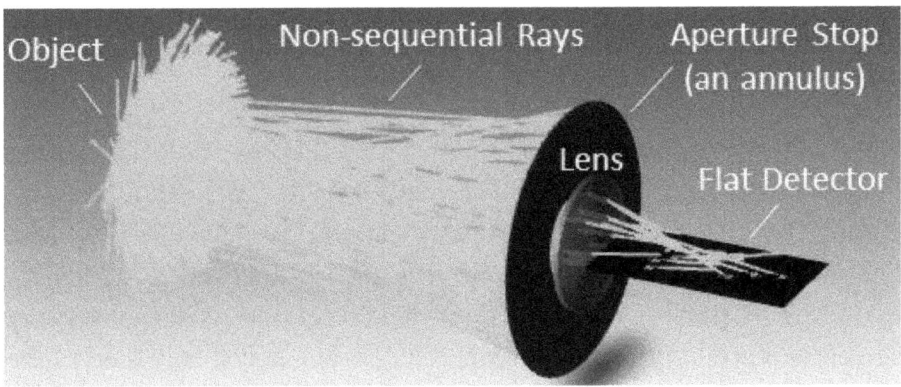

FIGURE 2.51 Flat detector for the lens in Fig. 2.47.

In Fig. 2.51, the length of the flat detector is equal to the distance from the lens's rear vertex to the paraxial image plane. Using a width x length array of 50 x 118 pixels and tracing 20 million rays yields the irradiance profile across the plane of the detector (smoothing value = 5) shown in Fig. 2.52. Note that there is a small region where irradiance seems quite high, and it is not at the image plane. At first glance, it looks as if it is located at the back focal length (BFL) of the lens, except that it really isn't. The BFL of this lens is 46.31 mm (at 587.56 nm), but the distance from the lens's rear vertex to the POP seems considerably lesser. In fact, by examining a plot of the irradiance profile for the central row of pixels on this detector (i.e., the irradiance profile along the optic axis) one obtains the plot shown in Fig. 2.53 (detector smoothing value = 5).

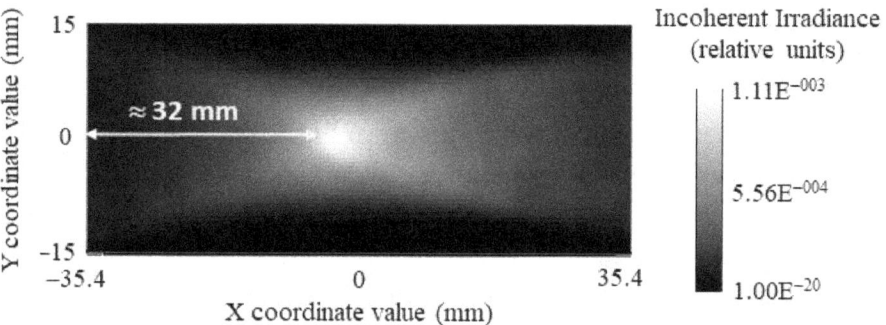

FIGURE 2.52 Irradiance profile across the flat detector of Fig. 2.51. The left side of the plot is where the lens's rear vertex is.

As can be seen from Figs. 2.52 – 253, the POP is not located at the image or focal planes. Let us take a look at a real-life example using a simple bi-convex lens (an ordinary magnifying glass) and a desk lamp with a round white LED bulb [Figs. 2.54 (left) and 2.54 (right)]. In Fig. 2.54 (left) a white sheet of paper serving as the observing screen is placed at the image plane. When the screen is placed at the POP (Fig. 2.54 right) the central distribution is clearly brighter than the image (I had ensured that the exposure level on my hand phone camera is the same in both cases when I performed this experiment).

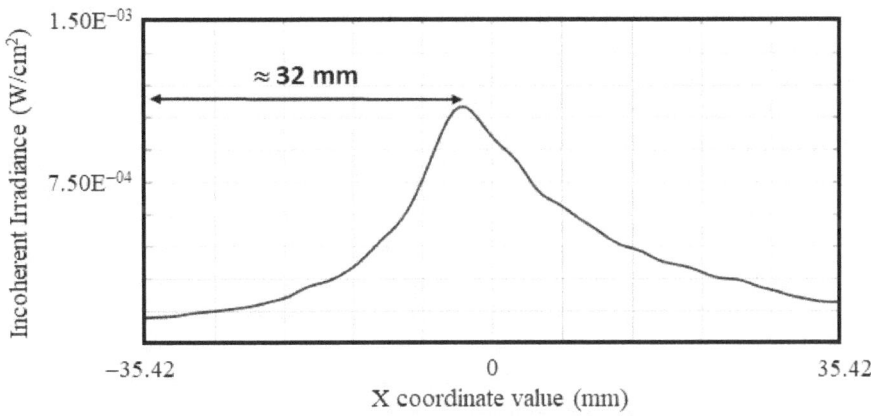

FIGURE 2.53 Irradiance profile for the central row of pixels on the flat detector in Fig. 2.52.

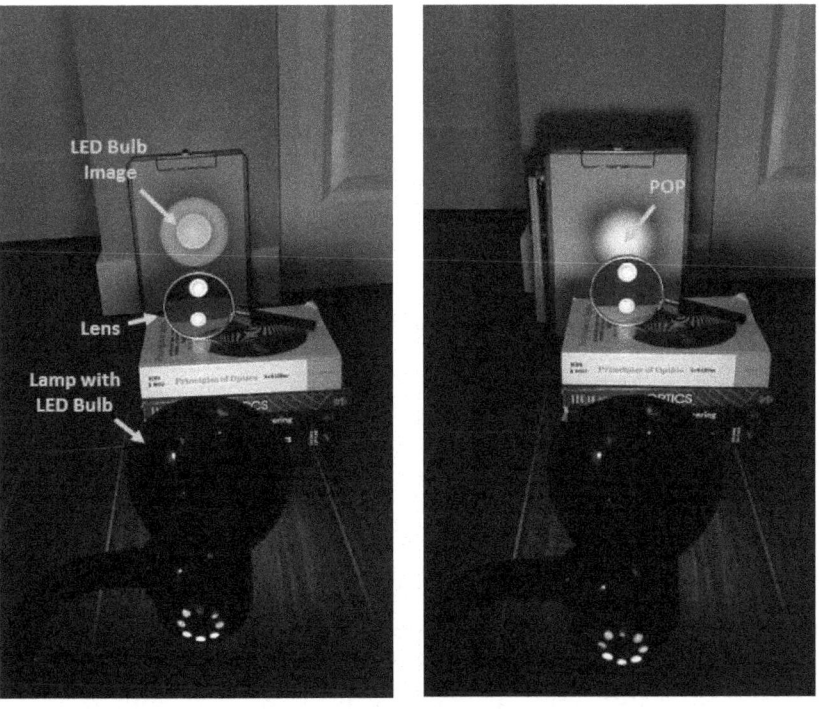

FIGURE 2.54 Simple experiment using a white LED lamp and a bi-convex lens. Screen at image plane (left), screen at the POP (right).

For systems that are imaging incoherent flat symmetric extended Lambertian sources (such as the lens example from Figs. 2.51 – 2.53), there is a simple way to determine the location of the POP: It is approximately located at the intersection between the farthest edge-ray from the source and the optic axis (Fig. 2.55).

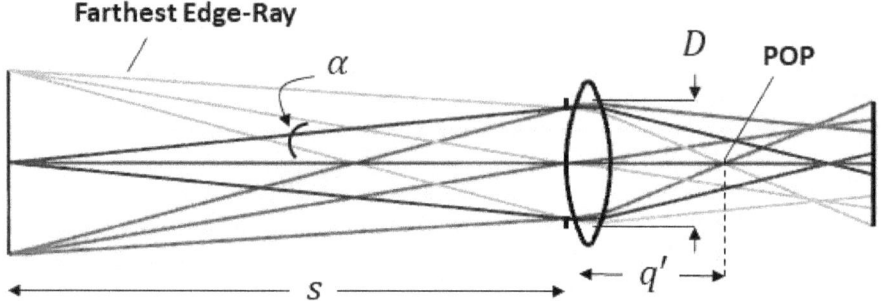

FIGURE 2.55 Determination of the POP for the lens in Fig. 2.51.

In Fig. 2.55, s is the distance between the source and the entrance pupil, D is the diameter of the exit pupil, and q' is the distance from the exit pupil to the POP. For a Paraxial Thin Lens Model (PTLM) of EFL f, an approximate formula [63] for estimating q' is given by

$$\frac{1}{q'} \approx \frac{1}{f} - \frac{1}{s} + \frac{2\tan\alpha}{D}. \tag{2.109}$$

Evidently, the POP is generally not located at a lens's infinite conjugate focal point, nor is it located at the image plane, except under limiting cases. In particular, in the limit that $s \to \infty$ and $\tan\alpha \ll D$, the POP is located at the infinite conjugate focus. And in the limit $\tan\alpha \ll D$ and s is finite, the POP's location is at the image. Let us perform some calculations using Eq. (2.109). For the lens in Fig. 2.51 (whose prescription is given by the same lens used in Fig. 2.46), $f = 50.4368$ mm at 587.56 nm, $s = 150$ mm, $\alpha = 9.46^0$ and $D \approx 30$ mm. Inserting these values into Eq. (2.109) yields $q' \approx 41.2$ mm. In a sequential ray trace, one finds the paraxial exit pupil to be located 79.47 mm from the paraxial image plane, which means that it is 79.47 − 70.85 = 8.62 mm to the left of the lens's rear vertex. Hence, in the absence of aberrations, the POP should be about 41.6 − 8.62 ≈ 32.6 mm to the right of the lens's rear vertex, which is close to the 32 mm distance estimated from the non-sequential plot in Figs. 2.52 and 2.53. In Zemax OS sequential mode, one finds the intersection between the farthest edge-ray with the optic axis for this lens to be at about 30.65 mm from the lens's rear vertex. There are several possible reasons for these differences: First, placing a detector lying flat in the manner shown in Fig. 2.51 actually measures a different irradiance from when it is orthogonal to the optic axis, which is its proper orientation (but we oriented the detector as shown just to perform a quick analysis). Secondly, Eq. (2.109) applies to a PTLM. And finally, in a Zemax OS sequential model for this lens, one notices the presence of significant spherical aberration. All of these effects serve to make our estimates of the POP to be within perhaps about +/- 2 mm accuracy.

It is also useful to have an analytical expression for the variation of axial irradiance with distance z' along the optic axis from a lens's exit pupil to the image. For a PTLM imaging a flat incoherent extended circular Lambertian source, it can be shown [63] that this variation of axial irradiance is given by

$$E'(z') \approx \frac{\pi L s'^2}{\left[m_p z' - (s' - z')\right]^2} \left[1 - \frac{1}{1 + \tan^2\alpha \left(\frac{m_p z'}{s' - z'} - 1\right)^2}\right]$$

(2.110)

(for $0 \leq z' \leq q'$),

$$E'(z') \approx \frac{\pi L s'^2}{\left[m_p z' - (s' - z')\right]^2}\left[1 - \frac{1}{1 + \tan^2\theta \left(\frac{s' - z'}{m_p z'} - 1\right)^2}\right]$$

(2.111)

(for $q' \leq z' \leq s'$).

In Eqs. (2.110) and (2.111), s' is the fixed distance between the exit pupil and paraxial image plane, m_p is the paraxial image magnification, L is the source radiance, θ is the half angle subtended by the entrance pupil from the source, and α is as defined in Fig. 2.55. Incidentally, note that in the limit $z' \to s'$, Eq. (2.111) becomes Eq. (2.84). A plot of Eqs. (2.110) and (2.111) (normalized to πL) is shown in Fig. 2.56, using the same values as before for estimating q' (note that we can have $m_p \approx s'/s$). Note that Figs. 2.56 and 2.53 are indeed similar.

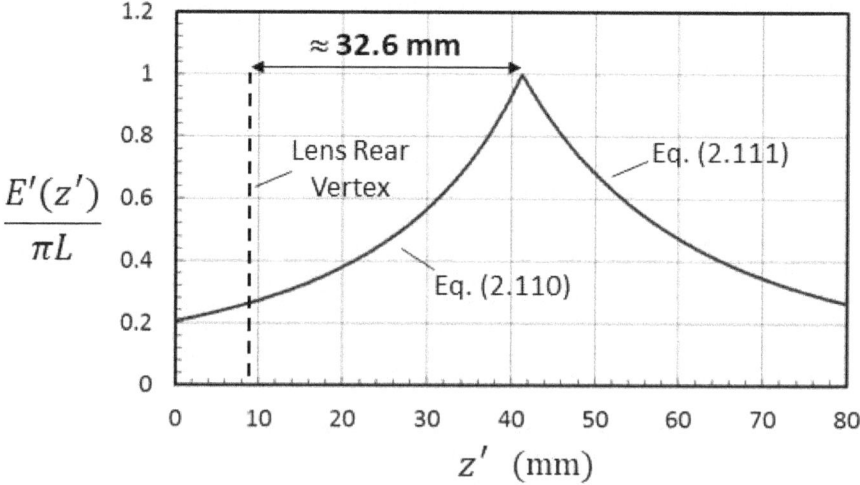

FIGURE 2.56 Normalized plot of Eqs. (2.110) and (2.111).

Note that the ray geometry and condition for observing the POP is not the same as the so-called "searchlight condition" that is discussed in some texts and references [e.g., 64 – 66]. However, a closely related analysis is provided by Stewart and Johnson [67]. Note also that the POP property of an imaging lens is not a defocusing characteristic due to diffraction and aberrations, such as those effects discussed by Vokinger *et al.* [68], Mahajan [69], and Rehn [70]. A POP for any imaging lens is a consequence of geometrical optics for incoherent extended Lambertian sources, and it is present even in the absence of diffraction and aberrations.

Evidently, it seems advantageous to account for the POP in applications involving signal detection using imaging systems. Even if a source were not large, but if it were comprised of an array of smaller but sufficiently extended sources (such as an array of LEDs), and if the array of sources were symmetrically placed about the optic axis at some finite distance from a lens, there will result a POP behind the lens [63], though it would not be as pronounced.

Because a POP is a position of high flux density and not necessarily where any image resides, it may be regarded as a *nonimaging* characteristic of an imaging lens. The subject known as *nonimaging optics* involves the design and analysis of optical systems that provide efficient concentration of optical flux within an area of illumination, generally with no requirement to form images (and in some cases, a preference not to form images) across surfaces [e.g., 71, 72]. A related field is *illumination engineering* [73], whose objective is to efficiently transfer and distribute a prescribed light pattern over a surface (or perhaps even over a volume of space). One generally regards a nonimaging optical problem as a flux concentration problem, while illumination engineering would be somewhat the reverse problem. In both cases, image formation is not a required result. However, that is not to say that nonimaging and illumination optical systems disregard imaging and aberrations completely. Laser beam shaping, for instance, may be regarded as a nonimaging/illumination problem for a coherent source. In this case, lens aberration is of interest. In fact, in beam shaping, spherical aberration enables the generation of a top-hat flux distribution at the plane of observation.

In some other cases, image formation could be a useful result of an illumination system. For instance, perhaps a design task requires a spot illumination distribution resembling third order coma. Or perhaps a design task requires a spread of optical flux across a surface whose distribution resembles spherical aberration. Moreover, all reflective or refractive optical components generally have some image forming property, no matter what we do to their surfaces. Even a grossly deformed refractive surface can form an image, albeit a grossly deformed one. Therefore, an illumination problem using an imaging lens can in some cases be regarded as a task of engineering a lens with high levels of aberrations to form a desired "image pattern" of the source. Further, even if a lens were well-corrected, shifting the plane of observation away from the image plane can produce some desired result. If, for example, the problem is simply to obtain a point of high flux density using an incoherent extended Lambertian source, then the solution is simply to place the plane of observation at the POP of an imaging lens. Generally, because the goal of nonimaging optics and illumination engineering is to efficiently transfer some prescribed distribution of flux over surfaces and volumes, "anything goes". Anything – almost literally. Any optic that can provide a required flux distribution is an optic of choice for solving the problem. Therefore, if an imaging lens yields the required irradiance distribution at any plane of observation (which may not necessarily be the image plane), then this imaging lens is a nonimaging optic.

2.3.6 F-number of a lens with non-circular pupils

In some cases, perhaps the aperture stop in an imaging system is not circular. Perhaps it is a rectangle, or perhaps it is some arbitrary shape. Sometimes, non-circular aperture stops may be required because of an asymmetrical distribution of stray light reaching the aperture. If the stop is not circular, then neither are the entrance and exit pupils. In such cases, if one is required to compute or specify the f-number of the system, then how would one go about doing this? After all, the infinite conjugate f-number [Eq. (2.8)] implicitly assumes that the stop is circular, for it is the ratio of the EFL to the entrance pupil's diameter. What is the "diameter" of a non-circular entrance pupil? While it seems reasonable (and actually, not a bad approximation) to convert a non-circular aperture into a circle of equivalent surface area to estimate the f-number, this approach is generally not consistent with conservation of energy and radiometry. Hence, what is needed is a physically meaningful way to formulate an "effective" f-number for an imaging system possessing non-circular pupils.

The problem of how to arrive at an expression for the effective f-number of an imaging system with non-circular pupils is equivalent to the problem of how to compute the axial image irradiance for such a system. Now, note in Eq. (2.81) that the axial image irradiance is essentially proportional to the square of the image numerical aperture (NA) given by Eq. (2.5). Hence, we can insert Eq. (2.5) into Eq. (2.81) and have

$$E' = \pi L (\text{NA})^2. \qquad (2.112)$$

If we return to Eq. (2.81), we note that it is obtained by the integration initiated in Eq. (2.79). That is, E' in Eq. (2.112) starts from performing the integration

$$E' = \frac{L}{z'^2} \int \cos^4 \theta' da'. \qquad (2.113)$$

In cross-referencing Eq. (2.112) with Eq. (2.113), one notes that the factor $\pi (\text{NA})^2$ in Eq. (2.112) must be equal to everything in Eq. (2.113) except L. Hence, it must be true that

$$(\text{NA})^2 = \frac{1}{\pi z'^2} \int \cos^4 \theta' da'. \qquad (2.114)$$

It follows that if the integration in Eq. (2.114) were performed over the area of an arbitrary shaped exit pupil, the NA must represent an "effective numerical aperture" for that exit pupil. It turns out that this is indeed the case [74]. Thus, the effective NA for a system with arbitrary stop shape may be defined by

$$(\text{NA})^2{}_{\text{eff}} = \frac{1}{\pi z'^2} \int\limits_{(\textit{Arbitrary Area})} \cos^4\theta' da'. \qquad (2.115)$$

The quantity z' in Eq. (2.115) is the distance from the exit pupil to the image plane. In theory, we're done. One needs to perform the integration in Eq. (2.115) over the exit pupil area for a system with some arbitrary shaped aperture stop. Upon substituting Eq. (2.115) into the definition of working f-number (WFN) provided by Eq. (2.6), one obtains the effective f-number of a system with arbitrary stop shape. But in practice, the form given by Eq. (2.115) isn't very practical. Also, there's nothing physically appealing about it. It's just mathematics and rather abstract. We can do better. Take a look back at the process of integration that led to the result in Eq. (2.80). The factor "2π" in that equation originates from integrating around a full circle. Suppose we didn't do that. Instead, suppose we divide the area of some arbitrary shaped exit pupil into pie slices (Fig. 2.57).

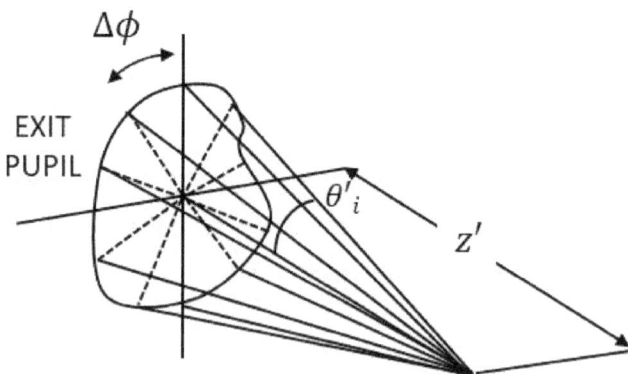

FIGURE 2.57 Pie slice samples in an arbitrary shaped exit pupil.

By dividing an arbitrary shaped exit pupil into pie slices, the integration in Eq. (2.80) may be expressed as a discrete sum given by

$$E' \approx L\Delta\phi \sum_{i=1}^{n} \left[\int_0^{\theta'_i} \sin\theta' d(\sin\theta') \right] \approx \frac{L\Delta\phi}{2} \sum_{i=1}^{n} \sin^2\theta'_i. \quad (2.116)$$

Multiplying the right side of Eq. (2.116) by π/π and noting that $n\Delta\phi = 2\pi$ yields

$$E' \approx \pi L \left(\frac{1}{n} \sum_{i=1}^{n} \sin^2\theta'_i \right). \quad (2.117)$$

Finally, cross-referencing Eq. (2.117) with Eq. (2.112) one notes that

$$(NA)^2_{\text{eff}} \approx \frac{1}{n} \sum_{i=1}^{n} \sin^2\theta'_i. \quad (2.118)$$

Eq. (2.118) states that the effective NA for a system with arbitrary stop shape is approximately given by the root mean square (RMS) average of NAs that sample around the shape of the exit pupil. This result seems far more physically appealing than Eq. (2.115), and it may be applied to numeric computations in any optical design program for a suitable choice of marginal rays. Substituting Eq. (2.118) into Eq. (2.6) yields an effective f-number for a system with non-circular aperture stop.

If both sides of Eq. (2.118) were multiplied by π, one obtains an "effective projected solid angle" Ω_P of the exit pupil in image space, which means that we have

$$\Omega_P = \pi(NA)^2_{\text{eff}}. \quad (2.119)$$

Moore [75] suggested that, since the projected solid angle is computed in the Zemax optical design program, substitution of Eq. (2.119) into

Eq. (2.6) yields a computationally convenient expression for the effective f-number:

$$\mathrm{EFNO} = \sqrt{\frac{\pi}{4\Omega_P}}. \tag{2.120}$$

Eq. (2.120) is the expression used by the "EFNO" operand in Zemax OS.

Pause for insight: $\rightarrow$ By the way, essentially anything that modifies image irradiance results in an effective f-number. You start by examining Eq. (2.112) and ask what possible things could change any of the quantities on the right side of that equation. In the above discussion, the shape of the aperture stop changes the way NA should be computed. Can anything change the source radiance? Absolutely. Radiance is invariant in a *lossless* medium. But optical glass is not lossless. It absorbs light. Consequently, multiplying the right side of Eq. (2.112) by a system's total transmittance yields the so-called "t-number" or "t-stop" [76, 77]. In theory, therefore, since diffraction and aberrations influence image irradiance when the size of the image is comparable to the blur radius of a system's PSF, one can define an effective f-number for a system whose image brightness is limited by such effects. Maybe call it a "d-number" if the system is diffraction-limited, or "a-number" if it is aberration-limited. And so on. So, if you discover something new that could influence image brightness in an optical system, slap that factor onto the right side of Eq. (2.112), and you've got yourself a new effective f-number.

2.3.7 Relative illumination (Part Two): Effect of angles, image distortion and pupil size

In reality, there is no such thing as a "cos^4th law" for the relative illumination in the image of a lens system. The relative illumination of an extended image – and we'll restrict our discussion to extended images, which is the condition most relevant to imaging – is a consequence of a number of factors that influence the irradiance across the plane of the image [e.g., 7, 61, 78 – 84]. One factor is vignetting, which we discussed in Sec. 2.1.4. In the current section, we examine other factors that are intrinsically present in the absence of vignetting. In particular, we will derive two simple and practical expressions for relative illumination that includes all of these factors, in the limit of low numerical aperture.

We shall take an approach similar to what I had done in an earlier study [84]. We start from examining Fig. 2.58.

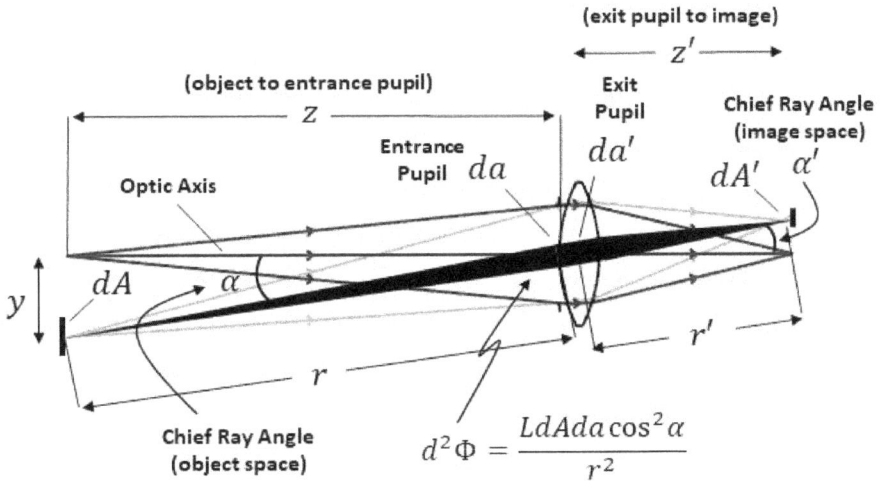

FIGURE 2.58 Flux through entrance and exit pupils.

In Fig. 2.58, the equation shown expresses the flux from an elemental area dA from a height y at the object plane, entering an elemental area da of a lens's entrance pupil. The angle α is between the optic axis and the central ray of this beam that's aimed towards da (i.e., the central ray is the chief ray in object space). Note that $\cos \alpha = z/r$, therefore

$$d^2\Phi = \frac{LdAda\cos^4\alpha}{z^2}. \tag{2.121}$$

By conservation of energy, this flux emerges from an elemental area da' at the exit pupil. As with Eq. (2.121), due to the fact that $\cos \alpha' = z'/r'$, we have

$$d^2\Phi = \frac{LdA'da'\cos^4\alpha'}{z'^2}. \tag{2.122}$$

Note that, by conservation of energy, Eqs. (2.121) and (2.122) are equal to one another. This equality tells us that anything that happens to the flux in Eq. (2.121) is expected to influence the flux in Eq. (2.122). This is interesting. It means that we can analyze the irradiance of an image either in terms of what happens to flux that enters the entrance pupil, or in terms of what happens to flux that emerges from the exit pupil. Let's first consider the entrance pupil. Dividing Eq. (2.121) by dA' yields

$$\frac{d^2\Phi}{dA'} = dE' = \frac{LdAda\cos^4\alpha}{dA'z^2}. \tag{2.123}$$

Eq. (2.123) expresses the differential image irradiance dE' as a function of variables that influence the flux through the entrance pupil. Note that the ratio dA'/dA is the magnification of the irradiance, defined by Eq. (2.107). Hence, we have

$$dE' = \frac{Lda\cos^4\alpha}{m^2z^2}. \tag{2.124}$$

The quantities $da, \alpha,$ and m are functions of field height y. Hence, for an axial point at the image, $y = 0$ and $\alpha = 0$, so the axial differential image irradiance is

$$dE'(0) = \frac{L[da(0)]}{[m(0)]^2z^2}. \tag{2.125}$$

The relative illumination R for the image is a function of field height y, and it is the ratio of the off-axis irradiance to the axial irradiance. Therefore, dividing Eq. (2.124) by Eq. (2.125) we have

$$R(y) = \frac{dE'(y)}{dE'(0)} = \frac{[m(0)]^2}{[m(y)]^2} \frac{[da(y)]}{[da(0)]} \cos^4\alpha. \qquad (2.126)$$

In Eq. (2.126), the ratio $[da(y)]/[da(0)]$ accounts for the possibility that the entrance pupil's area could have a different apparent size for rays entering it from an off-axis field height y relative to rays entering from the object's axial point. In the limit of low numerical aperture, we can approximate the differential area element da of the entrance pupil with its area a. Also, in this limit, the magnification of the irradiance is equal to the paraxial image magnification m_p [59], but the off-axis magnification $m(y)$ is influenced by off-axis aberrations, in particular, image distortion. Hence, leaving $m(y)$ in Eq. (2.126) and taking the low numerical aperture approximation, we may express Eq. (2.126) as

$$R(y) \approx \frac{[m_p]^2}{[m(y)]^2} \frac{[a(y)]}{[a(0)]} \cos^4\alpha. \qquad (2.127)$$

Now, we expect $m(y)$ to be influenced by image distortion, because image distortion affects the size of the area in which flux concentrates, thereby influencing the irradiance. To account for this, we first note that $[m(y)]^2 = dA'(y)/dA(y) = (dx'dy')/(dxdy)$, where $dxdy$ is the off-axis differential area at the object, and $dx'dy'$ is the off-axis differential area element that is influenced by image distortion. Substituting these into Eq. (2.127) we have

$$R(y) \approx \frac{[m_p]^2 dxdy}{dx'dy'} \frac{[a(y)]}{[a(0)]} \cos^4\alpha. \qquad (2.128)$$

It remains to determine what dx' and dy' are in the presence of image distortion. The image distortion D is often defined by the expression

$$D = \frac{y' - y'_p}{y'_p} = \frac{y' - m_p y}{m_p y}. \qquad (2.129)$$

Here, y' and y'_p are the heights of the real and paraxial chief ray at the image, respectively. Solving for y' in Eq. (2.129) yields

$$y' = m_p y [1 + D]. \qquad (2.130)$$

Therefore, the differential dy' is

$$dy' = m_p [1 + D + y(dD/dy)] dy. \qquad (2.131)$$

Now, at any height y' at the image, if one shifts slightly horizontally by an amount x', then it must be true that $y'/x' \approx y/x$. Applying this into Eq. (2.130) and solving for x' we have

$$x' \approx x m_p [1 + D]. \qquad (2.132)$$

Since D is not an explicit function of the orthogonal dimension x, its differential dx' is given by

$$dx' \approx m_p [1 + D] dx. \qquad (2.133)$$

Substituting Eqs. (2.133) and (2.131) into Eq. (2.128) we have

$$R(y) \approx \frac{1}{[1 + D][1 + D + y(dD/dy)]} \frac{[a(y)]}{[a(0)]} \cos^4 \alpha. \qquad (2.134)$$

(Entrance Pupil Formulation)

In Eq. (2.134), D is given by Eq. (2.129). The quantity dD/dy has been called "differential distortion" [84] and it is the instantaneous rate of change of image distortion with field height. As such, dD/dy is not necessarily zero when D is zero!

Eq. (2.134) expresses the effect flux has on relative illumination from the perspective of the entrance pupil. From this perspective, at low numerical aperture, the "cos^4th" effect of the chief ray angle comes from the angle α in object space rather than image space. However, this is not to say that there isn't *any* effect at all from the chief ray angle in image space. Recall that, by conservation of energy, Eqs. (2.121) and (2.122) are equal. In fact, dividing Eq. (2.122) by dA' yields the image irradiance as a function of flux emerging from the exit pupil. Just as we have done for deriving Eq. (2.134), applying Eq. (2.122) to provide an expression for relative illumination (in the low numerical limit) we have

$$R(y) \approx \frac{[a'(y)]}{[a'(0)]} \cos^4 \alpha'. \qquad (2.135)$$

(Exit Pupil Formulation)

Eq. (2.135) expresses the effect flux has on relative illumination from the perspective of the exit pupil. By conservation of energy, Eqs. (2.134) and (2.135) are equal. Any effect happening at the entrance pupil results in something happening at the exit pupil, and vice versa. These effects are consistent with one another, and cannot necessarily be said to be the "cause" of one another. Hence, there is no such thing as a "cos^4th law" for relative illumination. At most, it is a "rule", and a very "soft" one at best. This means that it is possible for a lens system to be telecentric in object or image space (i.e., $\alpha = 0$ or $\alpha' = 0$), and yet the relative illumination may not be uniform, due to effects from image distortion, differential distortion, and variations in the sizes of the entrance and exit pupils with field height. Conversely, it is possible to achieve uniform relative illumination with an appropriate balance of image distortion and pupil aberrations. The utility of Eqs. (2.134) and (2.135) to optical design is in noting what variables may be optimized without over-constraining a lens system. That is, even if an operand exists for controlling relative illumination directly, it would be useful to know if constraints on image distortion and chief ray angles are working against the constraint on relative illumination. We will revisit these ideas in Secs. 3.2.3.3 – 3.2.3.4 where we analyze and optimize a lens system by way of a practical example using Zemax OS.

References

1. G. H. Smith, *Practical Computer Aided Lens Design*, (William Bell, Inc., 1998).
2. J. M. Geary, *Introduction to Lens Design With Practical Zemax® Examples*, (William Bell, Inc., 2002).
3. M. Born and E. Wolf, *Principles of Optics*, 6th ed., (Cambridge University Press, 1980), p. 131.
4. A. Walther, *The Ray and Wave Theory of Lenses*, (Cambridge University Press, 1995), p. 13.
5. E. Hecht, *Optics*, 4th ed., (Addison Wesley, 2002).
6. F. L. Pedrotti, S. J., and L. S. Pedrotti, *Introduction to Optics*, 2nd ed., (Prentice Hall, 1993).
7. M. P. Rimmer, "Relative Illumination Calculations," in *Optical System Design, Analysis, Production for Advanced Technology Systems*, R. E. Fischer and P. J. Rogers (editors), *Proc. SPIE* **0655** (1986).
8. W. T. Welford, *Aberrations of Optical Systems*, (IOP Publishing, 1986), p. 139.
9. M. J. Kidger, *Fundamental Optical Design*, (SPIE Press, 2002), pp. 63 – 137.
10. J. Sasian, *Introduction to Aberrations in Optical Imaging Systems*, (Cambridge University Press, 2012), pp. 67 – 74.
11. R. Kingslake and R. B. Johnson, *Lens Design Fundamentals*, 2nd ed., (SPIE Press, 2010), pp. 101 – 113.
12. S. J. Boege, J. A. Hoshizaki, M. F. Oldham, and L. Ilkova, "Fluorescence Detector With Automatic Changing Filters," United States Patent Application Publication No. **US 2005/0151972 A1** (July 14, 2005).
13. See Ref. 2, pp. 370 – 373.
14. P. P. Clark, "Lens Design and Advanced Function for Mobile Cameras," in *Smart Mini-Cameras*, edited by T. V. Galstian, (CRC Press, 2014), pp. 32 – 33.
15. M. Laikin, *Lens Design*, 4th ed., (CRC Press, 2007), pp. 33 – 34.
16. W. J. Smith, *Modern Lens Design*, 2nd ed., (McGraw-Hill, 2005), pp. 96 – 98.
17. M. Born and E. Wolf, *Principles of Optics*, 6th ed., (Cambridge University Press, 1980), pp. 449 – 453.
18. J. W. Goodman, *Introduction to Fourier Optics*, 2nd ed., (McGraw-Hill, 1996), p. 67.

19. See, for example, Ref. 18, pp. 96 – 114.
20. See, for example, Ref. 18, pp. 145 – 151.
21. D. C. O'Shea, T. J. Suleski, A. D. Kathman, and D. W. Prather, *Diffractive Optics: Design, Fabrication, and Test*, (SPIE Press, 2004), pp. 67 – 70.
22. See Ref. 21, p.197 – 199.
23. See Ref. 18, pp. 151 – 154.
24. N-H. Kim, "How to Design Diffractive Optics Using the Binary 2 Surface," available online at: http://www.zemax.com/os/resources/learn/knowledgebase/how-to-design-diffractive-optics-using-the-binary
25. M. Nicholson, "What is the Normalization Radius?" Available online at: http://www.zemax.com/os/resources/learn/knowledgebase/what-is-the-normalization-radius
26. J. E. Hernandez, "Using Diffractive Surfaces to Model Intraocular Lenses," available online at: http://www.zemax.com/os/resources/learn/knowledgebase/using-diffractive-surfaces-to-model-intraocular-le
27. See Ref. 18, p. 143.
28. I. S. Gradshteyn and I. M. Ryzhik, *Table of Integrals, Series, and Products*, (Academic Press, 2000), p. 333.
29. See Ref. 17, pp. 518 – 523.
30. J. W. Goodman, *Statistical Optics*, (Wiley, 1985).
31. E. Wolf, *Introduction to the Theory of Coherence and Polarization of Light*, (Cambridge University Press, 2007).
32. A. Walther, "Radiometry and Coherence," *J. Opt. Soc. Am.* **58**(9), pp. 1256 – 1259 (1968).
33. A. Walther, "Radiometry and Coherence," *J. Opt. Soc. Am.* **63**(12), pp. 1622 – 1623 (1973).
34. E. W. Marchand and E. Wolf, "Walther's definitions of generalized radiance," *J. Opt. Soc. Am.* **64**(9), pp. 1273 – 1274 (1974).
35. A. Walther, "Reply to Marchand and Wolf," *J. Opt. Soc. Am.* **64**(9), p. 1274 (1974).
36. E. Wolf, "Radiometry with sources of any state of coherence," *J. Opt. Soc. Am.* **64**(9), pp. 1219 – 1226 (1974).
37. W. H. Carter and E. Wolf, "Coherence and radiometry with quasihomogeneous planar sources," *J. Opt. Soc. Am.* **67**(6), pp. 785 – 796 (1977).
38. E. Wolf, "Coherence and radiometry," *J. Opt. Soc. Am.* **68**(1), pp. 6 – 17 (1978).
39. R. Martinez-Herrero and P. M. Mejias, "Radiometric definitions for partially coherent sources," *J. Opt. Soc. Am. A.* **1**(5), pp. 556 – 558 (1984).
40. W. Welford and R. Winston, "Generalized radiance and practical radiometry," *J. Opt. Soc. Am. A.* **4**(3), pp. 545 – 547 (1987).
41. K. Kim and E. Wolf, "Propagation law for Walther's first generalized radiance function and its short-wavelength limit with quasi-homogeneous sources," *J. Opt. Soc. Am. A.* **4**(7), pp. 1233 – 1236 (1987).

42. R. Winston and X. Ning, "Generalized radiance of uniform Lambertian sources," *J. Opt. Soc. Am. A.* **5**(4), pp. 516 – 519 (1988).

43. K. Yoshimori and K. Itoh, "Interferometry and radiometry," *J. Opt. Soc. Am. A.* **14**(12), pp. 3379 – 3387 (1997).

44. K. Yoshimori and K. Itoh, "On the generalized radiance function for a polychromatic field," *J. Opt. Soc. Am. A.* **15**(10), pp. 2786 – 2787 (1998).

45. M. A. Alonso, "Radiometry and wide-angle wave fields. I. Coherent fields in two dimensions," *J. Opt. Soc. Am. A.* **18**(4), pp. 902 – 909 (2001).

46. M. A. Alonso, "Radiometry and wide-angle wave fields. II. Coherent fields in three dimensions," *J. Opt. Soc. Am. A.* **18**(4), pp. 910 – 918 (2001).

47. M. A. Alonso, "Radiometry and wide-angle wave fields III: partial coherence," *J. Opt. Soc. Am. A.* **18**(10), pp. 2502 – 2511 (2001).

48. R. W. Boyd, *Radiometry and the Detection of Optical Radiation*, (Wiley, 1983), pp. 18 – 20.

49. See, for example, Ref. 48, pp. 75 – 81.

50. R. J. Koshel, *Illumination Engineering: Design with Nonimaging Optics*, (Wiley, 2013), p. 68.

51. CIE Publication No. 127, *Measurement of LEDs*, Central Bureau of the CIE, Vienna, Austria (2007).

52. C. Say and R. Young, "The New CIE 127 Standard for LED Measurement," *Optics and Photonics News*, October (2008), pp. 12 – 13.

53. See Ref. 48, p. 22.

54. R. Siew, "Corrections to classical radiometry and the brightness of stars," *Eur. J. Phys.* **29**, pp. 1105 – 1114 (2008).

55. B. H. Walker, *Optical Design for Visual Systems*, (SPIE Press, 2000), pp. 2 – 7.

56. M. Zeilik, *Astronomy: The Evolving Universe*, 9th ed., (Cambridge University Press, 2002), p. 254.

57. See, for example, Ref. 48, pp. 84 – 85.

58. See, for example, Ref. 3, pp. 188 – 189.

59. R. Siew, "Distinction between image magnification and irradiance magnification: a commentary," *Opt. Eng.* **56**(2), 029701 (2017), DOI:10.1117/1.OE.56.2.029701.

60. See, for example, Ref. 10, pp. 173 – 186.

61. D. Reshidko and J. Sasian, "Role of aberrations in the relative illumination of a lens system," *Opt. Eng.* **55**(11), 115105 (2016).

62. M. R. Teague, "Image formation in terms of the transport equation," *J. Opt. Soc. Am. A.* **2**(11), pp. 2019 – 2026 (1985).

63. R. Siew, "Axial nonimaging characteristics of imaging lenses: discussion," *J. Opt. Soc. Am. A.* **33**(5), pp. 970 – 977 (2016).

64. See, for example, Ref. 48, pp. 86 – 89.

65. M. V. Kline, *Optics*, (Wiley, 1986), pp. 219 – 221.

66. E. O. Hulburt, "Optics of searchlight illumination," *J. Opt. Soc. Am.* **36**(8), pp. 483 – 491 (1946).

67. S. M. Stewart and R. B. Johnson, *Blackbody Radiation: A History of Thermal Radiation Computational Aids and Numerical Methods*, (CRC Press, 2017), p. 150.

68. U. Vokinger, R. Dandliker, P. Blattner, and H. P. Herzig, "Unconventional treatment of focal shift," *Optics Communications* **157**, pp. 218 – 224 (1998).

69. V. N. Mahajan, "Axial irradiance of a focused beam," *J. Opt. Soc. Am. A.* **22**(9), pp. 1814 – 1823 (2005).

70. H. Rehn, "Elliptical reflector: efficiency gain by defocusing," in Nonimaging Optics and Efficient Illumination Systems IV, R. Winston and R. J. Koshel (editors), *Proc. of SPIE* **6670** (2007).

71. R. Winston, J. C. Minano, and P. Benitez (with contributions from N. Shatz and J. C. Bortz), *Nonimaging Optics*, (Elsevier, 2005).

72. J. Chavez, *Introduction to Nonimaging Optics*, 2nd ed., (CRC Press, 2016).

73. See, for example, Ref. 50.

74. R. Siew, "F-number and the radiometry of image forming optical systems with non-circular aperture stops," in Optical Modeling and Performance Predictions II, Mark A. Kahan (editor), *Proc. of SPIE* **5867** (2005).

75. K. E. Moore, Founder of Zemax, (personal communication during a discussion over the study that had led to Ref. 74 in July, 2004).

76. R. Kingslake, "The Effective Aperture of a Photographic Objective," *J. Opt. Soc. Am.* **35**(8), pp. 518 – 520 (1945).

77. W. J. Smith, *Modern Optical Engineering*, 2nd ed., (McGraw-Hill, 1990)), p. 144.

78. P. Foote, "Illumination from a radiating disk," in *Bulletin of the Bureau of Standards* **12**, pp. 583 – 586 (1915).

79. M. Reiss, "The Cos^4 Law of Illumination," *J. Opt. Soc. Am.* **35**(4), pp. 283 – 288 (1945).

80. G. Slussareff, "A Reply to Max Reiss," *J. Opt. Soc. Am.* **36**(12), p. 707 (1946).

81. I. C. Gardner, "Validity of the Cosine-Fourth-Power Law of Illumination," *J. Res. Nat. Bur. Stand.* **39**, pp. 213 – 219 (1947).

82. M. Reiss, "Notes on the Cos^4 Law of Illumination," *J. Opt. Soc. Am.* **38**(11), pp. 980 – 986 (1948).

83. F. Wachendorf, "The Condition of Equal Irradiance and the Distribution of Light in Images Formed by Optical Systems without Artificial Vignetting," *J. Opt. Soc. Am.* **43**(12), pp. 1205 – 1208 (1953).

84. R. Siew, "Relative illumination and image distortion," *Opt. Eng.* **56**(4), 049701 (2017).

3. Optics and Imaging in Modern Applications

In this section, we get acquainted with a number of selected modern applications and their operational principles. In those subsections where specific Zemax OS examples are discussed, lens prescriptions are provided so that readers who do not use Zemax OS may try the examples using any program of their choice. Unlike the organization of topics in Sec. 2, Sec. 3 does not necessarily require that the reader proceed sequentially throughout each subsection. However, *some* logical order is present. For instance, Sec. 3.1 first introduces a central theme concerning what, in my view, is "modern optical engineering", and the connection between this theme and the rest of the subsections. Next, Sec. 3.2 discusses the characteristics of machine vision optics, which in my view serves as a natural prerequisite for understanding modern imaging. As such, Sec. 3.2 is rather lengthy and applies much of the principles from Sec. 2 to practical imaging problems. Sections beyond machine vision touch on systems of somewhat increasing complexity, such as the impact of aberrations on the coherent and diffractive nature of imaging using the Grating Light Valve™ spatial light modulator developed by the company Silicon Light Machines.

3.1 Modern optical engineering: Pushing the limits of the three backbones

If the theoretical limit to some physical process can be pushed by breaking certain "laws", then chance would have it that those laws probably aren't "hard" laws. A hard law is unshakeable, such as the conservation of energy. But "soft" laws can be broken. The Rayleigh limit is an example of a soft law, because it is rather arbitrary how one wishes to define the limit or boundary between resolvable and unresolvable imagery. The Rayleigh limit also makes the assumption that lenses aren't apodized, and we have seen in Secs. 2.2.2.2 – 2.2.2.3 that apodization can increase the MTF of a lens system within certain spatial frequencies, thereby pushing the limits of imaging at those spatial frequencies. Another example of a soft law is the so-called "cos^4th law" of relative illumination, which, as we have seen in Sec. 2.3.7, is more of a "rule" that can easily be broken by the presence of image distortion and pupil aberrations (we will see a practical Zemax OS example of this in Sec. 3.2.3.4).

Modern imaging systems push the limits of the "soft" laws of optical science. Computer-aided optimization of multi-element lenses enable apochromatic lens designs with extended field of view, which pushes limits on field of view often imposed by secondary axial chromatic aberration (Sec. 3.2.4.4). Curved image sensors push "limits" imposed by astigmatism and Petzval field curvature (Sec. 3.2.5). Liquid lenses provide tunable focus without mechanical focus adjustment (Sec. 3.2.6). Miniaturization reduces aberrations (Secs. 3.2.2.6 and 3.2.8). Free-form lenses push limits on correcting distortion in mobile phone imaging (Sec. 3.3.3). Wave-front coding and plenoptic cameras push depth of field limits (Sec. 3.2.7.6). And so on. In this way, the methods of modern optical engineering altogether seem to be a part of some common program: To push the limits of the soft laws of optics and imaging. Therefore, while geometrical optics is the backbone of "classical" optical engineering, and while physical optics is the backbone of geometrical optics, and while radiometry is the backbone of illumination and nonimaging optics, the program of "modern optical engineering" is to push the limits of these three backbones. In this spirit, let us dive into the amazing world of optics and imaging in modern applications.

3.2 Machine vision imaging principles and practices

3.2.1 The world of machine vision optics

At a basic level, any system that comprises a lens and an electronic image sensor is a machine vision system. Therefore, everyone carries a machine vision system in their pockets and hands these days. The mobile phone camera is perhaps the most common machine vision device today, other than the camera on a laptop or tablet pc. Mobile phone cameras fall into a class of systems that are gaining significant interest in industry – the "smart mini camera" [1]. Broadly, it is an *embedded vision* system – a device that integrates system on chips (SoCs) and system on modules (SoMs) with optical imaging components. The embedded electronics are what makes devices "smart", thereby extending the range of capabilities and possibilities for modern practical applications. Hence, together with embedded electronics, an embedded vision system is today's modern imaging device. In the discussions that follow, we shall examine the *optical* characteristics of such machine vision systems rather than the computational and electronic aspects. That is, we shall focus on the "vision" rather than the "machine".

3.2.2 Optical characteristics of "stock" lenses

Perhaps the simplest approach to develop a prototype imaging system for machine vision is to pick commercially available "off-the-shelf" or "stock" lenses. However, it is often necessary to speak with a technical representative from the lens supplier to clarify any possible ambiguities in lens parameter specifications. Let us examine some of these lens parameters and their optical characteristics.

3.2.2.1 What exactly is a lens's "field of view"?

Unless a stock lens's full design prescription (or, if one is using Zemax OS, the lens's black box file) is made available, the definition of its field of view (FOV) can be somewhat ambiguous. Let us take, for example, the lens from the Zemax OS exercise discussed in Sec. 2.2.2.3. If one possesses no knowledge of this lens's prescription, then it is quite possible for one to assume that the half angle FOV is defined by either of the two angles shown in Fig. 3.1, where α_1 is the angle from the front mechanical vertex, and α_2 is the angle from the front lens vertex. Therein lies an ambiguity. Additionally, from a lens designer's perspective, it is actually the ***chief ray angle in object space*** that defines the half angle α for the half angular FOV (Fig. 3.2).

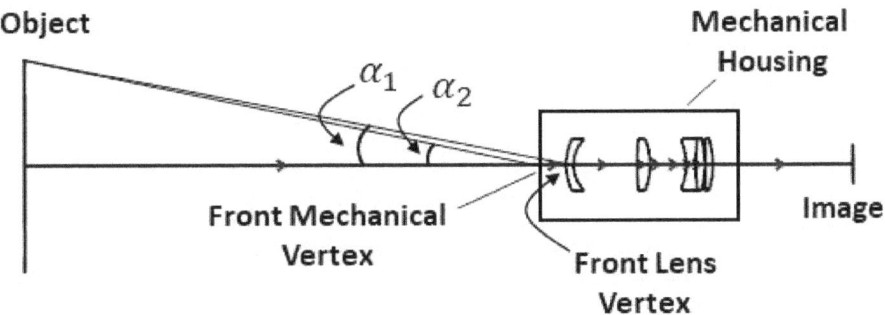

FIGURE 3.1 Ambiguous half FOV angles for a "stock lens".

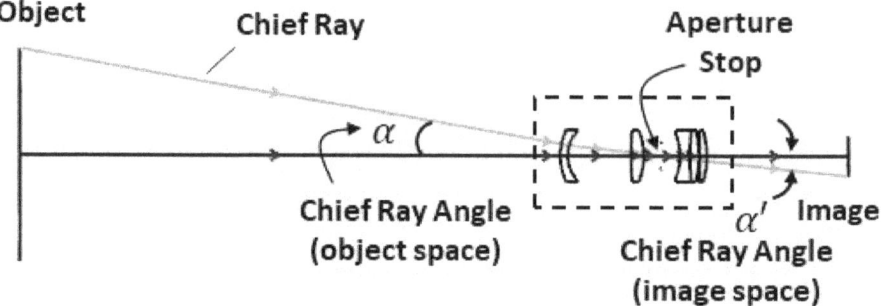

FIGURE 3.2 FOV based on the chief ray angle in object space.

During the process of lens design, in the absence of vignetting, the chief ray is aimed at the center of the entrance pupil, ensuring this ray's passage through the center of every pupil in the system. As such, the chief ray emerges from the center of the exit pupil and strikes precisely at the height of the image, which of course corresponds to the edge of the object, and hence, defining the FOV. Yet, perhaps this may not necessarily define the *edge of visibility* of the object. In the absence of vignetting, the relative illumination of the image can be rather smooth. Therefore, if an image sensor's size allows it, extending the chief ray beyond its design angle can yield visibility beyond the lens's intended design field height (Figs. 3.3 and 3.4).

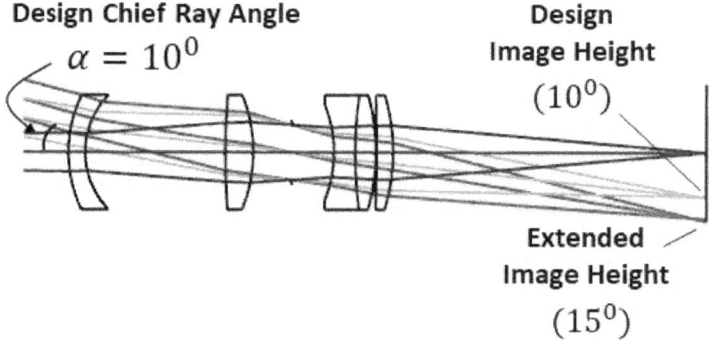

FIGURE 3.3 Design vs. extended image height for the lens in Fig. 3.2.

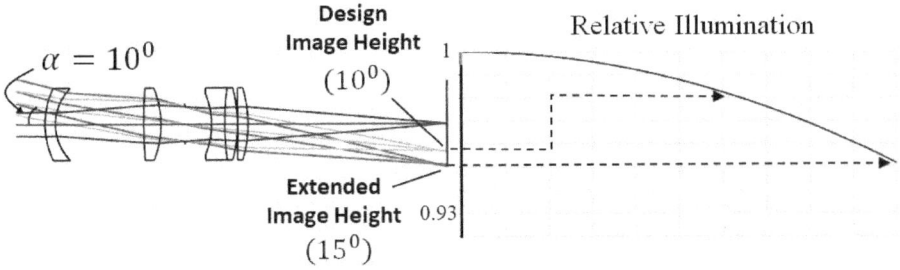

FIGURE 3.4 FOVs of the lens in Fig. 3.3 based on relative illumination.

To obtain the result shown in Figs. 3.3 and 3.4, use the prescription provided in Table 2.1 from Sec. 2.2.2.3, but set the aperture stop semi-diameter at 0.12 mm. Set wavelength = 587.56 nm. Set the thickness from the object to surface 1 at 5 mm. Set $\alpha = -10^0$ for the design chief ray angle, and set -15^0 for the extended angle. Set the thickness on surface 11 as a Solve for the zero paraxial marginal ray height, which should yield a distance of about 1.35141 mm. Note that the image is clearly visible beyond the design chief ray angle. However, the image quality beyond the design chief ray angle may perhaps be less than desirable as implied by the MTF curves in Fig. 3.5.

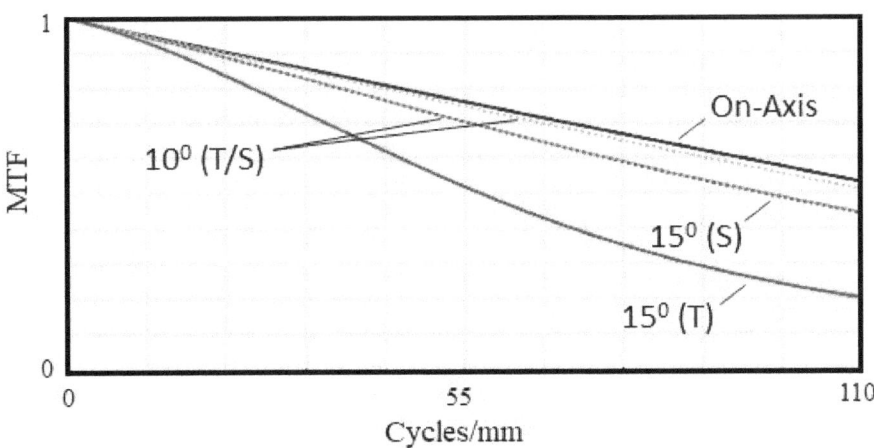

FIGURE 3.5 MTF of the lens from Fig. 3.4 at extended FOV angle.

Now, of course, a lens supplier might specify an image sensor format appropriate for this lens, which would certainly imply that the specified angular FOV corresponds to the maximum spatial extent of the sensor format (usually, the sensor's diagonal dimension). However, there would still be ambiguity in the field angle definition for the specified image sensor format. Is it the object space chief ray angle? Or is it any of the two angles illustrated in Fig. 3.1? The only way to be sure is to ask the lens supplier.

For a stock lens specified for imaging at finite conjugates, sometimes, a lens supplier provides the angular FOV, the sensor format, and the image magnification. In this case, dividing the sensor's half diagonal length by the magnification yields the implied object height. If the *working distance* (usually, this is the lens's specified distance between the object and the front mechanical flange of the lens housing) is provided, then one may check if the half angle FOV provided is the chief ray angle or the angles shown in Fig. 3.1. To do this, dividing the object height by the working distance yields the tangent of the angle that's referenced to the front mechanical flange [e.g., α_1 in Fig. 3.1], hence, its arc tangent yields an angle that one may compare against the specified half angle FOV. On the other hand, dividing the object height by the tangent of the specified half angle FOV would yield the distance between the object and the entrance pupil, and if this distance seems "reasonable" (relative to dimensions of the lens housing), then maybe the half angle FOV is the chief ray angle. Then again, maybe not. This assumes that vignetting is not present, which is a highly unlikely condition for a well corrected large FOV lens. Moreover, we have assumed that the working distance is the distance referenced to the front mechanical flange. In some cases, perhaps it is the distance from the object to the front lens vertex. Generally, there is no standard being followed by everyone, so I usually speak with a technical representative (such as an Applications Engineer) to get my answers.

Now, there is one optical characteristic of any lens that sort of naturally "generates" a FOV that actually looks like a real FOV in the sense that there is the appearance of a boundary between where the image lies and beyond where there is darkness. This characteristic is a consequence of vignetting, which cuts off visibility beyond a certain height at the image. For example, one will see that extending the half field angle for the lens in Fig. 3.4 beyond about 15 degrees results in vignetting. In particular, at about 22 degrees, the relative illumination falls to about 46% (Fig. 3.6).

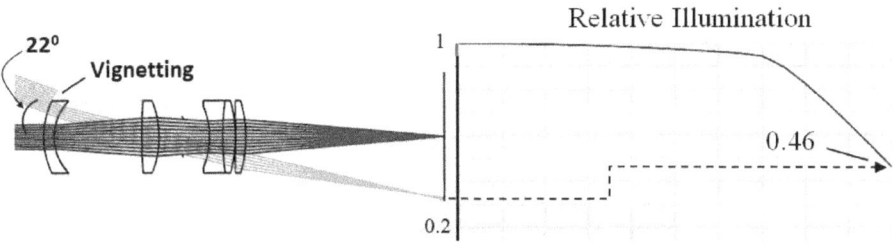

FIGURE 3.6 FOV based on vignetting at about 46% relative illumination.

What would an image look like in the presence of vignetting with the relative illumination illustrated in Fig. 3.6? Taking again the lens example we have been discussing, if the object were a grid of 6 horizontal lines by 6 vertical lines across a square of dimensions 5 mm x 5 mm. its simulated image under the vignetting conditions illustrated in Fig 3.6 would be as shown in Fig 3.7.

The image in Fig. 3.7 appears as if it were enclosed by a dark circle. This is known as the so-called ***image circle*** of a lens, and it is sometimes provided as a specification by lens suppliers. The appearance of the image circle would seem to provide a clear indication of the limits to the FOV of a lens because there is now a rather physical circular border enclosing the image. However, there is still some ambiguity for the image FOV. This is because one would not really know at what level of relative illumination an image circle's diameter refers to, unless the relative illumination is specified at full field. The designer of this lens can make this definition, but unless this definition is provided as a specification, the end user cannot really tell for sure what the relative illumination of this lens is at the stated image circle diameter. It is quite literally a gray area, as one can see from the relative illumination in Fig. 3.6. The reason why the image circle is not sharp is because vignetting occurs at lens surfaces rather than at the image plane. The only way to create a sharp circle boundary at the image plane is to have a physical aperture at or near the image, but this is unreasonable, because one would not be able to focus the lens. Sometimes, vignetting is introduced at the rear most element of the lens system, and in this case, if the distance between this element and image plane is close, then the image circle would appear somewhat sharper (but this is not the reason why vignetting is introduced at the rear lens element by the lens designer).

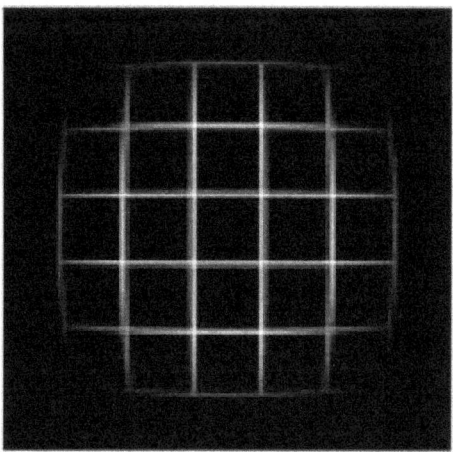

FIGURE 3.7 Simulated image of a square grid for the lens in Fig. 3.6.

In many high-performance lenses with large FOV, vignetting is introduced deliberately in order to remove off-axis rays that are afflicted with high aberrations. This is the primary reason for the use of vignetting. For the present lens example, if we aren't happy with the MTF performance at 15 degrees field angle (Fig. 3.5), then we can vignette a significant part of the light from this angle using the last lens element (Fig. 3.8). For instance, setting the last element's semi-diameter to 0.12 mm results in the MTF shown in Fig. 3.9 (compare this MTF plot with Fig. 3.5).

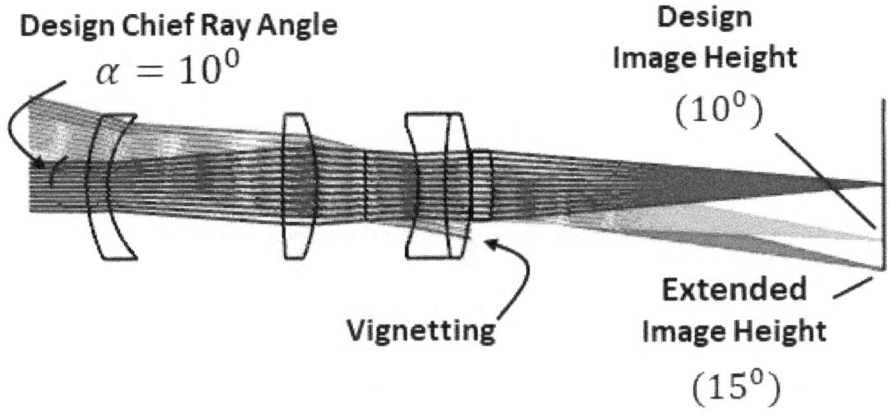

FIGURE 3.8 Vignetting of the extended field for the lens in Fig. 3.4.

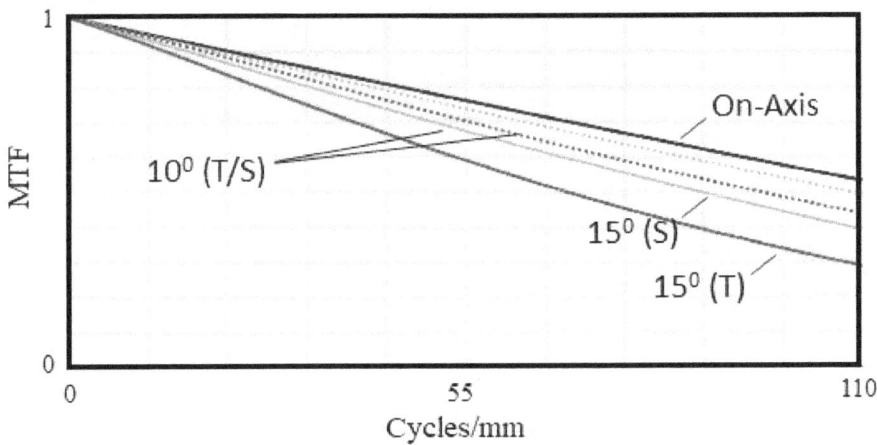

FIGURE 3.9 MTF for the lens in Fig. 3.8.

As a consequence of vignetting introduced by the last element, the image circle for the same object pattern from Fig. 3.7 is smaller (Fig. 3.10a). This image circle now defines the lens's usable or optimum FOV. But the ambiguity in what level of relative illumination this image circle is defined at is still present, so a good piece of information to have is a plot of the image's relative illumination (Fig. 3.10b). Although a lens's relative illumination is not a standard published specification for stock lenses, it is sometimes possible to request it from the lens supplier. In some cases, perhaps some non-disclosure agreement (NDA) might be required between parties involved in obtaining this plot.

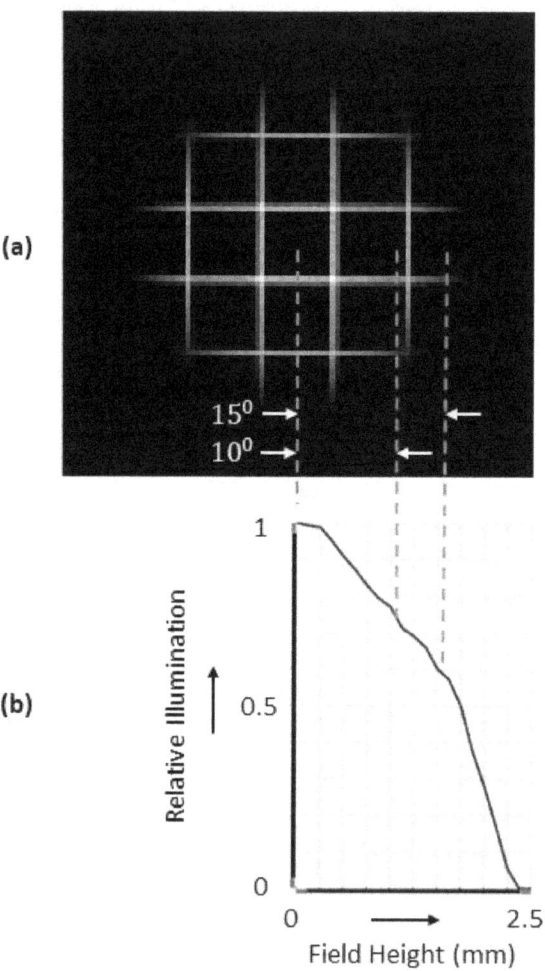

FIGURE 3.10 (a) Image simulation for the lens in Fig. 3.8. (b) Relative illumination.

3.2.2.2 What exactly is an "image circle"?

Let us re-examine Fig. 3.10, which shows the image circle and relative illumination for the lens in Fig. 3.8. Perhaps we may define the lens's image circle by the diameter corresponding to about 60% relative illumination, which is approximately at the extended field angle of 15 degrees for this lens. Now that vignetting has improved the MTF for this lens at 15 degree field (Fig. 3.9), we might as well regard this extended field as the design chief ray angle (i.e., let $\alpha = 15^0$) for the lens. So, let us say that we now have a stock lens whose half angle FOV is 15 degrees. During the process of stock lens selection, the real value in knowing the stock lens' half angle FOV is in being able to analytically determine the "view-ability" of the object by the lens over a range of working distances. In many practical situations, it may be desired to use a stock lens at more than a single conjugate, even if the lens may have been specified for a fixed working distance. At any working distance, the view-ability of the image lies within the image circle, which is defined by the lens's angular FOV. For example, Fig. 3.11 depicts objects at two working distances WD_1 and WD_2. The object heights h_1 and h_2 are both viewable, because they lie within the half angle FOV of 15 degrees. Hence, their image heights h'_1 and h'_2 (which define half image circles at the sensor) are also viewable.

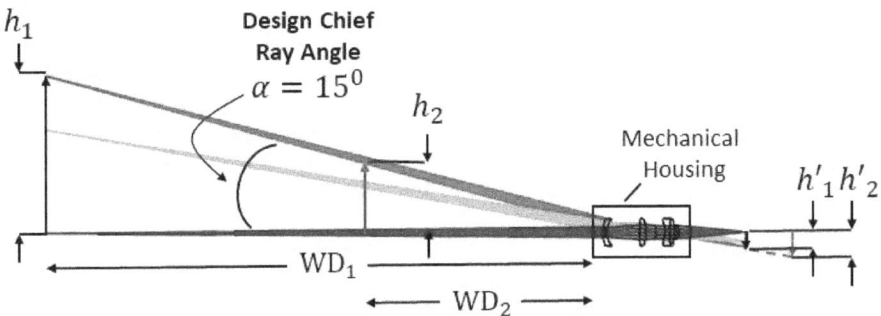

FIGURE 3.11 Object "view-ability" at different working distances.

From observing the variation of image heights in Fig. 3.11, it is evident that the image circle of a lens does not have a fixed diameter. Its size varies according to the working distance. That is, a lens's image circle actually 'breathes". It breathes in (gets *larger*) when the object is closer to the lens, and it breathes out (gets *smaller*) when the object is farther away, as illustrated by the image simulations for an object at the three working distances shown in Fig. 3.12. In this figure, the right plot shows simulated images for the grid pattern described in the previous section. The image sensor is assumed to be a square of fixed dimensions 1.2 mm x 1.2 mm. At each working distance, the grid's size is scaled proportionately according to the 15 degree chief ray angle, but they are each extended slightly larger than the viewable object height in order to reveal the image circles.

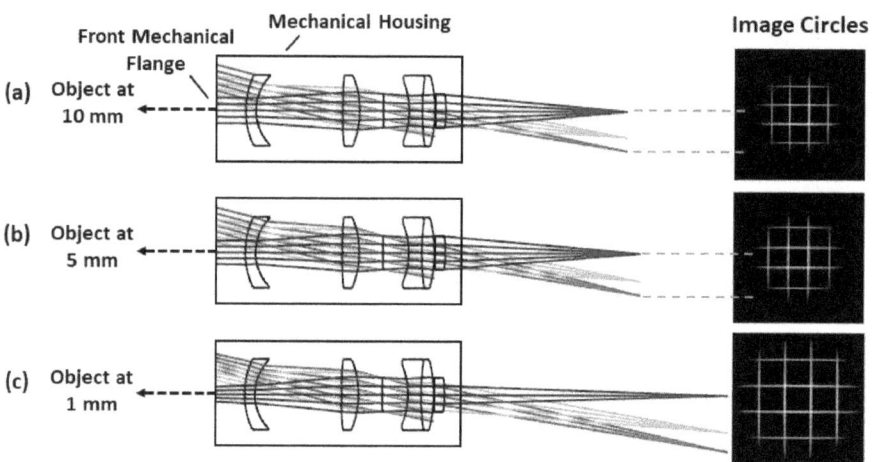

FIGURE 3.12 Image circles at three working distances. (a) 10 mm. (b) 5 mm. (c) 1 mm.

As a consequence of the breathing of image circles as a function of working distance, for a fixed image sensor size, it may not always be possible to have view-ability of the object at all desired working distances. For example, Fig. 3.13 shows a simulation of the same grid pattern object at the three working distances from Fig. 3.12. This time, at each working distance, the object height has been adjusted to fit approximately within the viewable half angle of 15 degrees (there is a small effect from image distortion). At each working distance, the sensor size is fixed at dimensions 0.8 mm x 0.8 mm. But if the sensor had instead been fixed to accommodate for the object at 10 mm working distance, then it surely would not see the entire object at 1 mm working distance even if the object had been of a height that's within the 15 degree half FOV.

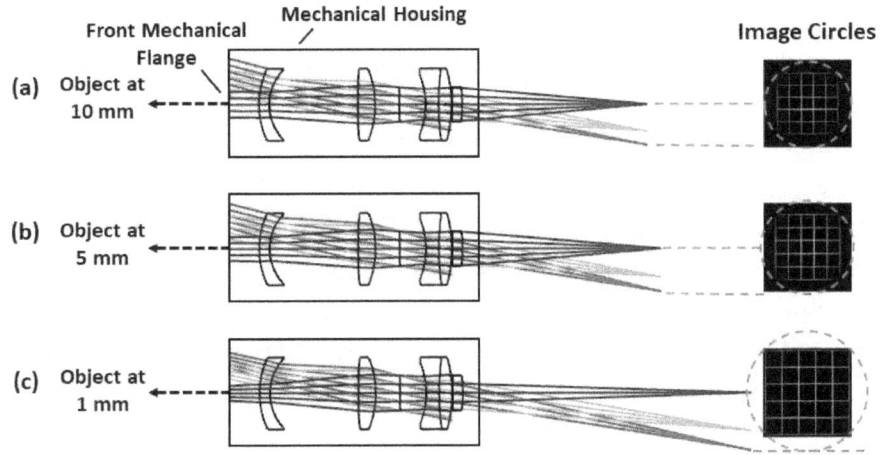

FIGURE 3.13 Image simulations for adjusted viewable object heights at three working distances. (a) 10 mm. (b) 5 mm. (c) 1 mm.

Clearly, if one wishes to use a stock lens at a variety of working distances, one requires image sensors of a variety of sizes. Fortunately, image circles don't breathe much when the working distances are large compared to the lens EFL. Still, suppose one wishes to use the stock lens over a wide working distance range, right down to distances on the order of the lens EFL. In order to determine the appropriate sensor size for any given working distance, one must know how the size of the image circle varies as a function of working distance. If the lens specifications provided by a supplier only consist of the EFL, angular FOV, image circle diameter at a specified working distance (or for an object at infinity), distortion, and sensor format, then they are actually not sufficient for determining the variation of image circle size with working distance. Let's discuss what's missing. We'll do this first for the case of infinite conjugate lenses, and then for the case of finite conjugate lenses.

Infinite conjugate lenses: → For a lens that has been specified for imaging at infinite conjugates or "infinity focus" (i.e., it has been designed and optimized for an object at infinity), the additional specification to request for would be the lens's infinite conjugate front focal length (FFL). This is the distance from some mechanical reference to the front focal point under the condition that the lens is used in reverse. The back focal length (BFL) would have already been provided (i.e., the distance from some mechanical referent to the rear infinite conjugate focal point). Knowledge of the FFL and BFL tells us the principal plane locations for the lens, relative to appropriate respective mechanical references. And with this information, we can determine the positions of the front and rear principal planes relative to each other. This is what we need, because with the front and rear principal plane locations, we may draw the geometry of rays illustrated in Fig. 3.14.

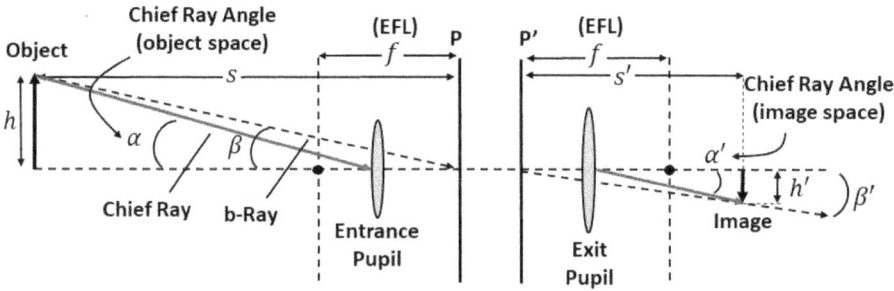

FIGURE 3.14 Geometry for determining image circle position and size.

Fig. 3.14 is reminiscent of the geometry shown in Fig. 2.12. In reality, only the chief ray is "physical" in that it passes through the center of the entrance and exit pupils and strikes the height of the image. In the presence of vignetting, although the pupils would be shifted laterally, the chief ray still passes through the centers of the shifted pupils. The "b-ray" in Fig. 3.14 refers to "Ray b" from Fig. 2.12, and it is a ray determined from first order optics. The b-ray may not necessarily make it through the system (it may be vignetted). However – and this the key point – for a focused image, the b-ray intersects the chief ray at the height of the image. Therefore, in the absence of image distortion, by applying Eqs. (2.1) and (2.2), the b-ray may be used for determining the semi-diameter h' of the image circle if we let h be the height of the viewable object within the design chief ray angle α. In the presence of a specified magnitude of image distortion D, one may apply Eq. (2.130) to account for distortion by noting that

$$h' = m_p h[1 + D].\tag{3.1}$$

Wow, this is our first equation of this section. Can't believe I managed to pull off some explanations without formulas until now. Anyway, in Eq. (3.1), h and h' refer to the heights in Fig. 3.14. For an infinite conjugate lens, h' is the image circle's semi-diameter when $s' = f$. If one wishes to use this lens for finite conjugate imaging, placing an object at some finite distance s from the front principal plane results in an extended location s' of the image circle, which may now be determined using Eq. (2.1). To determine the size of the image circle, we first determine m_p using Eq. (2.2). We then plug m_p into Eq. (3.1) to determine h', the semi-diameter of the image circle. The discussed method for determining the image circle location and semi-diameter is rather useful because it is valid with or without vignetting, and with or without the use of optical design programs (if you don't have access to such programs that is). Additionally, one does not require any knowledge of the locations of the entrance and exit pupils.

Finite conjugate lenses: $\rightarrow$ When a stock lens is specified for some finite working distance, the additional information to look out for (or to request) is the image magnification. If distortion is specified, you need only request for the real magnification h'/h rather than the paraxial one, because you can use Eq. (3.1) to determine the paraxial magnification m_p. If the lens supplier provides the paraxial magnification, then ask for the image distortion so that you can determine the real image height h'. Then, note that combining Eqs. (2.1) and (2.2) allows determination of s and s'. Finally, use the methods described for the infinite conjugate case to determine image circle position and size as a function of object distance s.

Some other important notes:

1. Generally, note in Fig. 3.14 that α is not necessarily equal to α', and β is not necessarily equal to β'. However, if there is no distortion, then $\beta = \beta'$, but α is still not necessarily equal to α' unless there is complete lens symmetry.
2. The image distortion D is not necessarily the same for all object and image distances. Therefore, one either accepts that using a single distortion for all working distances results in some error in estimating image circle sizes and locations, or one could request several distortion values at different working distances from the lens supplier, and then fit a curve.
3. Finally, note that the front and rear principal planes labelled P and P' in Fig. 3.14 may in general be located in reversed order. That is, P' can sometimes be in front of P. But Eqs. (2.1) and (2.2) remain valid. One just needs to be sure that the object and image distances s and s' respectively are measured from P and P' respectively.

3.2.2.3 What exactly is "working distance"?

In this book, the object distance s is strictly defined as the distance between the object and the front principal plane of an optical system, which is a virtual plane. In many practical situations, the optical engineer works with mechanical engineers to develop a physical system comprised of lens elements, mounts, and a mechanical housing. Mechanical engineers generally need a physical reference or "datum" to work with. The simplest reference is the front and back flanges of the lens housing. Generally, the "flange" refers to the outermost face of the mechanical housing. To a mechanical engineer, image distance must therefore be referred to the rear mechanical flange, and object distance must be referred to the front mechanical flange. As mentioned previously, the term ***working distance*** refers to the distance between the object and the front mechanical flange. But this term can often be confusing. Sometimes, working distance is referred to the distance between the object and the front vertex of the first lens element of a lens system. This is indeed, ambiguous, so the best thing to do is usually to ask a technical representative from the lens supplier. Whatever mechanical reference is used, one can perform the type of calculations discussed in Sec. 3.2.2.2 in terms of front and rear working distances WD and WD' respectively by inclusion of the differences Δ and Δ' as illustrated in Fig. 3.15. Accounting for this geometry, we may re-express Eq. (2.1) as

$$\frac{1}{f} = \frac{1}{\text{WD} + \Delta} + \frac{1}{\text{WD}' + \Delta'} . \tag{3.2}$$

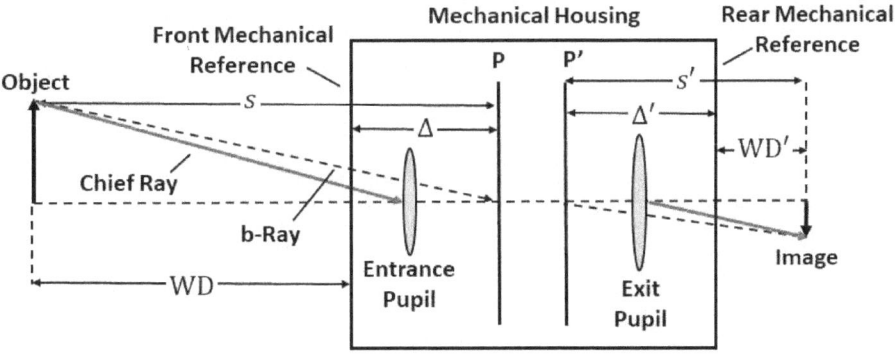

FIGURE 3.15 Working distance "WD" of a lens with mechanical housing.

3.2.2.4 Determining principal plane locations in real life

In the event that only a stock lens's EFL is known, and you've already purchased the lens and have it physically with you, then determining the lens's principal plane locations is a matter of first measuring the paraxial magnification m_p in Eq. (3.1) on an optics lab bench. To do this, set up a lens and image sensor in the lab for imaging at finite conjugates. Use a small object as a target (i.e., a target with small height h). For small heights, we may assume zero distortion in Eq. (3.1). At the image sensor, measure h' (e.g., by counting the number of pixels that make up the image height and multiplying this by the pixel size, interpolating as needed). Determine the paraxial magnification by taking the ratio h'/h. Upon determining the paraxial magnification, we note from combining Eqs. (2.1) and (2.2) that

$$\frac{1}{f} = \frac{1}{s} + \frac{1}{m_p s} .$$

$$(3.3)$$

Thus, one may insert the paraxial magnification into Eq. (3.3) to solve for s. From this, determine the front principal plane location relative to a front mechanical reference by noting from Fig. 3.15 that $\Delta = s -$ WD (note that you would determine WD from a lab measurement). Then, since s is known, you may determine s' by applying Eq. (2.1). Finally, determine the rear principal plane location relative to a rear mechanical reference by noting from Fig. 3.15 that $\Delta' = s' - \text{WD}'$.

3.2.2.5 The chief ray angle in image space

In Figs. 2.58 and 3.14, the angle α' is the chief ray angle in image space. Nowadays, it would be unheard of for a lens supplier not to provide this angle as a specification, due to the rather widespread availability of image sensors with microlens arrays. If this is not provided, then one should request it from the lens supplier. As a consequence of the microlenses, the relative illumination for the image formed onto an image sensor drops more rapidly than is described by Eqs. (2.134) and (2.135). Suppliers of image sensors with microlens arrays usually provide a plot of the image's sensor's quantum efficiency η as a function of the image space chief ray angle α'. To account for this effect in the analysis of the total relative illumination in an image, a reasonable approximation is to multiply the lens's relative illumination by $\eta(\alpha')$.

In some applications, knowledge of the exit pupil location may be required (e.g., see Sec. 3.2.7.2). If a stock lens is used for such applications, then knowledge of α' may be used for analytically determining the location of the exit pupil. For an unvignetted lens, the location of the exit pupil is at a position along the optic axis, relative to the image plane. For infinite conjugate lenses, locating this position is just a matter of extending the chief ray from the height of the image (i.e., the edge of the image circle) towards the optic axis, in the direction away from image plane. For a finite conjugate lens, one must first locate the rear principal plane using the methods discussed in Sec. 3.2.2.2. From this, one may determine the image distance s', and from this, one may then extend the chief ray from the height of the image towards the optic axis. For a vignetted lens, none of the above works because the pupils in such a lens are shifted laterally to positions above or below the optic axis. In this case, one may perhaps determine the axial locations of the entrance and exit pupils through some experimental means. And in this case, it seems best to simply request all information about the stock lens from the supplier.

3.2.2.6 Sensor formats, image circles, and resolution

It is natural that one would select an image sensor to integrate with an imaging lens. If one were using a stock lens for machine vision, in most cases, it is a matter of selecting a sensor whose format matches the stock lens's image circle diameter. This generally means that one would have the sensor's diagonal length to be equal to the diameter of the lens's image circle. In this way, there would be optimized field coverage without the visual presence of the image circle on the sensor. There are a variety of sensor formats available in industry [2], with a variety of aspect ratios. Figs. 3.16a – 3.16c depict three sensors with aspect ratios (defined in this book as the ratio of the width w to the height h of the sensor) ranging from 1:1 to 16:9, whose diagonal lengths D match a lens's image circle diameter. If R is the aspect ratio of the sensor, then the sensors in Figs. 3.16a – 3.16b have $R = 1$, $R = 4:3 \approx 1.33$ and $R = 16:9 \approx 1.78$ respectively.

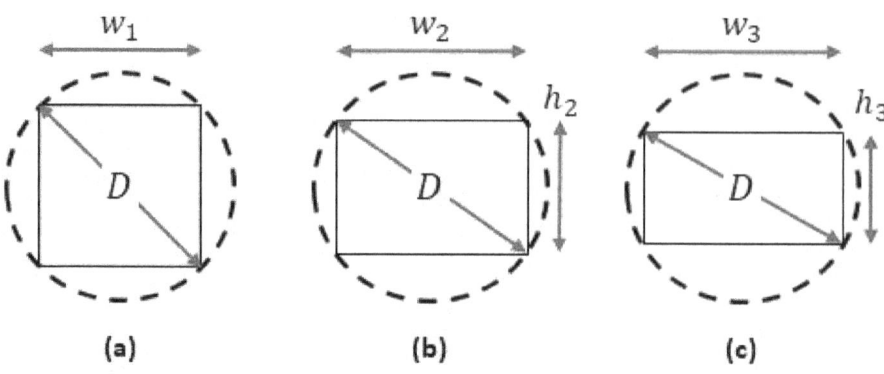

FIGURE 3.16 Sensors matched to image circles. (a) $R = 1{:}1$. (b) $R = 4{:}3$. (c) $R = 16{:}9$.

Now, it is not necessary to restrict one's choice of stock lenses to the range of lenses from a catalog that have been specifically designed to match available sensor formats. If I wanted to, I could take an existing eyepiece from an old telescope that nobody wants, and, provided that the eyepiece's *field stop* is accessible, I can use this lens as my machine vision lens. Why not? If the eyepiece's field stop is accessible, then the field stop's plane is the image plane when the eyepiece is used in reverse. I would determine the eyepiece's image circle diameter by measuring the field stop diameter (we'll talk about field stops in Sec. 3.3.2), and then search for a suitable sensor. But in this case, there may be no matching sensor format. For a fixed working distance, there is no possibility of scaling the eyepiece's image circle diameter D_c to match the diagonal length D_s of any sensor in order to optimize field coverage. At the image plane, one therefore ends up with any of the two possible scenarios depicted in Figs. 3.17a and 3.17b.

However, there is still hope, depending on application constraints. In particular, if I remove the constraint of a fixed working distance and fixed object size when using my eyepiece as a machine vision lens, then the image circle may be varied by changing both the working distance and object size proportionately in the manner discussed in Sec. 3.2.2.2. As long as the object height is within the half angular FOV of the eyepiece, its complete image would be visible within the lens's image circle. This would enable scaling of the image circle to suit the diagonal length of any image sensor. If one further removes the constraint of using a stock lens for machine vision, then lens customization provides far greater flexibility. One may scale any lens design to match its image circle to the diagonal length of any sensor.

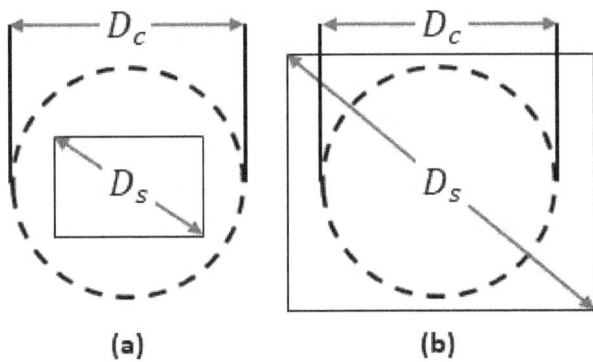

(a)   (b)

FIGURE 3.17 Sensor diagonal vs. image circle diameter. (a) $D_s < D_c$. (b) $D_s > D_c$.

In addition to matching image circles to sensor formats, one would also usually wish to select a sensor with high resolution (i.e., small pixel size). However, a complete description of the visual quality of a digital image involves a number of factors, not just pixel size [3 – 6]. For instance, as pixel size decreases, signal to noise ratio (SNR) may decrease as well, due to the lower light gathering area of smaller pixels. Suppose that SNR were maintained by varying the integration time or by varying the level of illumination at the object. And suppose that all other factors that influence image quality were maintained, other than pixel size. If we then regard the resolution of an image sensor as its ability to adequately resolve spatial frequencies through spatial sampling, then all other things constant, image resolution would be associated with pixel size. However, there is a tendency to associate image resolution with total pixel count, which is debatable.

For instance, under the conditions illustrated in Figs. 3.17a and 3.17b, increasing the total number of pixels whilst maintaining pixel size does nothing to improve spatial sampling. However, when it is assumed that lenses (and therefore, image circle diameters) scale and match with sensor diagonal lengths, then total pixel count is generally a fair indicator of image resolution.

However, consider a situation where sensor aspect ratios are different. In Figs. 3.16a – 3.16c, each sensor has a different aspect ratio, but their diagonals match the diameter of the image circle, which may either be held constant or may scale to any size. Assuming square pixels and a fixed total pixel count for each sensor, linear spatial sampling increases with increasing aspect ratio, because surface area decreases with aspect ratio (which results in reduced pixel sizes). But there are subtle points to note. Defining the aspect ratio R as the ratio of a sensor's width w to its height h, we have

$$R = \frac{w}{h}.$$
(3.4)

Since the area A of a sensor is the product of its width and height, we have

$$A = \frac{w^2}{R}.$$
(3.5)

Assuming only square pixels, the length (or pitch) p across a pixel for a total number of pixels n across a sensor is

$$p = \sqrt{\frac{A}{n}} = \sqrt{\frac{w^2}{nR}} = \frac{w}{\sqrt{nR}}.$$
(3.6)

If N_H is the number of horizontal pixels per unit width at the sensor, then we have

$$N_H = \frac{w}{p} = \sqrt{nR}. \tag{3.7}$$

If N_V is the number of vertical pixels per unit height at the sensor, then we have

$$N_V = \frac{h}{p} = \frac{w}{Rp} = \frac{w\sqrt{nR}}{Rw} = \sqrt{\frac{n}{R}}. \tag{3.8}$$

And if N_D is the number of diagonal pixels per unit diagonal at the sensor, then

$$N_D = \sqrt{N_H{}^2 + N_V{}^2} = \sqrt{n[R + (1/R)]}. \tag{3.9}$$

Finally, let us express Eq. (3.6) in terms of the diagonal D of the sensor, which is our reference (since image circle diameters should match sensor diagonals). To do this, we note that $N_D = D/p$. Therefore,

$$p = \frac{D}{\sqrt{n[R + (1/R)]}}. \tag{3.10}$$

Suppose we fix the total pixel count in Eqs. (3.7) – (3.10). Under this condition, Eq. (3.7) tells us that horizontal sampling increases as aspect ratio increases. This is not only due to the increase in sensor width. As Eq. (3.10) indicates, pixel size decreases with increasing aspect ratio. Eq. (3.8) tells us that vertical sampling decreases as aspect ratio increases. Eq. (3.9) tells us that diagonal sampling is generally higher as aspect ratio increases (though what this means physically is unclear because pixels aren't diagonally aligned). So, let us take the ratio of diagonal sampling N_D between sensors with 16:9 aspect ratio to those with 1:1 aspect ratio at fixed total pixel count. Applying Eq. (3.9), we have $\sqrt{1.78 + (1/1.78)}/\sqrt{2} \approx 1.08$. Let us also take the ratio of horizontal sampling N_H between sensors.

Applying Eq. (3.7), we have $\sqrt{1.78}/\sqrt{1} \approx 1.33$. Both ratios yield somewhat higher linear spatial sampling at fixed pixel count. On the other hand, it may be argued that the spatial information capacity within an image should be quantified in terms of a so-called "space-bandwidth product", which considers the image's area, not just sampling in a linear dimension [7]. Eqs. (3.7) – (3.10) also tell us that, for a fixed pixel size, increasing the sensor diagonal by some factor, say, N and increasing pixel count by N^2 (and scaling a lens to match the resulting larger sensor diagonal) increases linear sampling overall. This is the assumption that one implicitly makes when it is stated that image resolution (i.e., spatial sampling) increases with sensor size. On the other hand – and in the completely opposite direction – Eq. (3.10) also implies that if SNR and other factors that influence image quality are somehow maintained at any scale, then smaller image sensors (i.e., smaller D) with a disproportionately higher pixel count (i.e., larger n) and higher aspect ratios (i.e., larger R) may also yield higher spatial sampling, which is great for miniaturizing lenses. So, should we miniaturize all imaging systems? After all, scaling down the size of a lens reduces certain aberrations and increases depth of field (see Sec. 2.1.6). Yet it is difficult to imagine that everything would function well at small scales. Lens manufacturability becomes a challenge. Moreover, there has to be a limit at which geometrical optics fails at small scales. Finally, note also that the Airy diffraction blur radius does not scale when f-number is fixed. We discuss these considerations briefly in Sec. 3.2.8.

3.2.2.7 Illumination in machine vision

Suppliers of sources and illumination subsystems for machine vision provide a variety of illumination solutions, each with some special arrangement of bulbs. But if you do not have immediate access to these, then a rather practical and cost-effective way to start is to perform simple experiments using cheap handheld LED flashlights from the local hardware or convenience store. Hold one in your hand (tape others to lab mounts) and illuminate the object at a variety of positions and angles. Check to see how the image looks from your integrated lens and image sensor machine vision system. If it looks "good" (either through visual inspection or some quantitative analysis), then select a "real" source whose specs match as close as possible to the characteristics of the cheap source you had experimented with.

Every application has different requirements for illumination because not all target objects have the same physical properties. If you're imaging circuit boards, you may wish to discern between capacitors and weird looking wire bonds. If you're imaging food, you may wish to discern between ripe and rotten colored regions of interest (ROI). In the former, you may find that improving contrast requires creating some shadowing effect with sources at oblique angled illumination. In the latter, you may not want shadows and just need appropriate colored or dichroic filters. In either case, it is rather difficult to tell how the image would turn out without actually trying it out in real life.

Quite often, complex illumination problems will require some time to set up an accurate model in an optical design program. So, it seems reasonable to approach the problem by making use of observations and data gained from your flashlight experiments to fine-tune your model. Use your experiments to get immediate answers, and in the meantime, model your system during "free" periods. Do it when you're waiting for lab results. Do it when you're having your lunch. Do it whenever you get the chance, and let it be your little project, because when you finally get your model to work, it would serve you well in the long run. Simpler illumination problems are worth analyzing through basic hand sketches and calculations prior to setting up a basic optical model. Cross-check the hand calculations with your model to gain confidence. Let us consider two common examples of basic illumination problems that can be analyzed without complex models: 1. Illumination onto smooth or "glossy" surfaces. 2. Illumination onto diffuse or "matte" surfaces.

Glossy Surfaces: → The problem that concerns illuminating objects with features that lie on glossy surfaces is glare. By "glare" I mean the specular reflection of the source from the surface where the features to be imaged reside. A good example is the imaging of features printed onto glossy photographic film (e.g., if one were taking a photo of a glossy photograph). There is a particularly simple geometry for mounting a source to avoid such glare (Fig. 3.18).

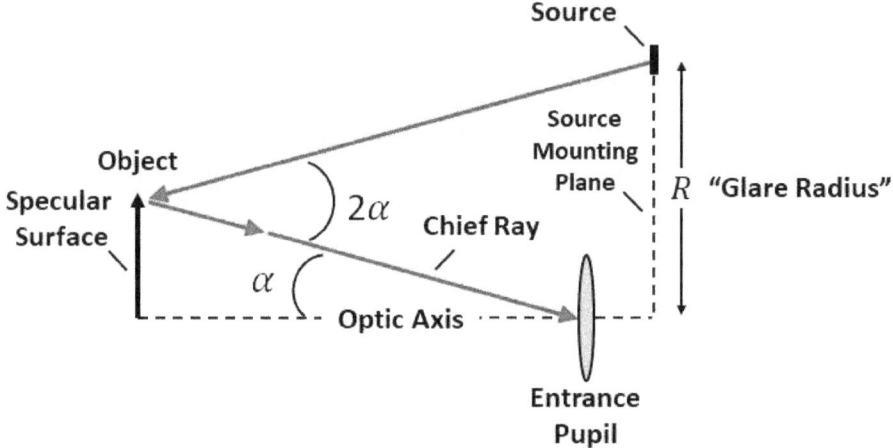

FIGURE 3.18 "Glare radius" for a source illuminating a glossy object surface.

In Fig. 3.18, the "Source Mounting Plane" refers to a plane where one has decided to mount a source. On this plane, light from a source placed at any height that's equal to or less than the "glare radius" R would reflect off the specular surface at the object plane and make its way into the entrance pupil, resulting in glare – the image of the source. Therefore, glare is avoided if a source's position on the mounting plane is $\geq R$. This would be a design consideration if, for example, one is selecting a so-called "ring light" for a machine vision system. To determine the glare radius R, one simply reverses the path of the chief ray and let it reflect off the object surface toward the plane where one wishes to mount the source. If the source's mounting plane is shifted along the optic axis, then retrace the chief ray to the new glare radial position. Evidently, lenses with larger angular FOVs result in larger glare radii, which takes up real estate in the mechanical design of the machine vision system. If mounting a source at a position $\leq R$ is unavoidable, then glare may be reduced through the use of crossed linear polarizers: one at the source, and the other in front of the lens. However, this of course significantly reduces brightness of the image. In other situations, if the source happens to illuminate a specular surface at the Brewster angle, then using a single polarizer in front of the lens suffices.

Matte Surfaces: → The problem in illuminating objects with features that lie on diffuse scattering or "matte" surfaces is often the uniformity of the irradiance distribution. For a small extended Lambertian source (such as a small LED whose output is Lambertian), the irradiance non-uniformity arises from the so-called cos^4th effect from an oblique angle. Fig. 3.19 shows the geometry for an oblique ray (dotted line) that contributes to this effect.

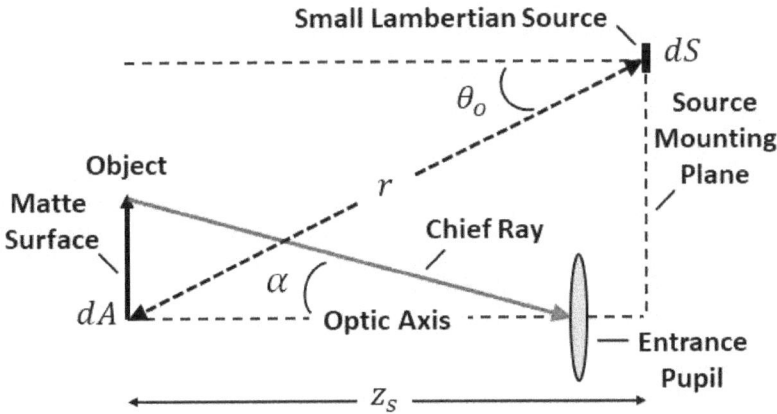

FIGURE 3.19 Illumination on a matte surface from a small Lambertian source.

From the geometry in Fig. 3.19 and applying Eq. (2.58), the flux $d^2\Phi$ from an elemental source area dS striking an elemental area dA at the object may be expressed as

$$d^2\Phi = LdS\cos\theta_o(dA\cos\theta_o/r^2). \qquad (3.11)$$

Since $\cos\theta_o = z_s/r$ and the irradiance dE at dA is $d\Phi/dA$, we have

$$dE = (LdS\cos^4\theta_o)/z_s{}^2. \qquad (3.12)$$

One can see that the irradiance along the vertical direction at the object varies as the fourth power of the cosine factor in Eq. (3.12), where the peak irradiance is at $\theta_o = 0$, which is not the center of the field. Placing a ring of small Lambertian sources about the optic axis can improve the brightness of illumination at the center of the field, but the non-uniformity persists. If one tilts the source by an angle θ_s relative to the optic axis (Fig. 3.20), then Eq. (3.12) is modified by this tilt and may be expressed as

$$dE = [LdS\cos^3\theta_o\cos(\theta_o - \theta_s)]/z_s{}^2. \qquad (3.13)$$

In Fig. 3.20, if $\theta_s > \theta_o$, the peak irradiance can be made to shift towards the center of the field. However, because this would just be the peak from a single source, the resulting irradiance distribution would not have circular symmetry, which is usually not preferred. In this case, mounting a ring of small tilted Lambertian sources about the optic axis (and adjusting the tilt angle) can result in improved circular symmetry for the irradiance distribution at the object. Additionally, since glare is minimal for matte surfaces, one need not be overly concerned about ensuring that the source's height above the optic axis satisfies the glare radius criterion illustrated in Fig. 3.19. Therefore, besides tilting, the source's mounting height is another variable for optimizing uniformity. The final irradiance distribution at the image is of course not the distribution at the object. As a reasonable approximation, the final distribution at the image may be taken as the product between the image's relative illumination and the irradiance distribution at the object plane. Hence, in optical design software, one need not model a complete system of imaging lens and illumination. Rather, one need only to place a detector at the object plane to record the irradiance distribution. Then, multiplying the recorded distribution with the lens's relative illumination yields the final output.

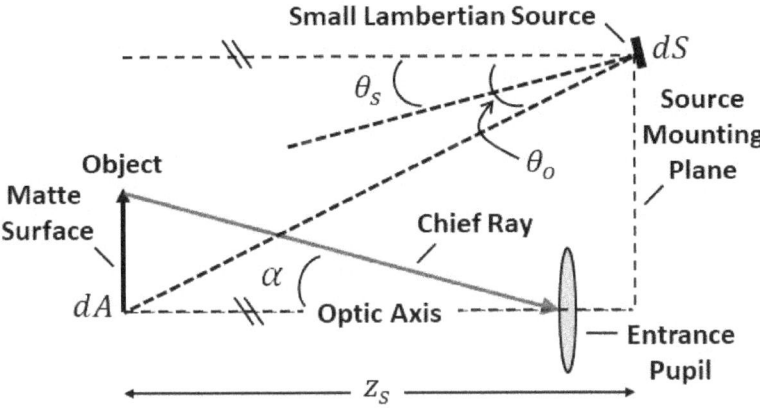

FIGURE 3.20 Tilted small Lambertian source illuminating a matte surface.

3.2.3 Customization and optical design for machine vision

3.2.3.1 Finding a starting point and scaling a lens

Modern lens design often begins with modifying an existing prescription, and modification often begins with scaling. Let us take, for example, the lens in Fig. 3.8, whose prescription is provided in Table 2.1 (but with the last element's semi-diameter at 0.12 mm). Scaling this lens by a factor of 5.18 makes it bigger and yields a lens suitable for a 1/6" image sensor, whose half-diagonal length is about 1.45 mm. Thus, if you have tried scaling this lens, note that the chief ray's height at the image plane should be about 1.45 mm (it does not have to be exact for what we're going to do). We are going to modify and optimize this lens further, but not for the purpose of arriving at a fully optimized machine vision lens design. This is not a book about how to design a lens (you can refer to many other fine books for that). Rather, we are going to do something interesting, whose result is still highly relevant to machine vision (or any imaging application for that matter). We are going to examine a claim I made in Sec. 2.3.7, where I said that in reality, there is no cos^4th law in a lens system. So, I'm going to show you that we can design a lens whose chief ray angle in image space is zero degrees, yet the resulting relative illumination is not uniform. And that's not all. We're going to apply our analysis from Sec. 2.3.7 to understand what controls relative illumination, and then we're going to control it. But to do all that, we're going to need to perform one more step in optimizing this lens.

3.2.3.2 Optimizing in steps

In order to apply the analysis from Sec. 2.3.7 for relative illumination, we need to work at low numerical aperture. So, let's make the aperture stop semi-diameter to 0.12 mm (make sure you use the "Float by stop size" setting for the aperture if you're using Zemax OS). Also, float all lens semi-diameters by removing all the "U" symbols under the Semi-Diameter column of the lens data editor. Set the field heights to 0, 10, and 14.2 mm. Change the thickness on surface 0 to 49.5 mm, and the thickness on surface 1 to 0.5 mm. Thus, the total working distance (WD) in object space should now be 50 mm (we're defining WD here as the distance from the object to the vertex of the first element). Ensure that ray aiming is "on" (aimed at the paraxial entrance pupil). We're going to "encourage" this lens to become telecentric in image space, so make the thickness of surface 6 to 4 mm (increasing the distance between the stop and the lens group after the stop helps to make the angle of incidence smaller at the image plane). A good machine vision lens is ordinarily telecentric in object space rather than image space, but we are merely performing an exercise to study relative illumination. Put a marginal ray height solve on surface 11 and set the ray height to zero at the zero-pupil zone. This sets the image plane to b at the paraxial focus. Set the wavelengths to the F, d, and C wavelengths (i.e., 486.13 nm, 587.56 nm, and 656.27 nm respectively), and set the d wavelength to be the primary wavelength (we'll assume that near infrared light is blocked by some surface coatings). All wavelength weights = 1. Note that there should be no more vignetting of any rays, so re-set vignetting factors to zero if you haven't already done so. In Zemax OS, the lens data editor should now look as shown in Fig. 3.21.

Lens Data

Update: All Windows

Surface 0 Properties — Configuration 1/1

	Surf:Type		Radius	Thickness		Material	Semi-Diameter	
0	OBJECT	Standard	Infinity	49.500			14.200	
1		Standard	Infinity	0.500			1.128	
2		Standard	3.024	0.370		N-LAK9	0.955	
3	(aper)	Standard	1.841	3.151			0.834	
4		Standard	17.660	0.622		N-BAF52	0.380	
5		Standard	-3.459	0.854			0.318	
6	STOP	Standard	Infinity	4.000			0.120	U
7		Standard	-2.898	0.470		N-SF11	0.883	
8		Standard	9.293	0.461		N-LAF2	1.050	
9		Standard	-4.908	0.012			1.120	
10		Standard	-54.466	0.371		N-SSK5	1.156	
11	(aper)	Standard	-4.512	4.214	M		1.195	
12	IMAGE	Standard	Infinity	-			1.695	

FIGURE 3.21 Print-screen from Zemax OS of the prescription for the current scaled and modified lens.

In Fig. 3.21, note that surfaces 3 and 11 have "(aper)" labelled for their surface type. That's because I left them as floating apertures. This is not important. Note the "M" label on the Thickness cell on surface 11, which indicates that a paraxial focus solve is used. Now, build your merit function so that it looks as shown in Fig. 3.22. In this figure, note that the default merit function for RMS spot size is used with an overall weight of 1 (rows > 24 aren't displayed). Axial color is unconstrained (it's not important for our present exercise). For those not using Zemax OS, build a merit function that constrains EFL = 6 mm, image height = 1.45 mm (use your program's sign convention), chief ray angle in image space = 0 degrees, total lens thickness < 15 mm, 0.1 mm < glass thicknesses < 4 mm, glass edge thicknesses > 0.1 mm, and air center and edge thicknesses (between surfaces 3 and 10) > 0.01 mm. Note that these dimensions may not be manufacture-friendly, but again, this is just an exercise. Image distortion is not constrained. Set variables as shown in Fig. 3.23. The lens layout prior to optimization should look as shown in Fig. 3.24. Upon clicking on the automatic local damped least squares optimization button, I obtained the lens shown in Fig. 3.25, whose prescription is given in Fig. 3.26.

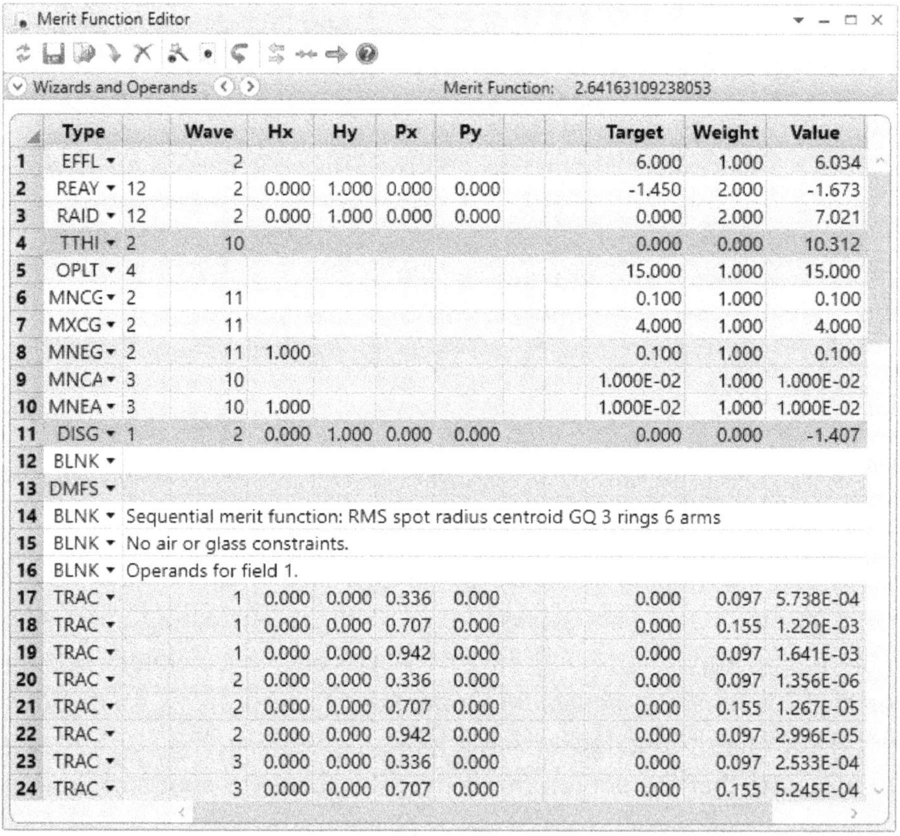

	Type	Wave	Hx	Hy	Px	Py		Target	Weight	Value
1	EFFL ▼	2						6.000	1.000	6.034
2	REAY ▼ 12	2	0.000	1.000	0.000	0.000		-1.450	2.000	-1.673
3	RAID ▼ 12	2	0.000	1.000	0.000	0.000		0.000	2.000	7.021
4	TTHI ▼ 2	10						0.000	0.000	10.312
5	OPLT ▼ 4							15.000	1.000	15.000
6	MNCG ▼ 2	11						0.100	1.000	0.100
7	MXCG ▼ 2	11						4.000	1.000	4.000
8	MNEG ▼ 2	11	1.000					0.100	1.000	0.100
9	MNCA ▼ 3	10						1.000E-02	1.000	1.000E-02
10	MNEA ▼ 3	10	1.000					1.000E-02	1.000	1.000E-02
11	DISG ▼ 1	2	0.000	1.000	0.000	0.000		0.000	0.000	-1.407
12	BLNK ▼									
13	DMFS ▼									
14	BLNK ▼ Sequential merit function: RMS spot radius centroid GQ 3 rings 6 arms									
15	BLNK ▼ No air or glass constraints.									
16	BLNK ▼ Operands for field 1.									
17	TRAC ▼	1	0.000	0.000	0.336	0.000		0.000	0.097	5.738E-04
18	TRAC ▼	1	0.000	0.000	0.707	0.000		0.000	0.155	1.220E-03
19	TRAC ▼	1	0.000	0.000	0.942	0.000		0.000	0.097	1.641E-03
20	TRAC ▼	2	0.000	0.000	0.336	0.000		0.000	0.097	1.356E-06
21	TRAC ▼	2	0.000	0.000	0.707	0.000		0.000	0.155	1.267E-05
22	TRAC ▼	2	0.000	0.000	0.942	0.000		0.000	0.097	2.996E-05
23	TRAC ▼	3	0.000	0.000	0.336	0.000		0.000	0.097	2.533E-04
24	TRAC ▼	3	0.000	0.000	0.707	0.000		0.000	0.155	5.245E-04

Merit Function Editor

Wizards and Operands   Merit Function: 2.64163109238053

FIGURE 3.22 Print-screen of the Zemax OS merit function editor with operands for the current lens (default RMS spot size operands beyond no. 24 not shown).

	Surf:Type		Radius	Thickness	Material	Semi-Diameter
0	OBJECT	Standard ▾	Infinity	49.500		14.200
1		Standard ▾	Infinity	0.500		1.128
2		Standard ▾	3.024 V	0.370 V	N-LAK9	0.955
3	(aper)	Standard ▾	1.841 V	3.151 V		0.834
4		Standard ▾	17.660 V	0.622 V	N-BAF52	0.380
5		Standard ▾	-3.459 V	0.854 V		0.318
6	STOP	Standard ▾	Infinity	4.000 V		0.120 U
7		Standard ▾	-2.898 V	0.470 V	N-SF11	0.883
8		Standard ▾	9.293 V	0.461 V	N-LAF2	1.050
9		Standard ▾	-4.908 V	0.012 V		1.120
10		Standard ▾	-54.466 V	0.371 V	N-SSK5	1.156
11	(aper)	Standard ▾	-4.512 V	4.214 V		1.195
12	IMAGE	Standard ▾	Infinity	-		1.695

FIGURE 3.23 Print-screen from Zemax OS of the prescription for the current lens with variables.

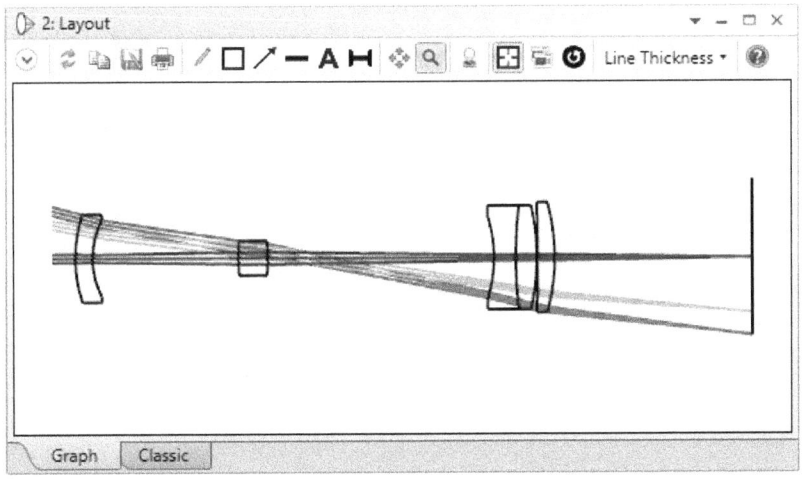

FIGURE 3.24 Print-screen from Zemax OS of the lens layout for the prescription in Fig. 3.23.

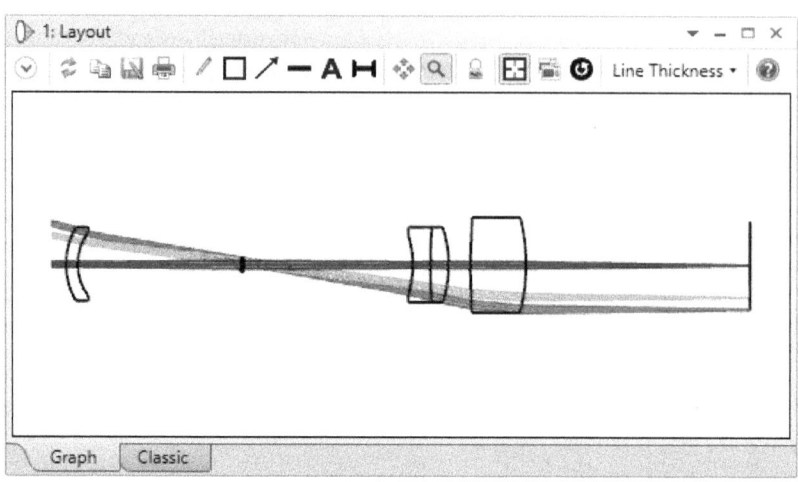

FIGURE 3.25 Print-screen from Zemax OS of the optimized lens layout from Fig. 3.24 with 0^0 chief ray angle at the image.

	Surf:Type		Radius	Thickness	Material	Semi-Diameter
0	OBJECT	Standard ▾	Infinity	49.500		14.200
1		Standard ▾	Infinity	0.500		1.421
2		Standard ▾	2.444 V	0.376 V	N-LAK9	1.209
3	(aper)	Standard ▾	1.778 V	5.388 V		1.056
4		Standard ▾	-2.948 V	0.104 V	N-BAF52	0.238
5		Standard ▾	-1.980 V	0.582 V		0.231
6	STOP	Standard ▾	Infinity	4.859 V		0.120 U
7		Standard ▾	-3.158 V	0.573 V	N-SF11	0.966
8		Standard ▾	18.938 V	0.583 V	N-LAF2	1.149
9		Standard ▾	-4.074 V	0.715 V		1.238
10		Standard ▾	20.639 V	1.817 V	N-SSK5	1.427
11	(aper)	Standard ▾	-6.464 V	7.277 V		1.569
12	IMAGE	Standard ▾	Infinity	-		1.451

FIGURE 3.26 Print-screen from Zemax OS of the prescription for the lens in Fig. 3.25.

3.2.3.3 Telecentricity does not imply uniform relative illumination

Let us take a look at the Zemax OS print-screen of the relative illumination plot (Fig. 3.27) for the lens from Fig. 3.25, whose prescription is given in Fig. 3.26.

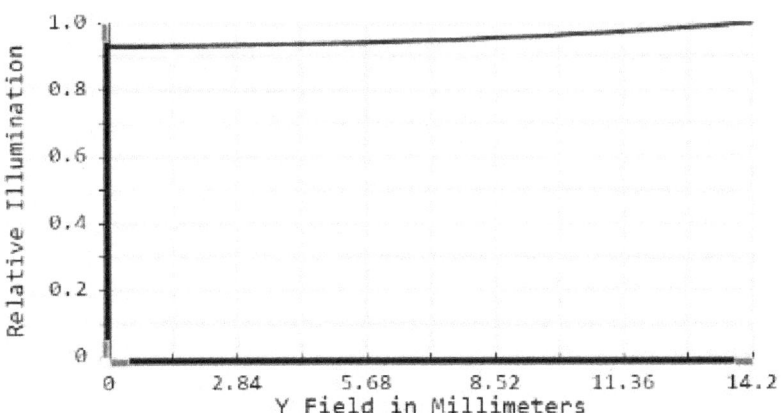

FIGURE 3.27 Relative illumination of the lens in Fig. 3.25.

Note that the relative illumination in Fig. 3.27 is not uniform. To understand the cause of this non-uniformity, we apply Eqs. (2.134) and (2.135). Let's start with Eq. (2.134), which describes relative illumination from the perspective of flux passing through the entrance pupil. Note that we can express Eq. (2.134) as the scalar product of three factors:

$$R(y) \approx \left\{ \frac{1}{[1+D][1+D+y(dD/dy)]} \right\} \times \left\{ \frac{[a(y)]}{[a(0)]} \right\} \times \{\cos^4 \alpha\}.$$

$$= \{R_D\} \times \{R_a\} \times \{R_\alpha\}. \tag{3.14}$$

In Eq. (3.14), R_D is the contribution of image distortion D and differential distortion dD/dy to relative illumination, R_a is the contribution from the variation of entrance pupil size (area) with field height, and R_α is the contribution from the chief ray angle in object space. The image distortion at full field is -7.01% (barrel). The differential distortion (i.e., the instantaneous rate of image distortion with field height) at full field is about -9.296E-03 mm^{-1}. Or, if you're using Zemax OS, you may compute this value by appending the merit function construction from Fig. 3.22 with the operands shown in Fig. 3.28 (see rows 12 – 19).

	Type	Wave	Hx	Hy	Px	Py		Target	Weight	Value
1	EFFL ▾	2						6.000	1.000	6.000
2	REAY ▾ 12	2	0.000	1.000	0.000	0.000		-1.450	2.000	-1.450
3	RAID ▾ 12	2	0.000	1.000	0.000	0.000		0.000	2.000	1.182E-08
4	TTHI ▾ 2	10						0.000	0.000	15.000
5	OPLT ▾ 4							15.000	1.000	15.000
6	MNCG ▾ 2	11						0.100	1.000	0.100
7	MXCG ▾ 2	11						4.000	1.000	4.000
8	MNEG ▾ 2	11	1.000					0.100	1.000	0.100
9	MNCA ▾ 3	10						1.000E-02	1.000	1.000E-02
10	MNEA ▾ 3	10	1.000					1.000E-02	1.000	1.000E-02
11	DISG ▾ 1	2	0.000	1.000	0.000	0.000		0.000	0.000	-7.010
12	BLNK ▾ dD/dy									
13	DISG ▾ 1	2	0.000	0.999	0.000	0.000		0.000	0.000	-6.997
14	REAY ▾ 0	2	0.000	1.000	0.000	0.000		0.000	0.000	14.200
15	REAY ▾ 0	2	0.000	0.999	0.000	0.000		0.000	0.000	14.186
16	DIFF ▾ 14	15						0.000	0.000	0.014
17	DIFF ▾ 11	13						0.000	0.000	-0.013
18	DIVI ▾ 17	16						0.000	0.000	-0.930
19	DIVB ▾ 18		100....					0.000	0.000	-9.296E-...
20	RELI ▾ 3	2	3	0				0.000	0.000	1.072

FIGURE 3.28 Partial print-screen of the Zemax OS merit function editor for the lens from Fig. 3.26 with operands to compute differential distortion.

Using the values mentioned above for image distortion and differential distortion, we have

$$R_D = \frac{1}{[1 - 0.0701][1 - 0.0701 - 14.2(-9.296 \times 10^{-3})]}$$

$$= 1.347770747. \tag{3.15}$$

The number of decimals may be important, so let's keep them. The chief ray angle in object space is 14.5565 degrees, so

$$R_\alpha = \cos^4(14.5565) = 0.877652186. \qquad (3.16)$$

What's left is the contribution from the entrance pupil area R_a. Also, by conservation of energy, we should be able to estimate total relative illumination from the perspective of the exit pupil given by Eq. (2.135). In the following section, we will do all of these.

3.2.3.4 Zemax OS example: Relative illumination and distortion

To obtain the contribution of the entrance pupil area R_a to relative illumination for the lens example from the previous section, let us literally "look into the entrance pupil". To do this, save your original lens file, and then save it as a new file. Delete all elements to the right of the stop, and reverse the remaining system such that the stop is now the object. Set up 3 field points and make them 0, 0.12, and -0.12 mm. Insert surfaces 6 and 7. Make surface 6 into a coordinate break, and surface 7 into a paraxial lens model (EFL = 12 mm, Mode = 1, semi-diameter = 0.01 mm). Enter 15.368 mm for the thickness from the paraxial lens to the image. Set surface 0 (the object) as the Global Coordinate reference. Ensure that ray aiming is "on" (aim to the real entrance pupil). At the coordinate break, set the object y decenter (parameter 2) to 14.1 mm (this places your paraxial lens at close to the height of the object full field). In Zemax OS, the lens editor should look as shown in Fig. 3.29. Not shown in this figure are the coordinate break parameters on surface 6, whose values shall be controlled using the multi-configuration editor. Set up the multi-configuration editor to appear as shown in the print-screen in Fig. 3.30, which should yield the lens layout shown by the print-screen from Zemax OS in Fig. 3.31. Now, open the "geometric image analysis" feature and set it up for analysis by applying the numbers shown in Fig. 3.32. The resulting geometric image analysis plot is as shown in Fig. 3.33 for configuration 1 (imaging the entrance pupil along the optic axis), and

Fig. 3.34 for configuration 2 (imaging the entrance pupil at the height of 14.1 mm above the optic axis.

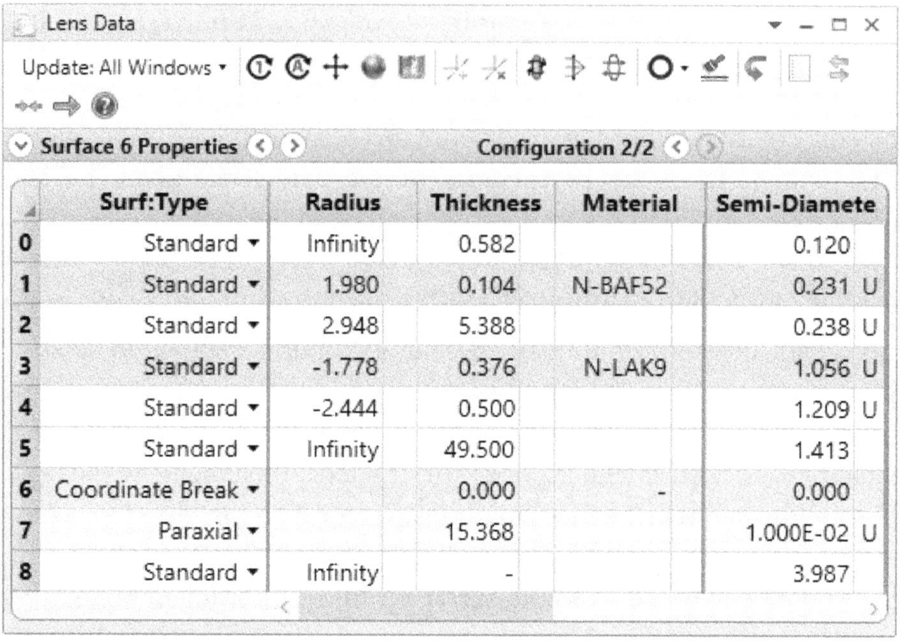

	Surf:Type	Radius	Thickness	Material	Semi-Diamete
0	Standard	Infinity	0.582		0.120
1	Standard	1.980	0.104	N-BAF52	0.231 U
2	Standard	2.948	5.388		0.238 U
3	Standard	-1.778	0.376	N-LAK9	1.056 U
4	Standard	-2.444	0.500		1.209 U
5	Standard	Infinity	49.500		1.413
6	Coordinate Break		0.000	-	0.000
7	Paraxial		15.368		1.000E-02 U
8	Standard	Infinity	-		3.987

FIGURE 3.29 Lens prescription for imaging the entrance pupil.

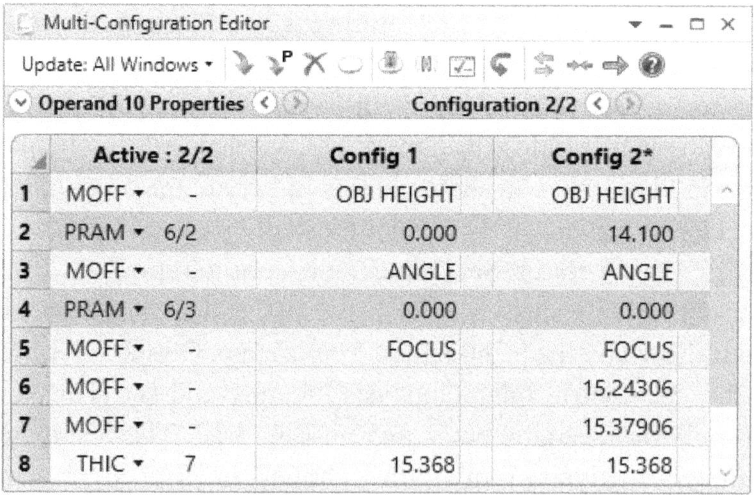

	Active : 2/2		Config 1	Config 2*
1	MOFF	-	OBJ HEIGHT	OBJ HEIGHT
2	PRAM	6/2	0.000	14.100
3	MOFF	-	ANGLE	ANGLE
4	PRAM	6/3	0.000	0.000
5	MOFF	-	FOCUS	FOCUS
6	MOFF	-		15.24306
7	MOFF	-		15.37906
8	THIC	7	15.368	15.368

FIGURE 3.30 Multi-configuration editor setup for lens in Fig. 3.31.

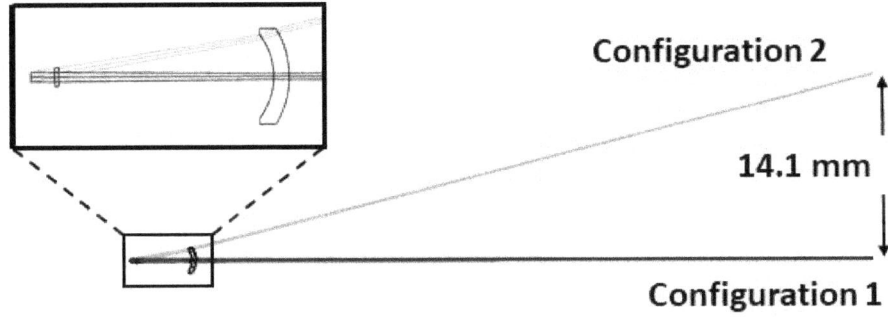

FIGURE 3.31 Lens layout using multi-configuration setup in Fig. 3.30.

Field Size:	0.24	Wavelength:	2
Image Size:	0.06	Field:	1
File:	CIRCLE.IMA		
Rotation:	0	Edit IMA File	
Rays x 1000:	500	Surface:	Image
Show:	Inverse Grey Scale		
Source:	Lambertian	# Pixels:	100
		NA	0
Use Polarization	☐	Total Watts:	1
Remove Vignetting Factors	☑	Plot Scale:	0
Scatter Rays	☐	Parity:	Even
Delete Vignetted	☑		
Use Pixel Interpolation	☐	Reference:	Chief Ray
Save as BIM File:			

☑ Auto Apply | Apply | OK | Cancel | Save | Load | Reset

FIGURE 3.32 Geometric image analysis settings for the lens in Fig. 3.31.

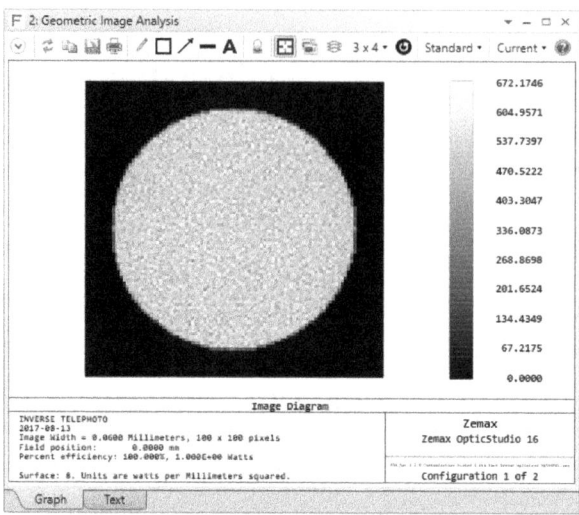

FIGURE 3.33 Entrance pupil at configuration 1 (Zemax OS print-screen of the output from the geometric image analysis feature using the settings in Fig. 3.32 at configuration 1).

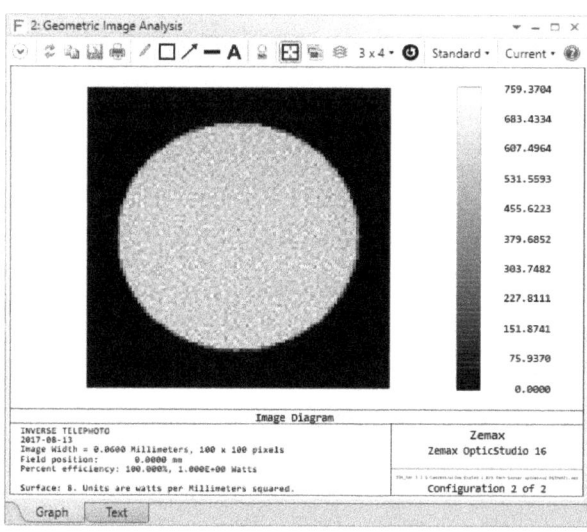

FIGURE 3.34 Entrance pupil at configuration 2 (Zemax OS print-screen of the output from the geometric image analysis feature using the settings in Fig. 3.32 at configuration 2).

Upon estimating the surface areas for these pupils, taking the ratio of the area from Fig. 3.34 to that in Fig. 3.33 I obtain

$$R_a = \frac{a(14.1)}{a(0)} = 0.914588069. \tag{3.17}$$

Applying the results from Eqs. (3.15) – (3.17) into Eq. (3.14) I obtain

$$Relative\ Illumination \approx R_D R_a R_\alpha \approx 1.082. \tag{3.18}$$

Note that Zemax OS's relative illumination plot shown in Fig. 3.27 is normalized to unity at full field. In Zemax OS, to obtain a list of numerical values for the relative illumination, click on where it says "text" at the lower left corner of the plot. Note that the on-axis magnitude is 0.932497. Taking the reciprocal, we have 1/0.932497 = 1.072, which is the relative illumination at full field. This is close to the value estimated in Eq. (3.18). Thus, we see that the total relative illumination in the image is a consequence of the three factors in Eq. (3.18). Now, let us estimate relative illumination from the perspective of the exit pupil by applying Eq. (2.135). As we did in Eq. (3.14), let's express Eq. (2.135) in terms of the product of its constituent parts:

$$R(y) \approx \frac{[a'(y)]}{[a'(0)]} \cos^4 \alpha' = R_{a\prime} R_{\alpha\prime}. \tag{3.19}$$

In image space, $\alpha' = 0$ so $R_{\alpha\prime} = 1$. To obtain $R_{a\prime}$, as we did for the entrance pupil, we will "look into the exit pupil" from the perspective of the image. To do this, return to your original optimized lens file (i.e., the one you saved prior to creating the layout in Fig. 3.31) and delete all elements to the left of the stop. Make the stop as the object. Set up a paraxial lens as we did for imaging the entrance pupil, and have your lens data editor to be as shown in Fig. 3.35. Set up your multi-configuration editor to appear as shown in Fig. 3.36. The lens layout should appear as shown in Fig. 3.37.

	Surf:Type	Radius	Thickness	Material	Semi-Diameter	
0	Standard ▾	Infinity	4.859		0.120	
1	Standard ▾	-3.158	0.573	N-SF11	0.966	U
2	Standard ▾	18.938	0.583	N-LAF2	1.149	U
3	Standard ▾	-4.074	0.715		1.238	U
4	Standard ▾	20.639	1.817	N-SSK5	1.427	U
5	Standard ▾	-6.464	7.277		1.569	U
6	Coordinate Break ▾		0.000	–	0.000	
7	Paraxial ▾		12.474		1.000E-02	U
8	Standard ▾	Infinity	–		0.198	

FIGURE 3.35 Lens prescription for imaging the exit pupil.

Active : 2/2		Config 1	Config 2*
1	PRAM ▾ 6/2	0.000	-1.440
2	THIC ▾ 7	12.474	12.474

FIGURE 3.36 Multi-configuration editor setup for the layout in Fig. 3.37.

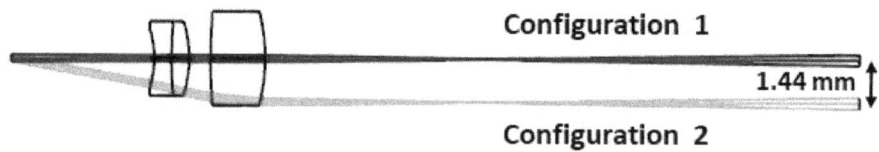

Configuration 1

1.44 mm

Configuration 2

FIGURE 3.37 Lens layout for multi-configuration setup in Fig. 3.36.

Now, set up the geometric image analysis feature with 0.24 for Field Size and 0.4 for Image Size. The resulting geometric image plots at configurations 1 and 2 should be as shown in Figs. 3.38 and 3.39.

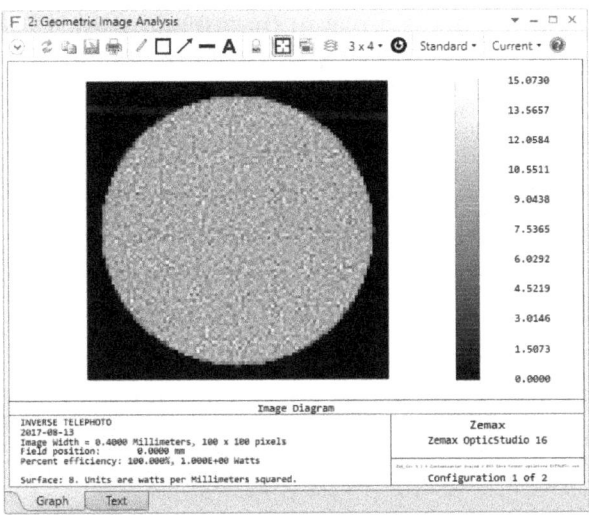

FIGURE 3.38 Exit pupil at configuration 1.

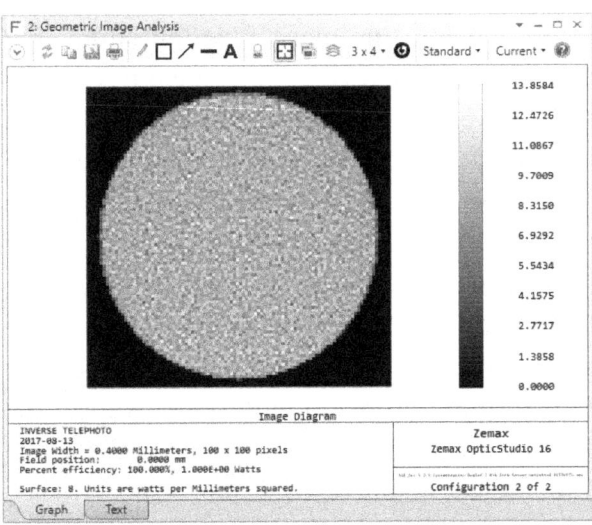

FIGURE 3.39 Exit pupil at configuration 2.

Estimating the surface areas in Figs. 3.39 and 3.40 and applying Eq. (3.19) with $\alpha' = 0$, I obtain

$$R(y) \approx R_{a\prime} = \frac{[a'(y)]}{[a'(0)]} = \frac{0.0341314}{0.03203205} \approx 1.066. \qquad (3.20)$$

Once again, the result in Eq. (3.20) is close to that obtain from Fig. 3.27, and it is also close to the value obtained in Eq. (3.18). Thus, due to conservation of energy, the larger relative size of the exit pupil at full field is consistent with the relative illumination in Fig. 3.27, and we see that the entrance pupil and exit pupil formulations for relative illumination given by Eqs. (2.134) and (2.135) are correct.

Note in Fig. 3.31 that we did not tilt the paraxial lens model to look towards the entrance pupil even though the chief ray angle in object space is not zero. The reason is because, in deriving Eq. (2.134), the elemental surface area at the object plane is not tilted. Note also that since elemental object and image areas were used in deriving Eqs. (2.134) and (2.135), the paraxial lens's semi-diameter in Figs. 3.31 and 3.37 is made small.

Now, suppose we wish to achieve uniform relative illumination for the lens in Fig. 3.25. In theory, one may simply apply the "RELI" operand for relative illumination into the merit function editor and target it to unity at full field. But our goal is to understand the effects of the contributing factors expressed in Eqs. (3.14) and (3.19) on relative illumination. To that end, suppose we wish to control relative illumination by varying image distortion without affecting the entrance pupil, and also without affecting the chief ray angle in object space. To do this, first return to the lens file relevant to Fig. 3.25, remove all variables on radii and thicknesses for surfaces prior to the stop, and convert all surfaces prior to the stop to user defined apertures. This ensures that further optimization does not impact the entrance pupil, nor would it affect the chief ray angle in object space (we are also maintaining the object height at 14.2 mm). Check to see if ray aiming is still "on" (aimed at the paraxial entrance pupil).

The question we now ask is, "what must image distortion be in order to achieve uniform relative illumination?" To answer this, we return to Eqs. (3.16) – (3.18) and note that the total relative illumination would be unity if we have

$$R_D = \frac{1}{R_a R_\alpha} = \frac{1}{(0.914588069)(0.877652186)} = 1.24581062.$$

$$(3.21)$$

So, the image distortion D must be such that $R_D = 1.24581062$. But there's something else: differential distortion dD/dy. To account for differential distortion, we recall from Seidel aberration theory [e.g., 8] that the ray intercept curve (i.e., the difference between the chief ray's real ray height y' and its paraxial ray height y'_p) for image distortion $\varepsilon_{y,D}$ as a function of field height y may be expressed as

$$\varepsilon_{y,D} = y' - y'_p = \sigma y^3. \tag{3.22}$$

The quantity σ in Eq. (3.22) is related to w_D in Eq. (2.13). Substituting Eq. (3.22) into Eq. (2.129) we have

$$D = \frac{\varepsilon_{y,D}}{y'_p} = \frac{\sigma y^3}{m_p y} = \frac{\sigma y^2}{m_p}. \tag{3.23}$$

Therefore,

$$\frac{dD}{dy} = \frac{2\sigma y}{m_p}. \tag{3.24}$$

Substituting Eq. (3.24) into the definition for R_D in Eq. (3.14) we have

$$R_D = \frac{1}{[1 + D][1 + D + y(dD/dy)]} \tag{3.25}$$

$$= \frac{1}{[1 + D][1 + D + y(2\sigma y/m_p)]}$$

$$= \frac{1}{[1 + D][1 + D + (2\sigma y^2/m_p)]}. \qquad (3.26)$$

Substituting D from Eq. (3.23) for the quantity $2\sigma y^2/m_p$ in Eq. (3.26) we have

$$R_D = \frac{1}{(1 + D)(1 + 3D)}. \qquad (3.27)$$

Solving for D in Eq. (3.27) we have

$$D = \frac{-4 \pm \sqrt{16 - 12[1 - (1/R_D)]}}{6}. \qquad (3.28)$$

Setting $R_D = 1.24581062$ [from Eq. (3.21)] in Eq. (3.28) and taking the negative root yields $D = -0.05130131$. Taking the positive root yields $D = -1.282$ which is absurd, because that's -128.2% barrel distortion, and we know that in comparing Eq. (3.21) with Eq. (3.15), the result in Eq. (3.21) implies that we need a lower amount of barrel distortion to reduce the relative illumination at full-field.

Now, return to the lens file associated with Fig. 3.25, and enter a target of -5.13% for the distortion operand in the merit function editor. Give it a weight of 2. Lift the constraint on EFL, because Eq. (3.22) tells us that y and y' are constrained, leaving only y'_p as a variable to effect $\varepsilon_{y,D}$, which is the distortion. Re-set the RMS spot size default merit function (as you'd normally do whenever you make changes to any operands and weights in the merit function editor) with an overall weight of 1. Click on the automatic damped least squares optimization. The resulting prescription I obtained after optimization is provided in Fig. 3.40, and the corresponding lens layout and relative illumination plots are shown in Figs. 3.41 and 3.42 respectively. Note that the relative illumination is quite uniform.

	Surf:Type		Radius	Thickness	Material	Semi-Diameter	
0	OBJECT	Standard ▾	Infinity	49.500		14.200	
1		Standard ▾	Infinity	0.500		1.420	
2	(aper)	Standard ▾	2.444	0.376	N-LAK9	1.208	U
3	(aper)	Standard ▾	1.778	5.388		1.055	U
4	(aper)	Standard ▾	-2.948	0.104	N-BAF52	0.238	U
5	(aper)	Standard ▾	-1.980	0.582		0.231	U
6	STOP	Standard ▾	Infinity	4.861 V		0.120	U
7		Standard ▾	-4.246 V	1.184 V	N-SF11	0.973	
8		Standard ▾	3.783 V	0.597 V	N-LAF2	1.283	
9		Standard ▾	-3.412 V	2.544 V		1.307	
10		Standard ▾	5465.425 V	0.210 V	N-SSK5	1.532	
11	(aper)	Standard ▾	-10.822 V	5.744 V		1.536	
12	IMAGE	Standard ▾	Infinity	-		1.450	

FIGURE 3.40 Resulting prescription of lens optimized for $D = -5.13\%$.

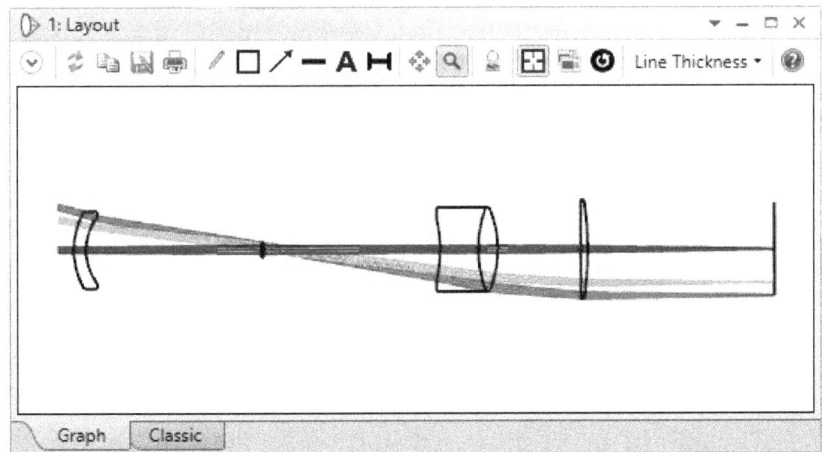

FIGURE 3.41 Layout of the lens given by the prescription in Fig. 3.40.

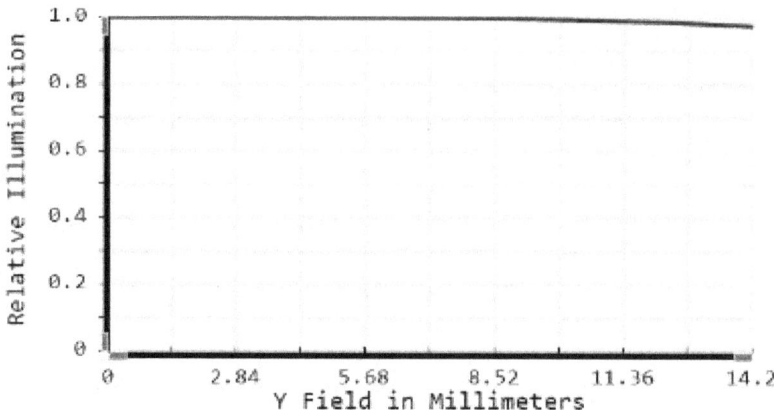

FIGURE 3.42 Relative illumination for the lens given by Fig. 3.40.

In Fig. 3.42, there is a small amount of residual non-uniformity, but that's mainly due to the approximations that we have been making in our calculations [note from the estimate in Eq. (3.18) that we did not obtain exactly the same full field total relative illumination as provided by Zemax OS in Fig. 3.27].

There is one more point to be made before closing this section. Recall that differential distortion is not necessarily zero when image distortion is zero. This being the case, it might seem odd that, in arriving at Eq. (3.28), we combined Eqs. (3.23) and (3.24) such that

$$\frac{dD}{dy} = \frac{2\sigma y}{m_p} = \frac{2\sigma y^2}{ym_p} = \frac{2D}{y}. \qquad (3.29)$$

Eq. (3.29) implies that differential distortion is zero when image distortion is zero, which contradicts my claim. However, Eq. (3.29) is true only when third order image distortion is the only order of distortion present. There are two ways in which a lens's image distortion can be zero. The first is when $\sigma = 0$, which is trivial. The second is when $\varepsilon_{y,D} = 0$ at some specific point along the field, such as at full field. This would occur if higher order aberrations that vary with field height are present to balance third order image distortion.

Under such a condition, $\sigma \neq 0$ and therefore, $dD/dy \neq 0$ at full field. In fact, under this condition, dD/dy is the instantaneous rate of change of the *total* balanced image distortion. When higher order field-dependent aberrations are present to "zero out" the third order distortion at full field, the image distortion curve never looks like the usual quadratic function given by Eq. (3.23). Take for example the lens we've just optimized (i.e., the one given by the prescription in Figs. 3.40 – 3.42.) A plot of this lens's image distortion is shown in Fig. 3.43, which is quadratic with field height [and therefore, is consistent with Eq. (3.23)]. Upon setting a target of zero for image distortion for this lens and re-optimizing, I obtained the results shown in Figs. 3.44 – 3.47.

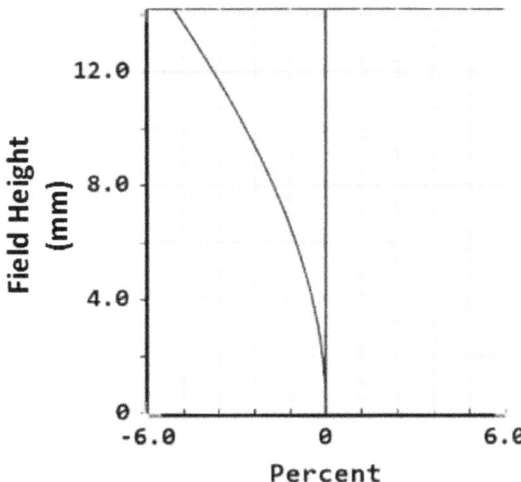

FIGURE 3.43 Image distortion (horizontal scale) for the lens in Fig. 3.41.

	Surf:Type		Radius	Thickness	Material	Semi-Diameter	
0	OBJECT	Standard ▾	Infinity	49.500		14.200	
1		Standard ▾	Infinity	0.500		1.420	
2	(aper)	Standard ▾	2.444	0.376	N-LAK9	1.208	U
3	(aper)	Standard ▾	1.778	5.388		1.055	U
4	(aper)	Standard ▾	-2.948	0.104	N-BAF52	0.238	U
5	(aper)	Standard ▾	-1.980	0.582		0.231	U
6	STOP	Standard ▾	Infinity	5.723 V		0.120	U
7		Standard ▾	-5.955 V	1.164 V	N-SF11	1.128	
8		Standard ▾	2.366 V	0.871 V	N-LAF2	1.457	
9		Standard ▾	-4.191 V	0.423 V		1.479	
10		Standard ▾	43.679 V	0.226 V	N-SSK5	1.548	
11	(aper)	Standard ▾	-12.279 V	6.773 V		1.553	
12	IMAGE	Standard ▾	Infinity	-		1.477	

FIGURE 3.44 Lens prescription after optimizing for $D = 0\%$.

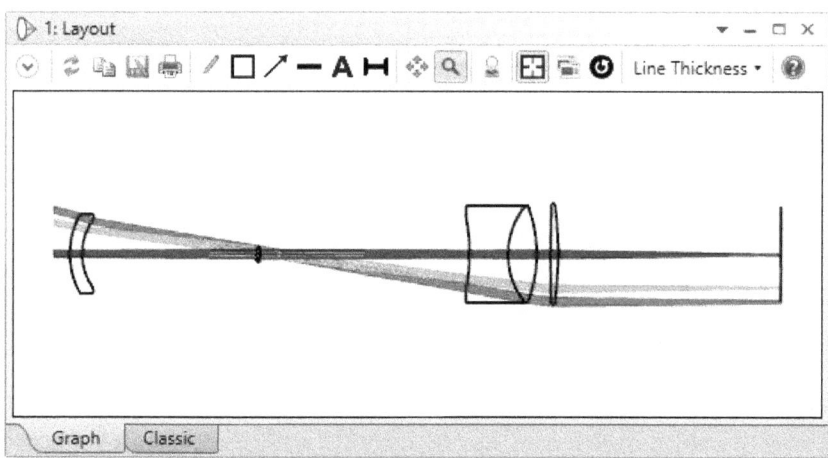

FIGURE 3.45 Layout for the lens given by the prescription in Fig. 3.44.

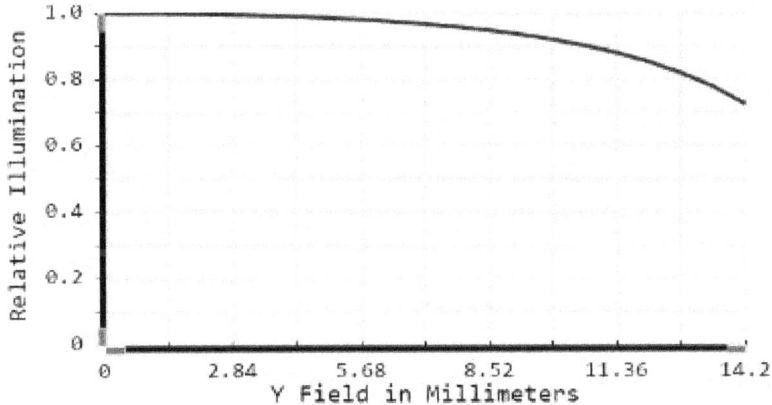

FIGURE 3.46 Relative illumination for the lens given by Fig. 3.44.

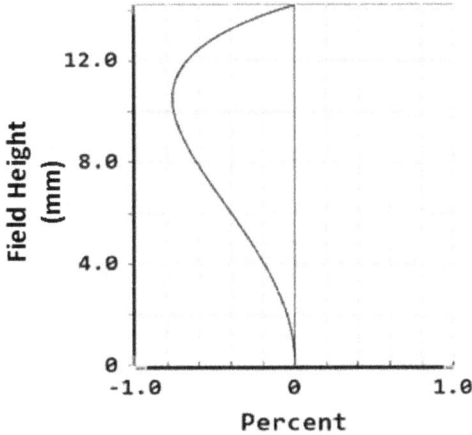

FIGURE 3.47 Image distortion for the lens given by the prescription in Fig. 3.44.

Comparing Fig. 3.47 with Fig. 3.43, we see that the image distortion curve in Fig. 3.47 is not quadratic. Note also that the slope of the distortion curve in Fig. 3.47 is not zero at full field. Hence, differential distortion is not zero there as well. In fact, computing for differential distortion at full field for this lens I obtain an estimate of $dD/dy \approx$ 0.00533271 mm^{-1} (note that it's a positive value this time). If we insert this value for differential distortion into Eq. (3.25) with $D = 0$ and $y = 14.2$ mm, we obtain $R_D = 0.929606062$. Applying this value for R_D along with the results from Eqs. (3.16) and (3.17) into Eq. (3.14) we obtain $R \approx 0.746$ for the image relative illumination at full field. Now, compare this value with the full field relative illumination obtained from Fig. 3.46. They are quite close. This is the role that differential distortion plays in the relative illumination of a lens system.

3.2.4 Chromatic aberration in machine vision

Chromatic aberration is a performance consideration for any optical imaging system used in broadband illumination, not just in machine vision. Hence, chromatic aberration is a design consideration for most imaging systems. Perhaps the simplest approach to correct it is to avoid dispersion altogether, which is achieved with the use of concave mirrors for imaging. However, mirrors often have limited field of view (which makes them more suited for telescopes). In contrast, lenses offer extended field coverage, but introduces dispersion. This brings about not just the question of how to correct chromatic aberration, but also how much correction is needed. The latter is difficult to answer. The total quality of a digital image is a function of a number of factors in the imaging chain, which adds complexity to an optical model. A practical approach is to test several existing lenses (whose chromatic aberrations are known) together with the intended image sensor for the end application. Even if the lenses being tested may not be the actual lenses used for the final application (perhaps because your intent is to have a custom designed lens), their data may be used to determine the level of color correction that is needed for the custom lens. That said, it is worth examining how to control and minimize chromatic aberration in lenses.

3.2.4.1 Achromatism

A *thin* lens is achromatic if it has the same EFL for two wavelengths. A doublet consisting of a cemented pair of thin lens elements has total EFL f given by

$$\frac{1}{f} = \frac{1}{f_1} + \frac{1}{f_2}.$$
(3.30)

Defining the optical power of a lens as $\phi = 1/f$, Eq. (3.30) may be expressed as

$$\phi = \phi_1 + \phi_2.$$
(3.31)

Ideally, we want the total optical power to be independent of wavelength, which means we want the rate of change of power with respect to wavelength to be zero. Differentiating Eq. (3.31) with respect to wavelength λ we have

$$\frac{d\phi}{d\lambda} = \frac{d\phi_1}{d\lambda} + \frac{d\phi_2}{d\lambda} = 0.$$
(3.32)

The power of any single element (in air) is proportional to $(n - 1)$, where n is the element's refractive index. If K is the constant of proportionality, we may express Eq. (3.32) as

$$K_1 \frac{dn_1}{d\lambda} + K_2 \frac{dn_2}{d\lambda} = 0.$$
(3.33)

In Eq. (3.33), because the sum has been set equal to zero, the denominators containing $d\lambda$ are not essential. Also, if it is only required that two wavelengths have the same focal length, then it is not required that the instantaneous rate of change of refractive index with wavelength be zero. If we let Δn_1 be the refractive index difference between the short and long wavelengths λ_S and λ_L

respectively for the first lens element, and if we apply the same representation for the second lens element, then we have

$$\Delta\phi = K_1\Delta n_1 + K_2\Delta n_2 = 0. \tag{3.34}$$

In theory, we're done. All that's needed is to find two different glasses, one for the first element, and the other for the second element, that satisfy Eqs. (3.34) and (3.31). But it is more convenient to express Eq. (3.34) in terms of ϕ_1 and ϕ_2. Since, as I mentioned earlier, the power of any single element (in air) is proportional to $(n-1)$, where the constant of proportionality is K, Eq. (3.34) may be expressed as

$$\frac{\phi_1}{(n_1-1)}\Delta n_1 + \frac{\phi_2}{(n_2-1)}\Delta n_2 = 0. \tag{3.35}$$

In Eq. (3.35), n_1 is the refractive index of the first lens element at some wavelength choice between the short wavelength λ_S and long wavelength λ_L. In other words, n_1 may lie somewhere between Δn_1, and n_2 may lie somewhere between Δn_2. In some cases, it is convenient to make the definition $\Delta n_1/(n_1-1) = D_1$, the dispersion of the first element, and $\Delta n_2/(n_2-1) = D_2$, the dispersion of the second element. Traditionally, their reciprocals are the so-called "Abbe dispersion" numbers $V_1 = 1/D_1$ and $V_2 = 1/D_2$ respectively. Using the Abbe number definitions for Eq. (3.35), we have

$$\frac{\phi_1}{V_1} + \frac{\phi_2}{V_2} = 0. \tag{3.36}$$

Applying Eq. (3.31) into (3.36), one obtains

$$\phi_1 = \frac{\phi}{1-(V_2/V_1)}. \tag{3.37}$$

And

$$\phi_2 = \frac{\phi}{1-(V_1/V_2)}. \tag{3.38}$$

In Eqs. (3.37) and (3.38), ϕ_1 is the optical power of the first element at n_1, and ϕ_2 is the optical power of the second element at n_2. These two equations suggest that achromatism is achieved for suitable choices of Abbe values. By "suitable" it is meant that the Abbe values should be as different as possible so that the denominators in Eqs. (3.37) and (3.38) are not zero.

3.2.4.2 Apochromatism

In some cases, one may require three wavelengths to have the same focus. For a thin doublet, suppose the third wavelength for the first element is the wavelength corresponding to n_1, and suppose we let this wavelength for the second element correspond to n_2. In other words, suppose that the third wavelength is between the short and long wavelengths λ_S and λ_L. If we let δn_1 be the refractive index difference between the third wavelength and λ_L for element 1, and if we let δn_2 be the refractive index difference between the third wavelength and λ_L for element 2, then we can impose a second condition in addition to Eq. (3.34) for the doublet, namely,

$$\delta\phi = K_1\delta n_1 + K_2\delta n_2 = 0. \tag{3.39}$$

In terms of the optical powers of the first and second elements, Eq. (3.39) may therefore be expressed as

$$\frac{\phi_1}{(n_1 - 1)}\delta n_1 + \frac{\phi_2}{(n_2 - 1)}\delta n_2 = 0. \tag{3.40}$$

Multiplying the first term in Eq. (3.40) by $\Delta n_1/\Delta n_1$ and the second term by $\Delta n_2/\Delta n_2$, we have

$$\frac{\phi_1}{V_1}\frac{\delta n_1}{\Delta n_1} + \frac{\phi_2}{V_2}\frac{\delta n_2}{\Delta n_2} = 0. \tag{3.41}$$

The quantities $\delta n_1/\Delta n_1$ and $\delta n_2/\Delta n_2$ are the so-called "relative partial dispersions" P_1 and P_2 of the glasses for the first and second elements respectively. In terms of these quantities, Eq. (3.41) may be written

$$\frac{\phi_1}{V_1} P_1 + \frac{\phi_2}{V_2} P_2 = 0. \qquad (3.42)$$

If $P_1 = P_2 = P$, then we have

$$P\left(\frac{\phi_1}{V_1} + \frac{\phi_2}{V_2}\right) = 0. \qquad (3.43)$$

Eq. (3.43) suggests that it is possible for a thin doublet to possess the same EFL for three wavelengths if the glasses for each element have the same relative partial dispersions, because the term $[(\phi_1/V_1) + (\phi_2/V_2)]$ can be made to satisfy Eqs. (3.36) – (3.38). Let us take a pause to think about something rather interesting.

Pause for insight: $\rightarrow$ Take a look at the ratios $\Delta n_1/(n_1 - 1)$ and $\Delta n_2/(n_2 - 1)$ in Eq. (3.35). Do they look like "relative dispersions"? They do, don't they? In fact, they look like they could be called "relative total dispersions" with respect to vacuum, whose refractive index is unity. Recall from Eq. (3.41) that we defined the quantities $\delta n_1/\Delta n_1$ and $\delta n_2/\Delta n_2$ as relative partial dispersions. In that case, couldn't we make the definition that the ratios $\Delta n_1/(n_1 - 1)$ and $\Delta n_2/(n_2 - 1)$ in Eq. (3.35) be relative total dispersions? This was why I suggested in Sec. 3.2.4.1 that, in some cases, it is convenient to define the ratios $\Delta n_1/(n_1 - 1)$ and $\Delta n_2/(n_2 - 1)$ in terms of the dispersions D_1 and D_2 respectively. In this case, we could re-write Eq. (3.42) as

$$\phi_1 D_1 P_1 + \phi_2 D_2 P_2 = 0. \qquad (3.44)$$

Applying Eq. (3.31) into (3.44) we obtain

$$\phi_1 = \frac{\phi}{1 - \left(\frac{D_1 P_1}{D_2 P_2}\right)}. \tag{3.45}$$

And

$$\phi_2 = \frac{\phi}{1 - \left(\frac{D_2 P_2}{D_1 P_1}\right)}. \tag{3.46}$$

Oddly, Eqs. (3.44) – (3.46) suggest that, for a thin apochromatic doublet, we may either choose two glasses where $D_1 \neq D_2$ and $P_1 = P_2$, or a pair of glasses where $D_1 = D_2$ but $P_1 \neq P_2$, if such glasses exist. I haven't found any.

3.2.4.3 Zemax OS example: Automated apochromat glass selection

Traditionally, one would browse glass supplier catalogs to select glasses with suitable Abbe values and relative partial dispersions to achieve either achromatism or apochromatism. However, the data in such catalogs are usually restricted to a number of wavelengths. For example, the common Abbe value V_d is often defined as $V_d = (n_d - 1)/(n_F - n_C)$ where n_d is the refractive index at the "d" wavelength (587.56 nm), n_F is the refractive index at the "F" wavelength (486.13 nm), and n_C is the refractive index at the "C" wavelength (656.27 nm). In the course of deriving the equations for achromatism and apochromatism in the previous two sections, there is no reason to restrict the computation of Abbe dispersions or relative partial dispersions to some agreed wavelengths. Any wavelengths may be used. Optical design programs usually contain glass libraries and dispersion data from a variety of glass suppliers. Therefore, one may simply construct a merit function that computes Abbe values and relative partial dispersions at any wavelength.[*†] In Zemax OS, one uses the INDX operand to obtain refractive index values at any wavelength for any glass that has been entered into the lens data editor. So, let's try a simple example using Zemax OS. We will design an apochromatic thin doublet.

Set up a doublet using the prescription provided in Fig. 3.48 (units in mm) and use a single field point at 0 degrees. Use "float by stop size" for the aperture setting. The lens layout is shown in Fig. 3.49. Surfaces 2 and 3 at this point actually do not need to have any powers, but I've applied radii and thickness values chosen at random. Note that this lens is not even achromatic at this stage. As shown in Fig. 3.48, place the "glass substitute" solves on the materials on surfaces 2 and 3.

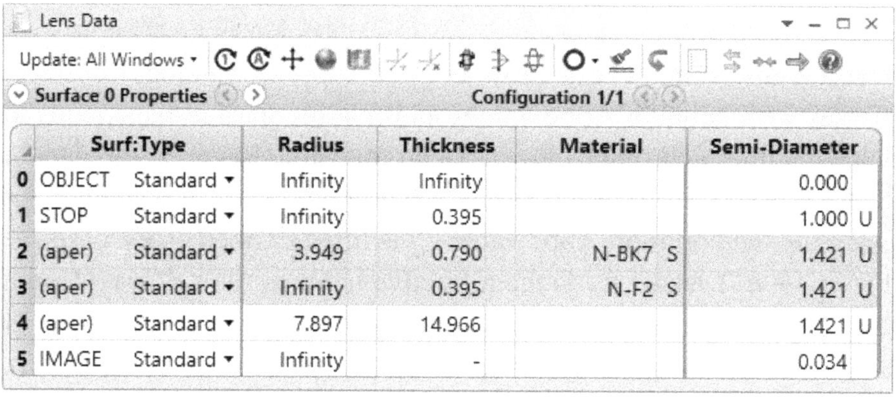

	Surf:Type		Radius	Thickness	Material		Semi-Diameter	
0	OBJECT	Standard ▾	Infinity	Infinity			0.000	
1	STOP	Standard ▾	Infinity	0.395			1.000	U
2	(aper)	Standard ▾	3.949	0.790	N-BK7	S	1.421	U
3	(aper)	Standard ▾	Infinity	0.395	N-F2	S	1.421	U
4	(aper)	Standard ▾	7.897	14.966			1.421	U
5	IMAGE	Standard ▾	Infinity	-			0.034	

FIGURE 3.48 Prescription to set up a glass search for thin doublet lens.

In the wavelengths dialogue box, use three wavelengths at 450, 550, and 650 nm, and set 550 nm as the primary wavelength. Set all three wavelength weights to 1. In some cases, it is effective to apply

---

[*] This seems somewhat similar to an analysis described by C. Alexay, "An achromatic lens for visible to far-IR light," *SPIE Newsroom* (10, November, 2010). DOI: 10.1117/2.1201011.003292.

[†] However, apochromatic single-glass dialyte designs clearly do not require much automation for glass selection (e.g., Ref. 8, pp. 465 – 468) and R. B. Johnson, "Very-broad spectrum afocal telescope," *Proc. SPIE* **3482** (1998).

the "Gaussian quadrature" spectral weights by clicking on the button that reads this at the lower right corner of the wavelength dialogue box in Zemax OS, but we'll stick to our weight of unity for now. Set up the operands in the merit function editor as shown in Fig. 3.50.

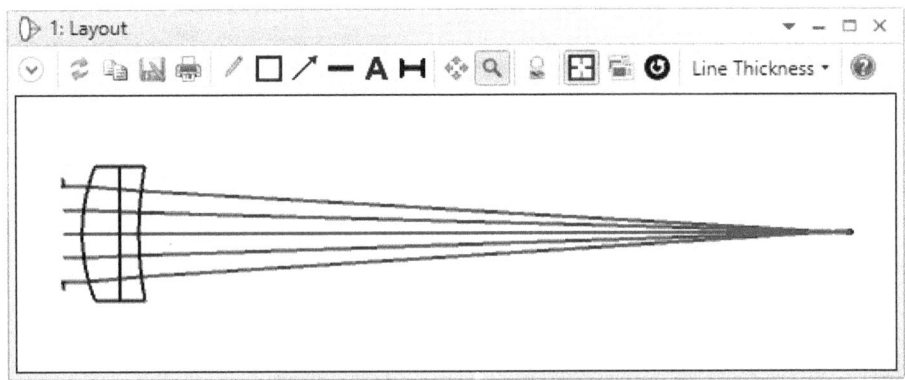

FIGURE 3.49 Layout for the doublet whose prescription is given in Fig. 3.48.

In Fig. 3.50, note that only the operands on lines 24, 27, and 28 have weights. These operands constrain the program to search for glasses that meet the criterion that the Abbe values are different and that the relative partial dispersions are close. For convenience, note on line 28 that I have forced the program to consider having a positive powered front element for this doublet. This is done by purposely constraining the ratio $V_2/V_1 < 0.8$. In "System Explorer", under "Material Catalogs", include the glass supplier Schott. Now, click on the "Hammer Current" button at the Optimize menu to bring up the "Hammer Optimization" dialogue box. Click the "Start" button to initiate hammer optimization using the damped least squares algorithm (do not click the "Automatic" button), which will perform an exhaustive search and substitution of glasses on surfaces 2 and 3 from the suppliers considered. After about 18 seconds, the prescription in my lens data editor has been altered to the state shown in Fig. 3.51.

Merit Function Editor

Wizards and Operands — Merit Function: 0.0337124016236206

	Type	Surf	Wave	Target	Weight	Value
1	CONS			1.000	0.000	1.000
2	INDX	2	1	0.000	0.000	1.525
3	INDX	2	2	0.000	0.000	1.519
4	INDX	2	3	0.000	0.000	1.515
5	DIFF	3	1	0.000	0.000	0.519
6	DIFF	2	4	0.000	0.000	0.011
7	DIFF	3	4	0.000	0.000	4.002E-03
8	BLNK	V1				
9	DIVI	5	6	0.000	0.000	48.015
10	BLNK	P1				
11	DIVI	7	6	0.000	0.000	0.371
12	INDX	3	1	0.000	0.000	1.639
13	INDX	3	2	0.000	0.000	1.624
14	INDX	3	3	0.000	0.000	1.615
15	DIFF	13	1	0.000	0.000	0.624
16	DIFF	12	14	0.000	0.000	0.023
17	DIFF	13	14	0.000	0.000	8.208E-03
18	BLNK	V2				
19	DIVI	15	16	0.000	0.000	26.656
20	BLNK	P2				
21	DIVI	17	16	0.000	0.000	0.351
22	BLNK	Achromatism				
23	DIVI	19	9	0.000	0.000	0.555
24	OPLT	23		0.800	1.000	0.800
25	BLNK	Apochromatism				
26	DIVI	11	21	0.000	0.000	1.056
27	OPLT	26		1.003	2.000	1.056
28	OPGT	26		0.998	2.000	0.998

FIGURE 3.50 Merit function construction to search glasses.

The reason why we're not clicking on the "Automatic" button for Hammer optimization is because we want to force the program to continue searching without "deciding" whether or not the merit function value is sufficiently minimized. In theory, one has to actually iterate several times between wavelength choices and Hammer optimization in order search the best glass choice, which may be performed through writing a script or code in Zemax Programming Language (ZPL). But my objective here is simply to show you the basic idea. In my exercise, the merit function value just prior to Hammer optimization began at 0.033712402, and I stopped the optimization at a merit function value of 0.000448224, which resulted in the glasses shown in Fig. 3.51.

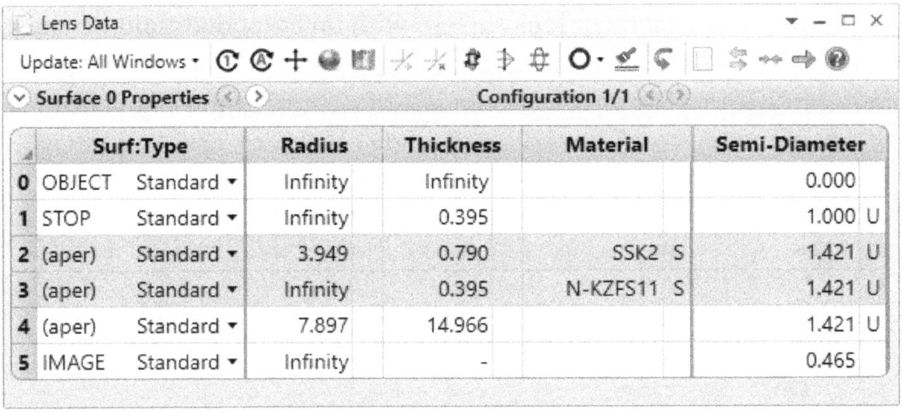

	Surf:Type		Radius	Thickness	Material	Semi-Diameter	
0	OBJECT	Standard ▾	Infinity	Infinity		0.000	
1	STOP	Standard ▾	Infinity	0.395		1.000	U
2	(aper)	Standard ▾	3.949	0.790	SSK2 S	1.421	U
3	(aper)	Standard ▾	Infinity	0.395	N-KZFS11 S	1.421	U
4	(aper)	Standard ▾	7.897	14.966		1.421	U
5	IMAGE	Standard ▾	Infinity	-		0.465	

FIGURE 3.51 Prescription with new glasses after hammer optimization.

Now, remove the Solves for the materials on surfaces 2 and 3. Set variables for the radii on surfaces 2, 3, and 4, as well as a variable for the thickness on surface 4. Delete all operands in the merit function (save your lens file first if you wish) and build a new merit function to appear as shown in Fig. 3.52. Note that I've applied the default merit function for RMS spot size with 3 rings and 6 arms (overall weight = 1). I've also included the EFL operand "EFFL|" with a target of 16 mm and weight of 1 at the primary wavelength of 550 nm. The operands "AXCL" target axial color to zero for the three wavelengths at a pupil zone of 0.7, and they each have weights of 2. After clicking on the automatic damped least squares optimization button once, I obtained the prescription shown in Fig. 3.53. One will note that the chromatic focal shift plot does not fully show apochomatic correction across the current chosen visible bandwidth 450 – 650 nm. To see the full apochromatism, open the wavelength settings and include one more wavelength at 1200 nm. The layout and chromatic focal shift plots for this optimized doublet are shown in Figs. 3.54 and 3.55

respectively. Note that there are three focal plane crossings across the bandwidth of 450 – 1200 nm for this lens. Now, this is just a simple doublet with low FOV. An apochromatic machine vision lens should have a larger FOV, which would naturally require more than two lens elements to achieve. In the following section, we will arrive at an apochromatic 10 element lens system with an extended field of view.

	Type	Wave	Hx	Hy	Px	Py		Target	Weight	Value
1	EFFL ▾	2						16.000	1.000	11.574
2	AXCL ▾ 1	3	0.700					0.000	2.000	0.234
3	AXCL ▾ 2	3	0.700					0.000	2.000	0.087
4	BLNK ▾									
5	DMFS ▾									
6	BLNK ▾ Sequential merit function: RMS spot radius centroid GQ 3 rings 6 arms									
7	BLNK ▾ No air or glass constraints.									

Merit Function Editor — Wizards and Operands — Merit Function: 1.56796500330791

FIGURE 3.52 Merit function to optimize lens given by Fig 3.51.

Lens Data — Update: All Windows ▾ — Surface 0 Properties — Configuration 1/1

	Surf:Type		Radius	Thickness	Material	Semi-Diameter
0	OBJECT	Standard ▾	Infinity	Infinity		0.000
1	STOP	Standard ▾	Infinity	0.395		1.000 U
2	(aper)	Standard ▾	3.949 V	0.790	SSK2	1.421 U
3	(aper)	Standard ▾	Infinity V	0.395	N-KZFS11	1.421 U
4	(aper)	Standard ▾	7.897 V	14.966 V		1.421 U
5	IMAGE	Standard ▾	Infinity	-		0.465

FIGURE 3.53 Prescription for the optimized apochromatic doublet.

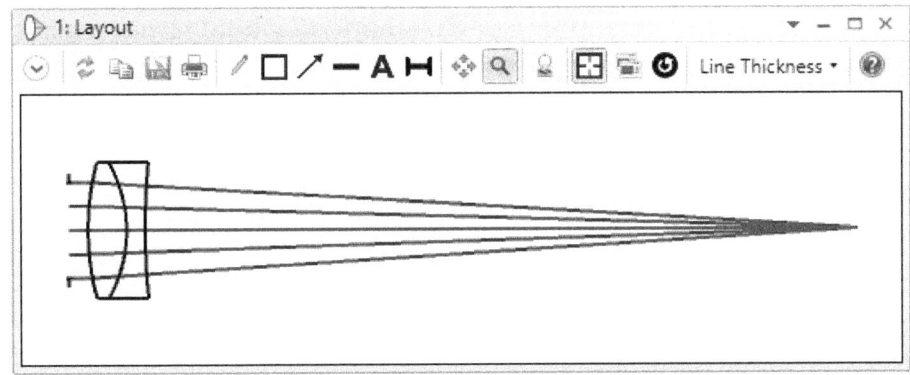

FIGURE 3.54 Apochromatic doublet optimized from Fig. 3.49.

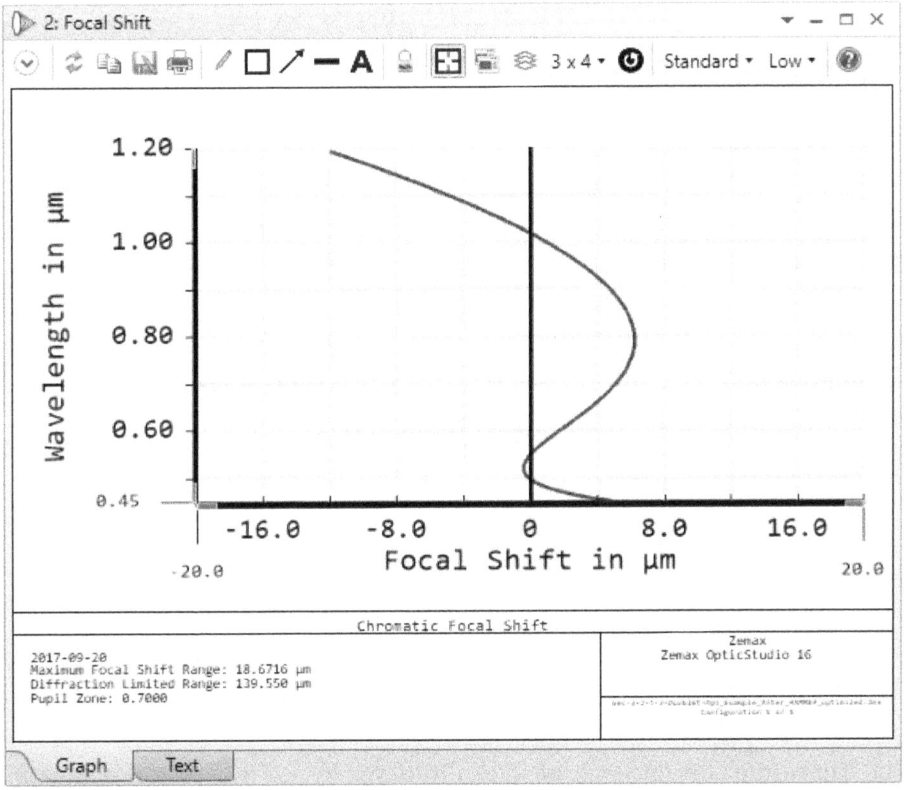

FIGURE 3.55 Chromatic focal shift plot (at the 0.7 pupil zone) for the lens in Figs. 3.53 – 3.54.

3.2.4.4 Zemax OS example: Apochromatic machine vision lens with extended field of view

The method of finding pairs of glasses with nearly the same relative partial dispersions (but different Abbe values) to achieve apochromatic designs follow closely the work of M. Herzberger [9, 10], and it is discussed in many books on optical design. In 1985, P. Robb [11] introduced a method of design and selection of optical glasses that does not apply the method above. Shortly after Robb's publication, R. D. Sigler [12] extended Robb's method to include systems of thin lenses.* More recently, A. Yang *et al.* successfully applied Sigler's method to the design of a 10-element apochromatic lens (corrected at the 0.7 pupil zone) with an extended FOV for machine vision [13]. Yang *et al.* referred to Sigler's method as "dispersion vector analysis" (DVA). Following from Yang *et al.*'s work, I showed [14] that one stage of the glass search and selection process may be simplified and automated in an optical design program, such as Zemax OS. The method is very rapid – taking seconds to minutes – to arrive at a selection of glass materials that, upon a final step in optimization, would tend to yield a completely apochromatic design. This is the method we shall apply in the current section.

For a system of thin lenses, the axial color Δ (i.e., the difference in focal positions along the optic axis between any two wavelengths) is given by [15]

$$\Delta = -\frac{1}{\phi_o Y_1} \sum_{i=1}^{n} \frac{\phi_i Y_i^2 P_i}{V_i}. \tag{3.47}$$

In Eq. (3.47), ϕ_o is the total system power, Y_1 is the marginal ray height at the first lens's surface, ϕ_i is the power of the i'th lens, Y_i is the marginal ray height at the i'th lens, P_i is the relative partial

---

* The study of apochromats is extensive, and a thorough list of references on this subject is in fact rather long. Thus, the references cited here focus only on those that are of immediate relevance to the discussion.

dispersion of the i'th lens, and V_i is the Abbe value of the i'th lens. In Sigler's DVA approach, one regards the sum in Eq. (3.47) as a vector sum, which interprets the products in the sum as vector components. As such, DVA enables one to plot "vectors" on a graph, which is appealing visually and is indeed helpful. Minimizing axial color in DVA is regarded as a process of minimizing the total vector on the graph. But modern lens design involves the use of computer programs that can calculate the axial color given by Eq. (3.47) through ray tracing. Hence, this eliminates the need to plot dispersion vectors on a graph. One simply needs to select optical glasses that would minimize the total system axial color.

Now, in section VII of his paper, Sigler [12] suggested that if a lens design is at a state where it is achromatic (but not necessarily apochromatic) with total system power ϕ_o, then improvements can be made on glass choices to improve the total system axial color Δ without significantly affecting the system power ϕ_o. Mathematically, it means that P_i and V_i in Eq. (3.47) may vary to reduce Δ whilst maintaining ϕ_o within some reasonable range. If ϕ_o deviates significantly in the process of substituting existing glasses with alternatives, then one need only to iterate between manually varying the choice of glasses (by applying DVA) and re-optimizing lens variables. Yang *et al.* recognized this in their paper [13], and they performed this iterative process successfully, alternating between applying DVA to select glasses and re-optimizing lens variables to maintain the system power.

By the way, this process of iteration is also implied by Eqs. (3.31), (3.36), and (3.42) for the case of a thin apochromatic doublet. In this case, if one has optimized a doublet to possess power $\phi = \phi_1 + \phi_2$, then one may continue to select different pairs of glasses to improve the conditions in Eqs. (3.36) and (3.42) with the hope that they do not significantly impact ϕ. If new pairs of glasses significantly impact ϕ, then re-optimization of lens radii and thicknesses would be required, and the process iterates. In our worked example involving the automatic glass selection process for a thin apochromatic doublet, we did not perform this iteration because a doublet's analytic solutions were sufficiently simple. We knew that we needed $P_1 = P_2$ and $V_1 \neq V_2$. So, the problem was simply to automate the search for glasses that satisfy these criteria. However, the problem is more complex for a system of thin lenses. In this case, it would appear to be simpler if the total system were first made to be approximately achromatic at some total system power.

In the paper that I wrote [14], the basic idea was that if a system of thin lenses has sufficiently large number of elements (admittedly, it has not been determined what "sufficient" means), then iteration should not be required. One should be able to let the optical design program automatically search and substitute existing glasses with alternatives that serve to reduce the axial color between pairs of desired wavelengths whilst constraining the system's total EFL to remain within some reasonable range through the use of control operands in the merit function editor. Upon completion of the automated glass selection process, a final optimization would yield the desired lens. So, let us try an example using Zemax OS.

As usual, the approach is to start with a design form that we know has the capacity to become what we want it to be. In this case, it makes sense to apply the form that Yang *et al.* had developed successfully, which was also the starting point in my paper [14]. Yang *et al.* referred to their design form as the "Preliminary Optimized Structure" (POS) [13]. However, subsequent to the publication of my paper [14], I had found that the choices of glasses used in Yang *et al.*'s POS were already so well chosen that, even if DVA weren't applied to select alternative glasses, the POS actually had the potential to become apochromatic. One reason perhaps was that at least two glasses had very low dispersion. Therefore, rather than to apply the same glass choices as Yang *et al.*'s for our POS, we shall use a different set of initial "more normal" glasses that cannot serve to make the total lens system apochromatic (thereby requiring the automated glass selection process we wish to try), but does provide achromatism. I have put together such a POS, and its layout and prescriptions are provided in Figs. 3.56 and 3.57 respectively.

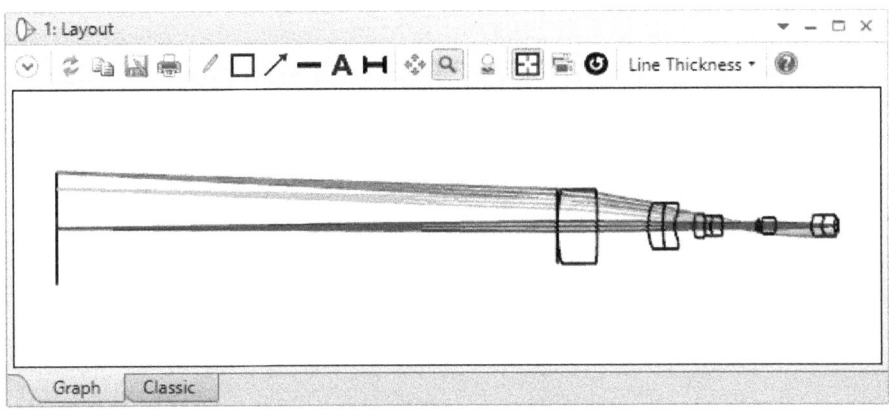

FIGURE 3.56 Machine vision lens starting design form, based on Yang *et al.*'s POS [12].

The lens system in Fig. 3.56 has field heights at 0, 8, and 11.33 mm. Wavelength settings are at 450, 550, and 650 nm (weights = 1). The aperture setting is "Image Space F-number" with a value of "5". Ray aiming is "on", and the chief ray is aimed at the paraxial entrance pupil. All lens semi-diameters are made to float (including the stop). The system has been scaled for a 1/6" image sensor. Note in Fig. 3.57 that "glass substitute" Solves have been placed on all glasses. The system is achromatic, and the image distortion is a little under 5% barrel. The MTF is good, near diffraction limited, at about 50% modulation at 110 cycles/mm. The merit function construction is shown in Fig. 3.58. Note that I am constraining the RMS average of the axial colors (at the 0.7 pupil zone) because at this stage, one ordinarily does not know if apochromatism is "minimizable" for the selected wavelength range. Note also that it is rather arbitrary what target to set for the operand on line 11, which constrains the RMS average axial color on line 10. I have arbitrarily chosen 1/10th of the present value. As for glass suppliers, go to the Materials Catalog menu under "System Explorer". I arbitrarily selected CDGM, HOYA, OHARA, and SCHOTT.

Lens Data

Update: All Windows

Surface 0 Properties Configuration 1/1

	Surf:Type		Radius	Thickness	Material		Semi-Diameter
0	OBJECT	Standard	Infinity	100.000			11.330
1		Standard	Infinity	0.000			7.531
2		Standard	33.079	8.000	N-BK7	S	7.499
3		Standard	-121.129	10.047			6.773
4		Standard	11.315	2.950	N-SF15	S	4.619
5		Standard	25.556	2.428	N-SK4	S	3.948
6		Standard	7.045	3.766			3.085
7		Standard	8.333	2.237	N-FK5	S	2.409
8		Standard	-12.738	0.020			2.010
9		Standard	-15.402	1.009	N-BASF2	S	1.986
10		Standard	3.369	2.230	N-BK7	S	1.671
11		Standard	13.735	4.224			1.370
12	STOP	Standard	Infinity	2.682			0.555
13		Standard	8.952	0.543	N-LASF41	S	1.137
14		Standard	4.759	0.487			1.168
15		Standard	5.400	2.880	N-FK5	S	1.344
16		Standard	-6.801	7.189			1.659
17		Standard	4.873	2.134	N-SK14	S	2.169
18		Standard	2.940	2.507	N-SF15	S	1.837
19		Standard	3.039	0.712			1.428
20		Standard	Infinity	4.587E-03			1.450
21	IMAGE	Standard	Infinity	-			1.451

FIGURE 3.57 Prescription for the POS lens in Fig. 3.56.

Now, go to the optimize menu and select "Hammer Current". You should see the dialogue box shown in Fig. 3.59. Use "Damped Least Squares" as the algorithm. Click "Start" to initiate the process of automatic glass selection (note that in my paper [14], I had used the

Global Search function instead, simply because I did not know at the time that Hammer Optimization can also be used to select glasses).

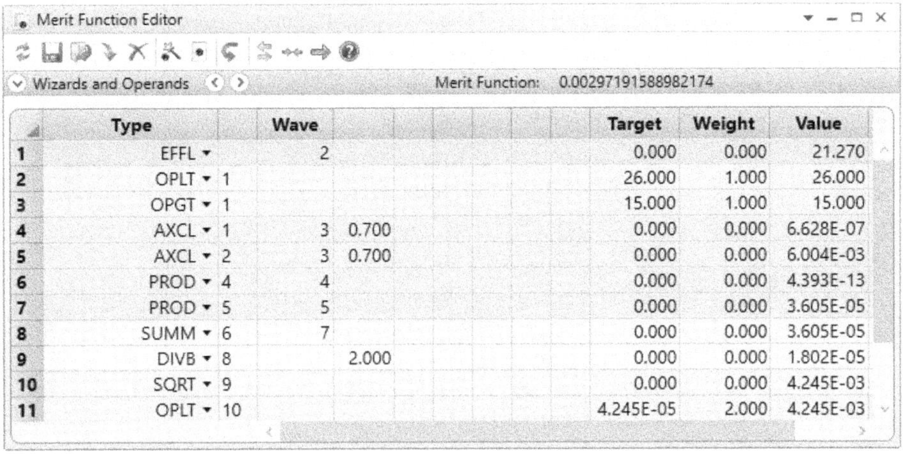

	Type	Wave				Target	Weight	Value
1	EFFL ▾	2				0.000	0.000	21.270
2	OPLT ▾ 1					26.000	1.000	26.000
3	OPGT ▾ 1					15.000	1.000	15.000
4	AXCL ▾ 1	3	0.700			0.000	0.000	6.628E-07
5	AXCL ▾ 2	3	0.700			0.000	0.000	6.004E-03
6	PROD ▾ 4	4				0.000	0.000	4.393E-13
7	PROD ▾ 5	5				0.000	0.000	3.605E-05
8	SUMM ▾ 6	7				0.000	0.000	3.605E-05
9	DIVB ▾ 8		2.000			0.000	0.000	1.802E-05
10	SQRT ▾ 9					0.000	0.000	4.245E-03
11	OPLT ▾ 10					4.245E-05	2.000	4.245E-03

Merit Function Editor — Merit Function: 0.00297191588982174

FIGURE 3.58 Merit function setup for the POS lens in Fig. 3.57, prior to auto glass selection.

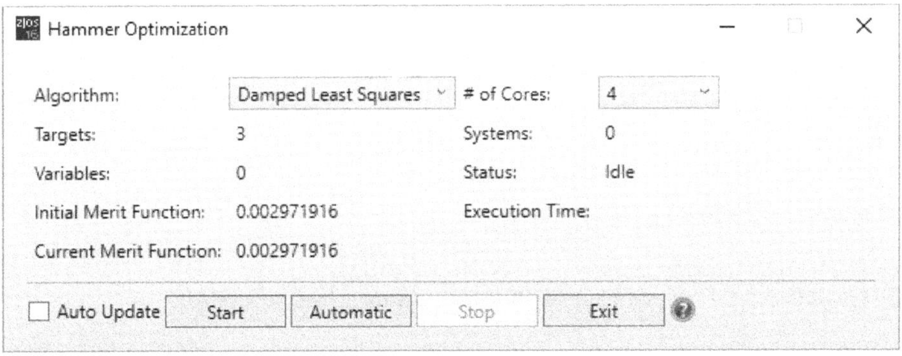

Hammer Optimization

Algorithm:	Damped Least Squares	# of Cores:	4
Targets:	3	Systems:	0
Variables:	0	Status:	Idle
Initial Merit Function:	0.002971916	Execution Time:	
Current Merit Function:	0.002971916		

☐ Auto Update | Start | Automatic | Stop | Exit

FIGURE 3.59 Hammer Optimization dialogue box.

After about 5 minutes, I terminated the optimization and obtained the resulting prescription shown in Fig. 3.61. Note that the optimization time depends on what target you set for the axial color. If you had set a larger value than what I had used on line 11 in the merit function editor (Fig. 3.58), then the optimization time is faster, and sometimes the "current merit function value" could be made zero.

		Surf:Type	Radius	Thickness	Material	Semi-Diameter
0	OBJECT	Standard ▾	Infinity	100.000		11.330
1		Standard ▾	Infinity	0.000		8.496
2		Standard ▾	33.079	8.000	BSL1 S	8.465
3		Standard ▾	-121.129	10.047		7.745
4		Standard ▾	11.315	2.950	TIH14 S	5.397
5		Standard ▾	25.556	2.428	M-BACD15 S	4.701
6		Standard ▾	7.045	3.766		3.619
7		Standard ▾	8.333	2.237	BAL12 S	2.879
8		Standard ▾	-12.738	0.020		2.472
9		Standard ▾	-15.402	1.009	ZF8 S	2.427
10		Standard ▾	3.369	2.230	L-PHL2 S	1.959
11		Standard ▾	13.735	4.224		1.574
12	STOP	Standard ▾	Infinity	2.682		0.359
13		Standard ▾	8.952	0.543	NBFD3 S	1.086
14		Standard ▾	4.759	0.487		1.137
15		Standard ▾	5.400	2.880	N-PSK57 S	1.345
16		Standard ▾	-6.801	7.189		1.662
17		Standard ▾	4.873	2.134	P-SK60 S	2.341
18		Standard ▾	2.940	2.507	SF18 S	2.063
19		Standard ▾	3.039	0.712		1.686
20		Standard ▾	Infinity	4.587E-03		1.726
21	IMAGE	Standard ▾	Infinity	-		1.727

FIGURE 3.61 Prescription of POS after auto glass selection.

Now, if you check the chromatic focal shift plot (Fig. 3.62), there is the classic "S" curve for apochromatism, though I do not always expect to see it at this stage. A final optimization is always required to ensure full apochromatism. Also, the final optimization would ensure the correction of monochromatic aberrations as well as meeting all other first order properties such as the system EFL. Now, remove all Glass Substitute solves. Place variables on all surface radii, and thickness variables on all surfaces except at the object and surface 1. Delete surface 20 (that was originally used for the lens in Fig. 3.57

to place the marginal ray at the paraxial focus, but it is now a redundant surface). The lens data editor should now look as shown in Fig. 3.63. Set up a merit function to appear as shown in Fig. 3.64. Note that the default merit function for RMS spot size is used starting at line 13 (lines beyond 13 are not displayed) with 3 rings and 6 arms, and overall weight = 1. Note that image distortion is not constrained. For those not using Zemax OS, build a merit function that constrains EFL = 21 mm, image height = 1.45 mm (use your program's sign convention), total lens thickness < 60 mm, 0.5 mm < glass thicknesses < 8 mm, glass edge thicknesses > 0.1 mm, and air center and edge thicknesses (between surfaces 3 and 17) > 0.01 mm. Leave the chief ray angles in object and image spaces unconstrained.

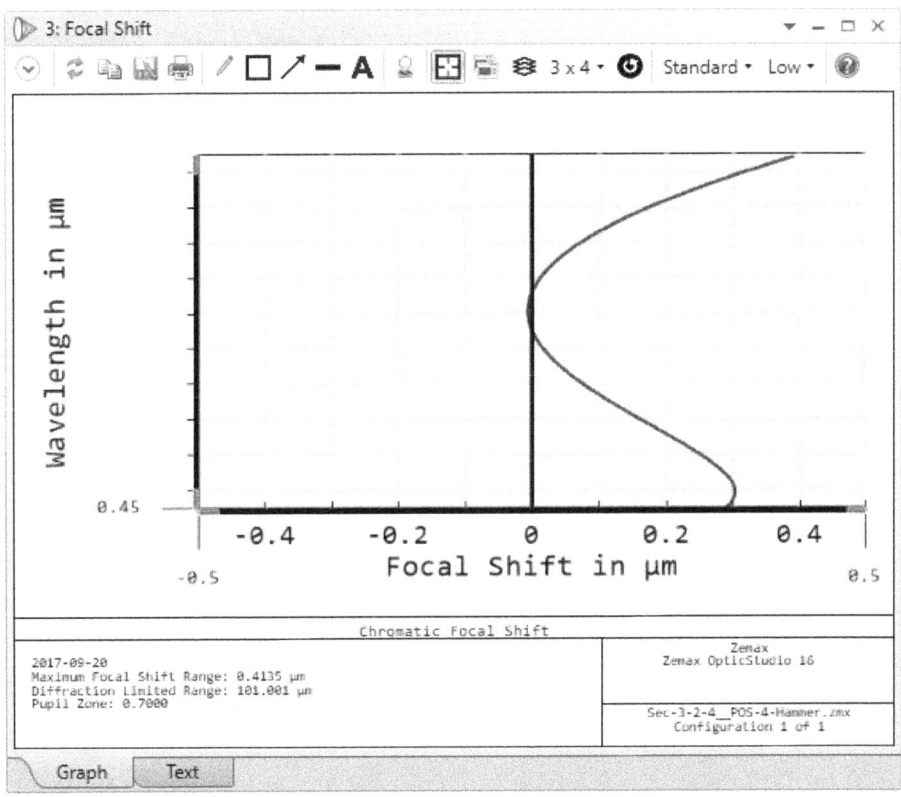

FIGURE 3.62 Chromatic focal shift (0.7 pupil zone) for the lens given by Fig. 3.61.

Lens Data

Update: All Windows

Surface 0 Properties | Configuration 1/1

	Surf:Type		Radius		Thickness		Material	Semi-Diameter
0	OBJECT	Standard	Infinity		100.000			11.330
1		Standard	Infinity		0.000			8.496
2		Standard	33.079	V	8.000	V	BSL1	8.465
3		Standard	-121.129	V	10.047	V		7.745
4		Standard	11.315	V	2.950	V	TIH14	5.397
5		Standard	25.556	V	2.428	V	M-BACD15	4.701
6		Standard	7.045	V	3.766	V		3.619
7		Standard	8.333	V	2.237	V	BAL12	2.879
8		Standard	-12.738	V	0.020	V		2.472
9		Standard	-15.402	V	1.009	V	ZF8	2.427
10		Standard	3.369	V	2.230	V	L-PHL2	1.959
11		Standard	13.735	V	4.224	V		1.574
12	STOP	Standard	Infinity		2.682	V		0.359
13		Standard	8.952	V	0.543	V	NBFD3	1.086
14		Standard	4.759	V	0.487	V		1.137
15		Standard	5.400	V	2.880	V	N-PSK57	1.345
16		Standard	-6.801	V	7.189	V		1.662
17		Standard	4.873	V	2.134	V	P-SK60	2.341
18		Standard	2.940	V	2.507	V	SF18	2.063
19		Standard	3.039	V	0.712	V		1.686
20	IMAGE	Standard	Infinity		-			1.726

FIGURE 3.63 Variables for prescription from Fig 3.61 set up for final optimization.

Clicking once on the local damped least squares optimization button, I obtain the results shown in Figs. 3.65 – 3.69 for the layout, prescription, chromatic focal shift, polychromatic MTF, and relative illumination respectively. The chromatic focal shift plot in Fig. 3.67 (at the 0.7 pupil zone) clearly shows the apochromatism within the 450 – 650 nm wavelength band. Ideally, for a machine vision lens, the object space chief ray angle should be near zero (telecentric), but the current lens's angle is reasonably small, and it may be improved through further optimization.

	Type	Wave	Hx	Hy	Px	Py	Target	Weight	Value
1	EFFL ▾	2					21.000	1.000	21.088
2	REAY ▾ 20	2	0.000	1.000	0.000	0.000	-1.450	1.000	-1.532
3	DISG ▾ 1	2	0.000	1.000	0.000	0.000	0.000	0.000	-5.947
4	AXCL ▾ 1	3	0.700				0.000	2.000	-1.596E-04
5	AXCL ▾ 2	3	0.700				0.000	2.000	3.152E-04
6	TTHI ▾ 2	18					0.000	0.000	55.334
7	OPLT ▾ 6						60.000	1.000	60.000
8	MNCG ▾ 2	19					0.500	1.000	0.500
9	MXCG ▾ 2	19					8.000	1.000	8.000
10	MNEG ▾ 2	19	1.000				0.100	1.000	0.100
11	MNCA ▾ 3	17					1.000E-02	1.000	1.000E-02
12	MNEA ▾ 3	17	1.000				1.000E-02	1.000	1.000E-02
13	DMFS ▾								
14	BLNK ▾ Sequential merit function: RMS spot radius centroid GQ 3 rings 6 arms								

Merit Function Editor — Wizards and Operands — Merit Function: 0.0604677166062578

FIGURE 3.64 Merit function for final optimization of lens from Fig. 3.63.

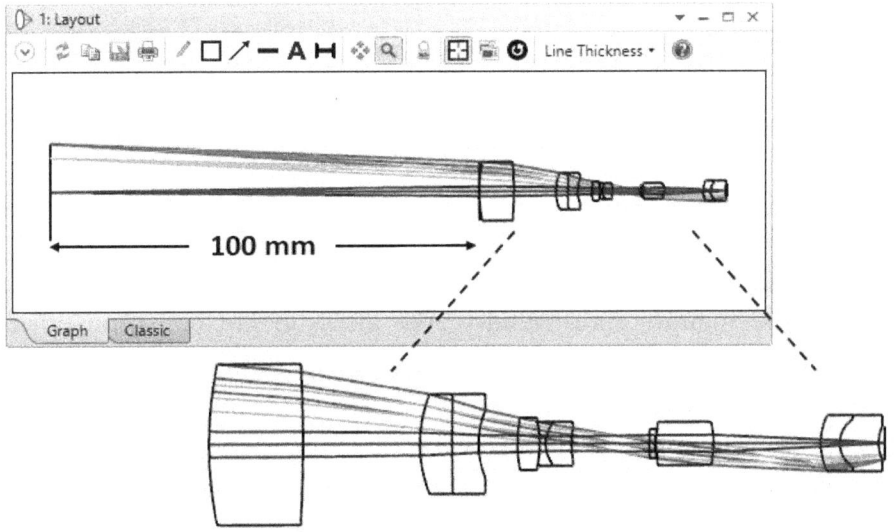

FIGURE 3.65 Lens layout after optimization.

Lens Data	▼ – □ ×

Update: All Windows ▾ ⓒ ⓐ ✛ ⬤ 🎴 ⚹ ⚹ ⬧ ⬧ ⬧ O· ✎ ↻ ☐ ⇅ ↔ ⇒ ❓

⌄ Surface 0 Properties ‹ › Configuration 1/1 ‹ ›

	Surf:Type		Radius		Thickness		Material	Semi-Diameter
0	OBJECT	Standard ▾	Infinity		100.000			11.330
1		Standard ▾	Infinity		0.000			6.927
2		Standard ▾	34.399	V	8.000	V	BSL1	6.896
3		Standard ▾	-129.090	V	9.914	V		6.192
4		Standard ▾	11.371	V	2.756	V	TIH14	4.266
5		Standard ▾	112.521	V	2.258	V	M-BACD15	3.720
6		Standard ▾	7.194	V	3.325	V		2.879
7		Standard ▾	11.282	V	1.899	V	BAL12	2.235
8		Standard ▾	-9.637	V	1.000E-02	V		1.919
9		Standard ▾	-11.537	V	0.500	V	ZF8	1.896
10		Standard ▾	2.763	V	2.242	V	L-PHL2	1.658
11		Standard ▾	11.078	V	3.898	V		1.370
12	STOP	Standard ▾	Infinity		2.651	V		0.667
13		Standard ▾	19.399	V	0.500	V	NBFD3	1.220
14		Standard ▾	5.853	V	0.062	V		1.269
15		Standard ▾	5.837	V	4.871	V	N-PSK57	1.294
16		Standard ▾	-7.991	V	9.110	V		1.851
17		Standard ▾	5.427	V	1.643	V	P-SK60	2.401
18		Standard ▾	2.427	V	3.351	V	SF18	2.063
19		Standard ▾	3.340	V	0.535	V		1.463
20	IMAGE	Standard ▾	Infinity		-			1.452

FIGURE 3.66 Prescription for the optimized lens in Fig. 3.65.

Of course, many of the lens mechanical characteristics at this state may not be manufacture-friendly. We also did not constrain other glass properties, such as climatic resistance, stain resistance, acid resistance, alkali resistance, and phosphate resistance, but in a real design, one may include such constraints in the merit function. What we have done was merely an exercise to illustrate the approach for rapidly arriving at an apochromatic design of a multi-element lens with extended FOV. There is one final note: image distortion and relative illumination (Fig. 3.69).

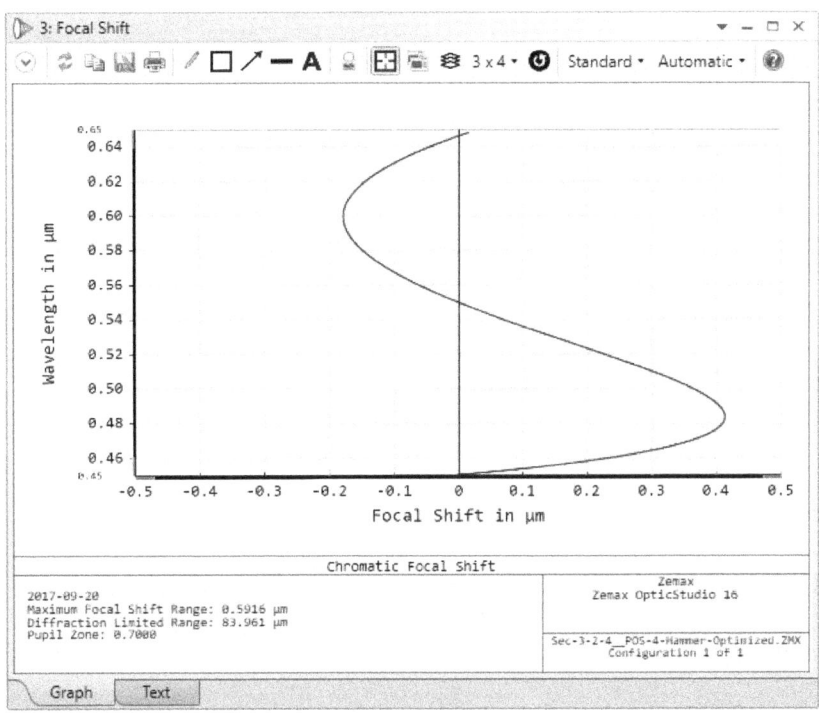

FIGURE 3.67 Zemax OS print-screen of the chromatic focal shift (at 0.7 pupil zone) plot for the optimized lens in Fig. 3.65.

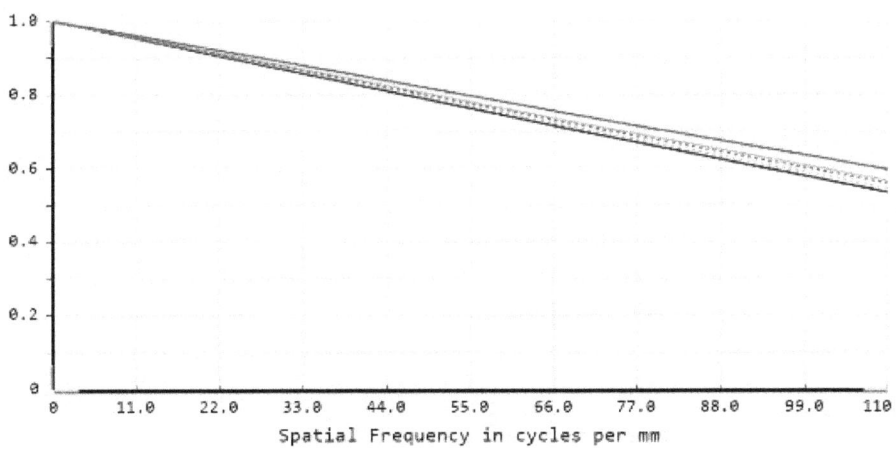

FIGURE 3.68 MTF plot for the optimized lens in Fig. 3.65.

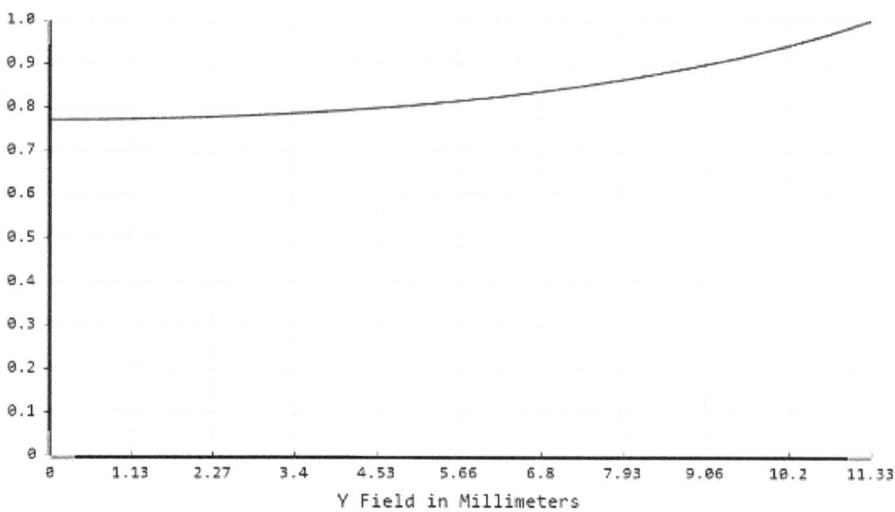

FIGURE 3.69 Relative illumination for the optimized lens in Fig. 3.65.

The image distortion is slightly under 5% barrel, which perhaps contributed partially to the increase in the illumination across the image plane as shown in Fig. 3.69. However, remember Fig. 3.20? For a machine vision system, one may install surrounding illumination with an appropriate tilt angle that yields a bright central area, which could be made to compensate the type of relative illumination at the image shown in Fig. 3.69. Furthermore, if an image sensor contains microlenses over its pixels, the drop-off in illumination with chief ray angle in image space at the sensor pixels may also be used to even-out the image relative illumination.

3.2.5 Curved image sensors

3.2.5.1 Where to put the best focus

Curved image sensors enable the possibility to reduce the number of lens elements needed in a design to correct field curvature [16 – 18]. However, note from Eq. (2.13) that, strictly, field curvature arises from the Petzval coefficient w_f, which is not associated with a physical wavefront (see also, e.g., Ref. 19). The physically converging wavefront is defined by the total effect from the astigmatism and Petzval curvature aberration terms such that the converging wave has different focal positions for sagittal and tangential rays across the image field. It is this "total" curvature that most refer to as field curvature. Therefore, the best focal surface across the image is generally the average between the astigmatic sagittal and tangential focal planes. In Zemax OS, plots of these astigmatic surfaces are shown in the "Field Curvature/Distortion" plots. Let's take a simple example. In Zemax OS, set up a simple bi-convex lens at a monochromatic wavelength of 550 nm, and field angles 0, 14, and 20 degrees using the prescription shown in Fig. 3.70. Set aperture to "float by stop size" and use the stop semi-diameter shown in Fig. 3.70. Note that a radius of -2.4 mm has been placed at the image. This is a curved image plane. The layout is shown in Fig. 3.71. The MTF for this lens is shown in Fig. 3.72.

	Surf:Type		Radius	Thickness	Material	Semi-Diamete
0	OBJECT	Standard ▼	Infinity	Infinity		Infinity
1	STOP	Standard ▼	Infinity	0.500		0.200 U
2	(aper)	Standard ▼	4.180	0.500	N-BK7	1.235 U
3	(aper)	Standard ▼	-4.180	3.947		1.235 U
4	IMAGE	Standard ▼	Infinity	-		1.487

FIGURE 3.70 Simple bi-convex lens prescription in Zemax OS.

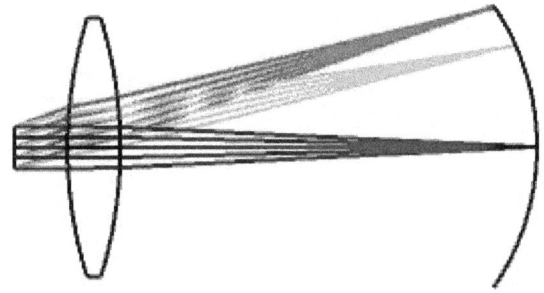

FIGURE 3.71 Layout for the lens given by the prescription in Fig. 3.70.

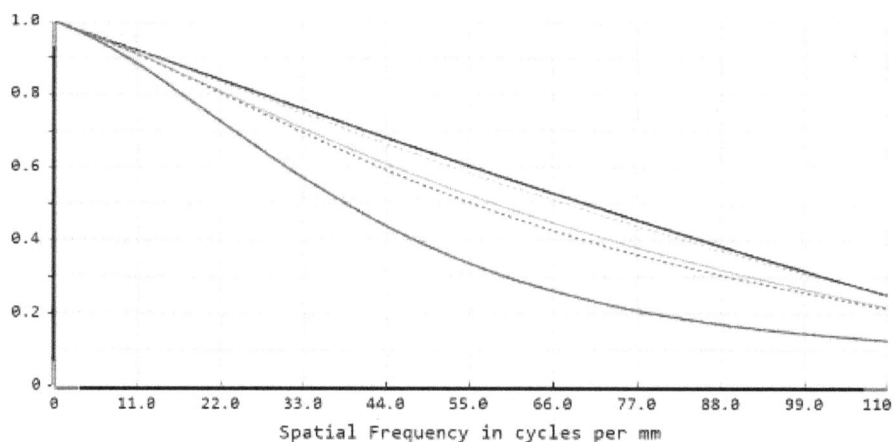

FIGURE 3.72 MTF for the lens in Fig. 3.71.

Now, bring up the "Field Curvature and Distortion" plots, whose results should appear as shown in Figs. 3.73a and 3.73b respectively.

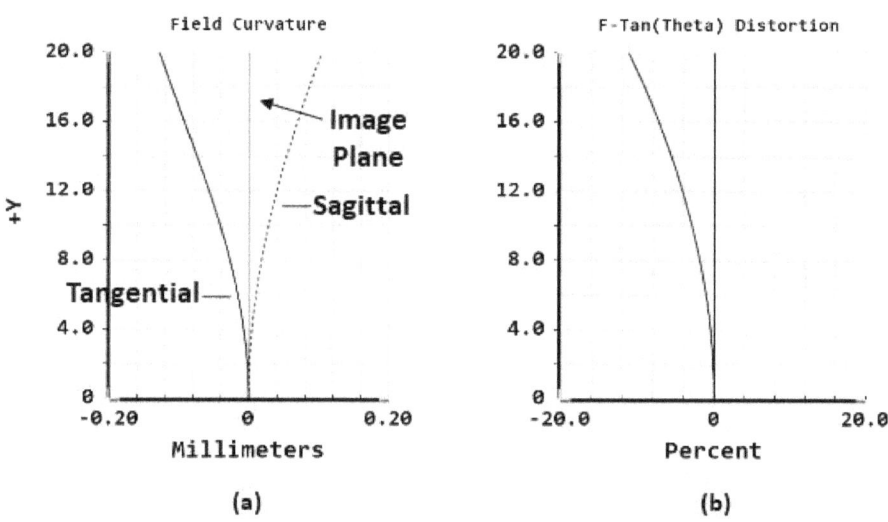

FIGURE 3.73 (a) Field curvature. (b) Distortion (for the lens in Fig. 3.71).

Note in Fig. 3.73a that the curved plane has been placed between the sagittal and tangential astigmatic surfaces, which yields the MTF curve shown in Fig. 3.72. It is important to note that the sagittal and tangential curves include the Petzval coefficient, because they are obtained from tracing real off-axis sagittal and tangential rays (at small angles).

3.2.5.2 Distortion on a curved surface

The interpretation of image distortion on a curved image plane is not obvious. However, for small field angles, it is approximately the same as the so-called "F – Tan (Theta)" distortion that you see in Fig. 3.73b. Generally, image distortion on a curved surface increases as the surface's radius of curvature decreases (i.e., as it gets more curved). To understand how this comes about, let us refer to the geometrical construction shown in Fig. 3.74, which is based on the lens in Fig. 3.71 (whose prescription is provided in Fig. 3.70). In Fig. 3.74, the vertical height (i.e., the height above the optic axis) of the chief ray at the tip of the curved surface is y'_c. Therefore, its arc length on the curved surface is γR, where γ is in radian units. To obtain the angle γ, one simply traces the real chief ray to the curved surface to obtain the height y'_c. The angle is therefore

$$\gamma = \sin^{-1}(y'_c/R). \qquad (3.48)$$

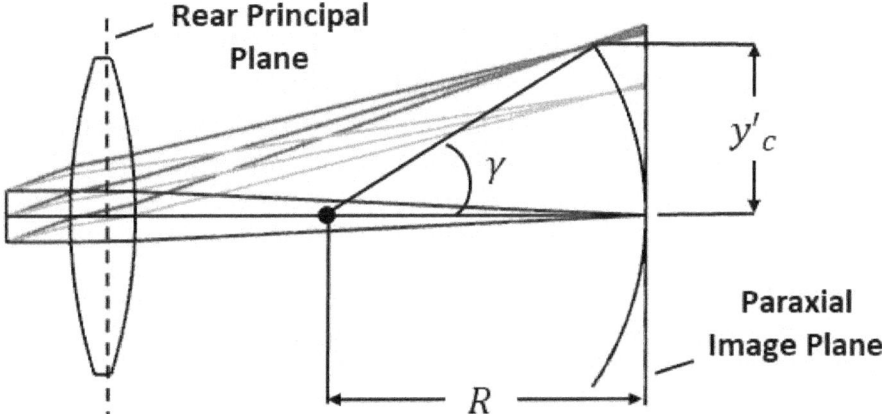

FIGURE 3.74 Geometrical construction to compute image distortion on a curved surface for the lens in Fig. 3.71. Angles are in radian units.

If the vertical height of the ideal chief ray at the curved surface is y'_{pc}, then it would subtend an angle β'' (in radian units) given by

$$\beta'' = \sin^{-1}(y'_{pc}/R). \qquad (3.49)$$

The image distortion D_c along the curved surface is therefore

$$D_c = \frac{\gamma R - \beta'' R}{\beta'' R} = \frac{\gamma}{\beta''} - 1. \qquad (3.50)$$

For the current lens example, $R = 2.4$ mm (the sign is not important here). Using the REAY operand with Hy $= 1$ and Py $=0$ yields $y'_c = 1.331$ mm at surface 4. Applying Eq. (3.48) with these values I obtain $\gamma \approx 0.588$ radians. Using the PARY operand with Hy $= 1$ and Py $= 0$ yields $y'_{pc} = 1.498$ mm at surface 4. Applying Eq. (3.49) yields $\beta'' \approx 0.674$ radians. Applying Eq. (3.50) we have $D_c = (0.588/0.674) - 1 \approx -0.12.7$, or in other words, -12.7% distortion. Compare this value with the F – Tan (Theta) plot in Fig. 3.73b. They are similar, though clicking on the "Text" button at the lower left corner of the plot in Zemax OS reveals that the F – Tan (Theta) distortion at full field is actually -11.1%. Reducing the full field angle to 5 degrees yields virtually the same results.

Now, upon setting the full field angle back to 20 degrees, note that reducing the radius of curvature of the image results in larger distortion. At a radius of -1.8 mm, I obtain -18.7% distortion using the equations above. Zemax OS's F – Tan (Theta) distortion yields -13.87%. Setting the image surface radius to infinity yields F – Tan (Theta) distortion at about -3.2%.

3.2.6 Focusing a lens in machine vision

3.2.6.1 Shifting the image plane versus shifting a lens

On an optics lab bench, the action of focusing the image is usually associated with shifting the image plane. This makes sense. Eq. (2.1) suggests that, for a fixed object distance s, the image is found at s'. Therefore, it is reasonable in the lab to maintain the object distance and to locate the image by shifting a screen along the optic axis until a sharp focus is achieved. However, this isn't done in any commercial imaging system, such as a digital single lens reflex (DSLR) or mobile phone camera. In such systems, it is the lens that shifts. There is actually a subtle difference between shifting the image plane and shifting the lens. In the former, only the image distance changes. In the latter, both the object and image distances change as the lens shifts to bring the image into focus. The result is that the image magnification is different between the two cases. In most practical cases, the difference is small, but it is rather difficult to prove this analytically even though the equations are solvable in closed-form. To see why, let us first consider the first case, which is shifting the image plane to focus the image. Applying Eq. (2.1) for a thin lens, if an object shifts away from the lens by an amount δ, then we expect the image plane to be required to shift towards the lens by some amount δ'. This means we have

$$\frac{1}{f} = \frac{1}{s + \delta} + \frac{1}{s' - \delta'}. \tag{3.51}$$

Solving Eq. (3.51) for δ' we have

$$\delta' = s' - \frac{f(s + \delta)}{s + \delta - f}. \tag{3.52}$$

Now, in the case where the lens shifts to focus the image, we expect the lens to shift towards the image plane by, say, some amount δ_L. But this also increases the distance between the lens and the object by δ_L, resulting in a total object distance of $s + \delta + \delta_L$. This means we have

$$\frac{1}{f} = \frac{1}{s + \delta + \delta_L} + \frac{1}{s' - \delta_L}. \tag{3.53}$$

Solving Eq. (3.53) for δ_L yields the formidable result

$$\delta_L = \frac{s' - s - \delta + \sqrt{[s' - s - \delta]^2 - 4\{f(s + \delta) + s'[f - s - \delta]\}}}{2}.$$

(3.54)

There is no easy way to compare the magnitude of δ_L with δ' using Eqs. (3.54) and (3.52). So, you may either plug in numbers (which I did) or model a paraxial thin lens in an optical design program (which I also did using Zemax OS). One will find that the differences are small. For instance, if $f = 8$ mm, $s = 40$ mm, and $\delta = 60$ mm, I obtain $s' = 10$ mm, $\delta' = 1.30435$ mm, and $\delta_L = 1.31414$ mm. The image magnification for the case of shifting the image plane to focus is $(s' - \delta')/(s + \delta) = 0.08696$. For the case of shifting the lens to focus, the magnification is $(s' - \delta_L)/(s + \delta + \delta_L) = 0.08573$. Thus, the differences in focal shift and magnification are both small for the lens being considered. However, perhaps this exercise brings to mind a need to check these differences for any application where one is designing an imaging system that includes focus adjustment.

3.2.6.2 How to maintain the working distance

As we found in the previous section, if focus adjustment is performed by shifting the lens by an amount δ_L, then the object distance changes (even if slightly) by that same amount. Therefore, the working distance (WD) is also altered by that amount. This would be true whether or not the WD is measured relative to the lens's front glass vertex or relative to a mechanical flange that's part of the lens housing (see Fig. 3.15). Moreover, the WD is subject to the lens's EFL tolerances. Hence, in practice, one needs to allow some room for a WD tolerance. One also needs to account for tolerances on the image magnification. At any conjugate, the paraxial image magnification and object distance is related through Eq. (3.3). Therefore, tolerances in the lens's EFL result in variations in both the magnification and object distance (and hence, WD too). In order to meet requirements on magnification and WD, a practical approach is to allow for some degree of adjustment to the imaging system such that one may iterate between setting WD and lens focus to achieve the desired magnification and WD.

But it is also entirely up to us to define what working distance means. In Fig. 3.15, the WD is measured relative to the lens's mechanical housing. If focus adjustment to the lens is provided, then it stands to reason that the lens's housing itself would be threaded and screwed into an image sensor's housing. The integrated lens and sensor could then be mounted inside a "system housing" (Fig. 3.75).

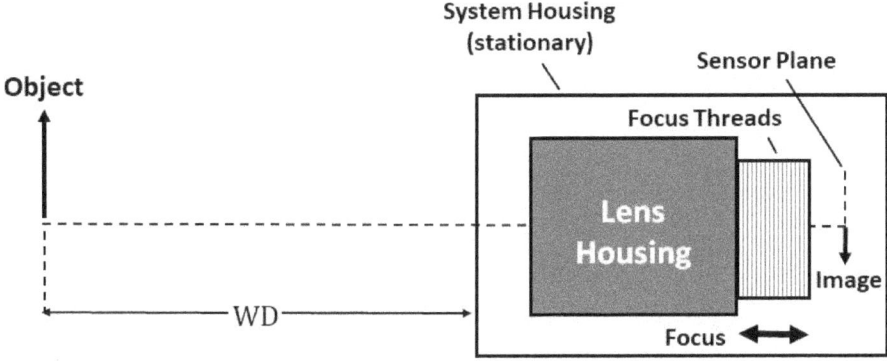

FIGURE 3.75 Working distance (WD) defined relative to system housing.

So, the WD may as well be defined by the distance between the object and the front plane of the system housing, which is stationary. This definition does not change any of the fundamental geometric optics equations for magnification, object distance, and image distance, and it provides a convenient datum for the mechanical engineer to design other mechanical related features related to mounting and assembly. There is of course also the question of how to allow sufficient room in the system housing for lens focus adjustment. A practical approach is to take note of the image distances required to focus at the object's far point (FP) and near point (NP). The difference $\Delta s'$ between the two conditions is the minimum allowable focus range (Fig. 3.76).

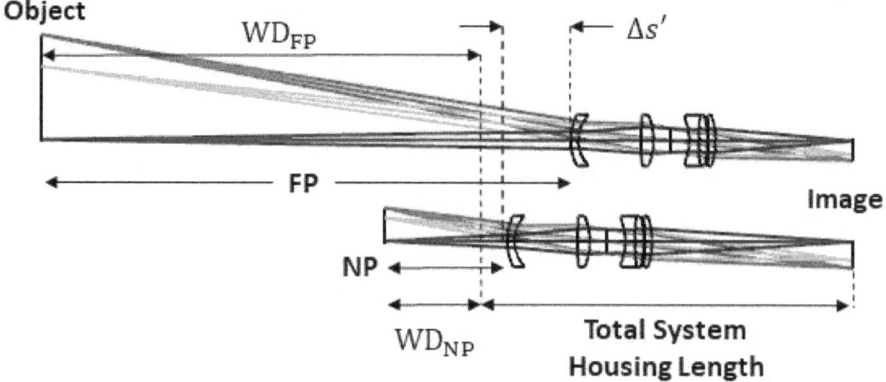

FIGURE 3.76 Determination of the required lens focusing space within system housing.

One may also add some mechanical space for the front end of the system housing in order to account for tolerances and "just in case we need it" situations. One may then define working distances for the near WD_{NP} and far WD_{FP} object points relative to the front of the system housing. In some cases, a "nominal WD" may also be defined for an object position somewhere between the NP and FP.

3.2.6.3 Liquid lenses in machine vision

Focus-tunable liquid lenses have come a long way and are available today as commercial off-the-shelf lenses [e.g., 20, 21]. Perhaps the simplest way to include such a component into an existing imaging system using a fixed focus lens (FFL) is to mount the liquid lens in front of the FFL [22, 23]. In this way, the liquid lens serves as a tunable spectacle lens or "eyeglass" for the FFL. But there are many other possibilities as well. In theory, one may mount a liquid lens virtually anywhere within the path of rays in an existing imaging system provided that one has considered vignetting effects and total aberrations of the liquid lens/FFL system.

3.2.6.4 Zemax OS example: Position of a liquid lens in a lens

As an example of examining where to place a focus-tunable liquid lens relative to a fixed focus lens (FFL), let us first consider the simple case of a liquid lens in front of an existing FFL. Applying the lens example from Sec. 2.2.2.3, let us insert a "model glass" with refractive index of 1.33 on surface 1 as shown in the prescription in Fig. 3.77. This model glass shall serve as our focus-tunable liquid lens, whose refractive index is that of water. Note that there is a "Pickup" on the radius of surface 2, which is set to take on the negative of the radius of surface 1. Evidently, we are modeling the liquid lens as a bi-convex lens. Set a single wavelength at 587.56 nm for this lens system, and ensure that ray aiming is "on" (aimed at the paraxial entrance pupil). Set the field points to 0, 7, and 10 degrees and clear all vignetting factors (all elements have user defined floating apertures).

Lens Data				▼ – ☐ ×
Update: All Windows ▾				
∨ Surface 0 Properties ⟨ ⟩			Configuration 3/3 ⟨ ⟩	

	Surf:Type	Radius	Thickness	Material	Semi-Diameter
0	OBJE(Standard ▾	Infinity	20.000		3.668
1	(aper) Standard ▾	16.810	0.100	1.33,0.0	0.300 U
2	(aper) Standard ▾	-16.810 P	0.050		0.300 U
3	(aper) Standard ▾	0.584	0.071	N-LAK9	0.250 U
4	(aper) Standard ▾	0.355	0.608		0.250 U
5	(aper) Standard ▾	3.409	0.120	N-BAF52	0.250 U
6	(aper) Standard ▾	-0.668	0.165		0.250 U
7	STOP Standard ▾	Infinity	0.181		0.118 U
8	(aper) Standard ▾	-0.559	0.091	N-SF11	0.210 U
9	(aper) Standard ▾	1.794	0.089	N-LAF2	0.250 U
10	(aper) Standard ▾	-0.947	2.392E-03		0.250 U
11	(aper) Standard ▾	-10.515	0.072	N-SSK5	0.250 U
12	(aper) Standard ▾	-0.871	1.160		0.250 U
13	IMAG Standard ▾	Infinity	-		0.181

FIGURE 3.77 Prescription for FFL with model "liquid lens" on surface 1 with index 1.33.

Open the multi-configuration editor and populate this editor with the multi-configuration operands shown in Fig. 3.78. This setup generates three configurations, each with the object set at distances 2, 5, and 20 mm for "NEAR", "MID" and "FAR" distances ranges. We may call these the working distances (WDs) for this lens system, defined as the distance between the object and the front vertex of the liquid lens (surface 0 in the lens data editor). The operand "CRVT" in the multi-configuration editor controls the "curvature" on surface 1 (i.e., the reciprocal of the liquid lens's radius). I have pre-optimized the curvature values for the three configurations. The resulting layouts are shown in Figs. 3.79a – 3.79c. Note that in order to display the lenses as shown in Fig. 3.79, click on the "3D Viewer" button under the "Setup" menu, and set the "Offset" to Y = -1 mm in its settings menu (right click on the window of the 3D layout). The resulting MTFs for the three configurations are shown in Figs. 3.80a – 3.80c.

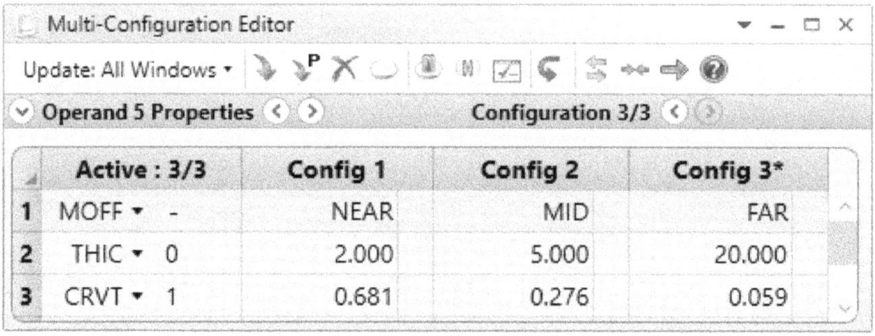

☐ Multi-Configuration Editor			▼ – ☐ ✕
Update: All Windows ▾ ⟩ ⟩P ✕ ⟳ ▦ ⊞ ⟲ ⟰ ⇢ ⇨ ❸			
⌄ **Operand 5 Properties** ‹ ›		**Configuration 3/3** ‹ ›	
Active : 3/3	**Config 1**	**Config 2**	**Config 3***
1 MOFF ▾ -	NEAR	MID	FAR
2 THIC ▾ 0	2.000	5.000	20.000
3 CRVT ▾ 1	0.681	0.276	0.059

FIGURE 3.78 Multi-configuration setup for the system given by Fig. 3.77.

In comparing Fig. 3.80 with Fig. 3.84, one can see that placing the modelled tunable liquid lens either at the front of the FFL or near the aperture stop of the FFL yields similar results under similar object distance conditions. However, one advantage of having the liquid lens mounted near the stop in the model given by Fig. 3.81 is that the rays near the stop are approximately parallel, which results in a reduction of their footprint over the surface area of the liquid lens. The lens elements before the stop are producing near collimated rays, and the lens elements after the stop focus these rays towards a plane whose distance from the rear principal plane is approximately the back focal plane. Also, rays from off-axis field points strike near the center of the liquid lens. This enables the liquid lens model at the stop not to perturb the FFL's performance too much. Also, it reduces the risk of vignetting. In some cases, a FFL may be designed to have the stop in front of all lens elements, thereby serving as the FFL's entrance pupil [23]. In such cases, this makes it very simple and practical to mount the liquid lens in front of the FFL, with virtually no vignetting and minimal perturbation to the FFL's imaging performance.

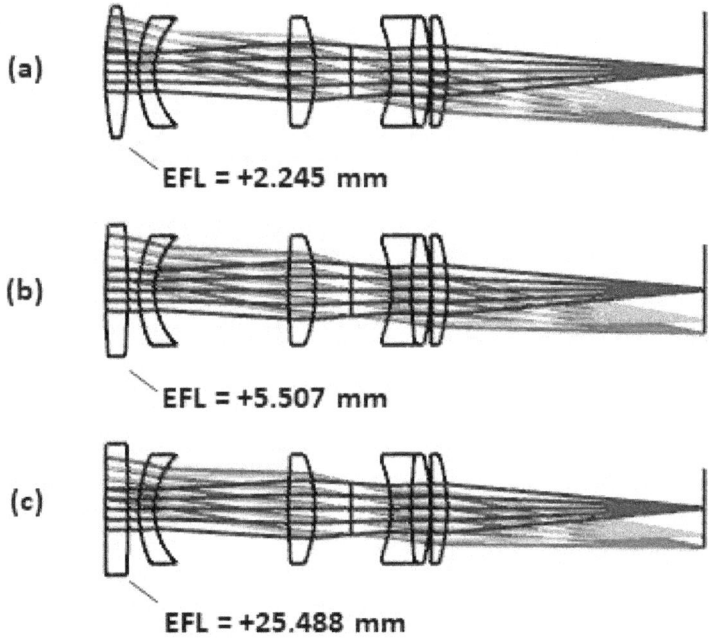

FIGURE 3.79 Layout for Fig. 3.78. (a) Configuration 1. (b) Configuration 2. (c) Configuration 3.

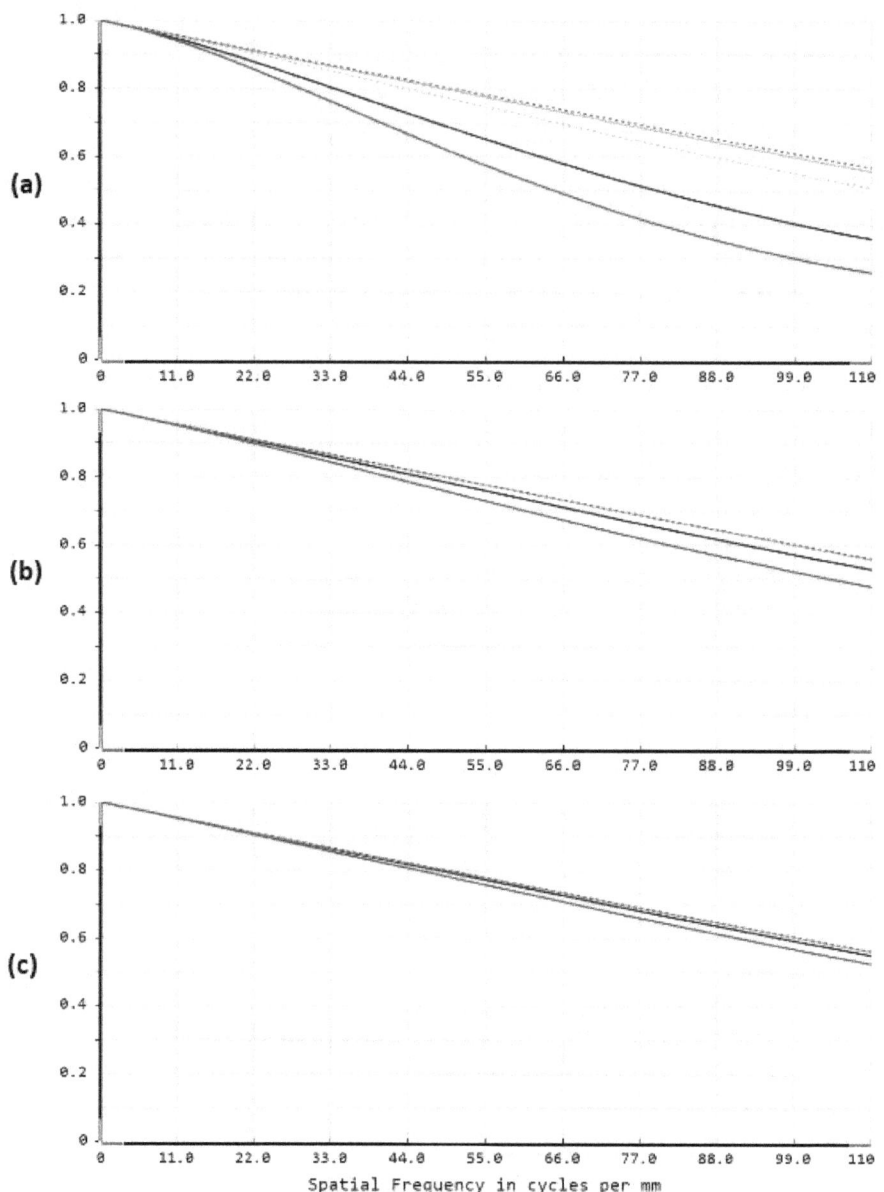

FIGURE 3.80 MTF. (a) Configuration 1. (b) Configuration 2. (c) Configuration. 3.

	Surf:Type	Radius	Thickness	Material	Semi-Diameter
0	OBJEC Standard ▾	Infinity	20.000		3.653
1	Standard ▾	Infinity	0.100		0.216
2	(aper) Standard ▾	0.584	0.071	N-LAK9	0.250 U
3	(aper) Standard ▾	0.355	0.608		0.250 U
4	(aper) Standard ▾	3.409	0.120	N-BAF52	0.250 U
5	(aper) Standard ▾	-0.668	0.109		0.250 U
6	STOP Standard ▾	Infinity	0.050		0.118 U
7	(aper) Standard ▾	20.722	0.100	1.33,0.0	0.300 U
8	(aper) Standard ▾	-20.722 P	0.100		0.300 U
9	(aper) Standard ▾	-0.559	0.091	N-SF11	0.210 U
10	(aper) Standard ▾	1.794	0.089	N-LAF2	0.250 U
11	(aper) Standard ▾	-0.947	2.392E-03		0.250 U
12	(aper) Standard ▾	-10.515	0.072	N-SSK5	0.250 U
13	(aper) Standard ▾	-0.871	1.160		0.250 U
14	IMAG Standard ▾	Infinity	-		0.174

FIGURE 3.81 Prescription for liquid lens model placed near the stop.

Active : 3/3		Config 1	Config 2	Config 3*
1	MOFF ▾ -	NEAR	MID	FAR
2	THIC ▾ 0	2.000	5.000	20.000
3	CRVT ▾ 7	0.362	0.174	0.048

FIGURE 3.82 Multi-configuration setup for prescription in Fig. 3.81.

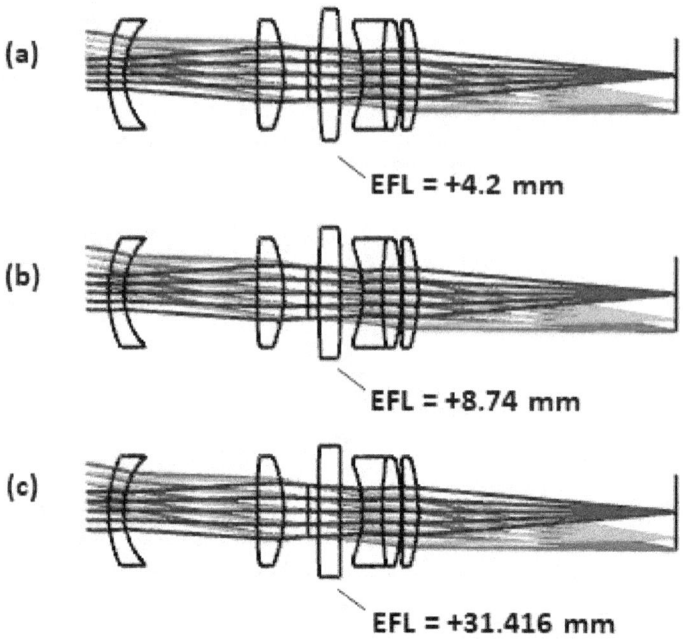

(a)

EFL = +4.2 mm

(b)

EFL = +8.74 mm

(c)

EFL = +31.416 mm

FIGURE 3.83 Layout for Fig. 3.82. (a) Configuration 1. (b) Configuration 2. (c) Configuration 3.

Note that in all of the examples provided above, the setup of the FFL is such that it has been made to image an object at infinity (one may check this in the present models by setting the radius of the liquid lens to infinity and note the resulting marginal ray focus). For example, in all of the configurations used in the model given by Fig. 3.77, the liquid lens essentially collimates rays from the object at all object distances. And even in the case of the model given by Fig. 3.81, rays from the front lens group of the FFL are not exactly collimated prior to entering the liquid lens; they diverge a little. Hence, the liquid lens must possess positive power to focus the rays towards the rear lens group. Note that all of these conditions have required that the liquid lens possesses positive power. In reality, there is usually a limit to the amount of positive power given to a focus-tunable liquid lens, which means that there is a limit to the near point (NP) of an object for the liquid lens/FFL system.

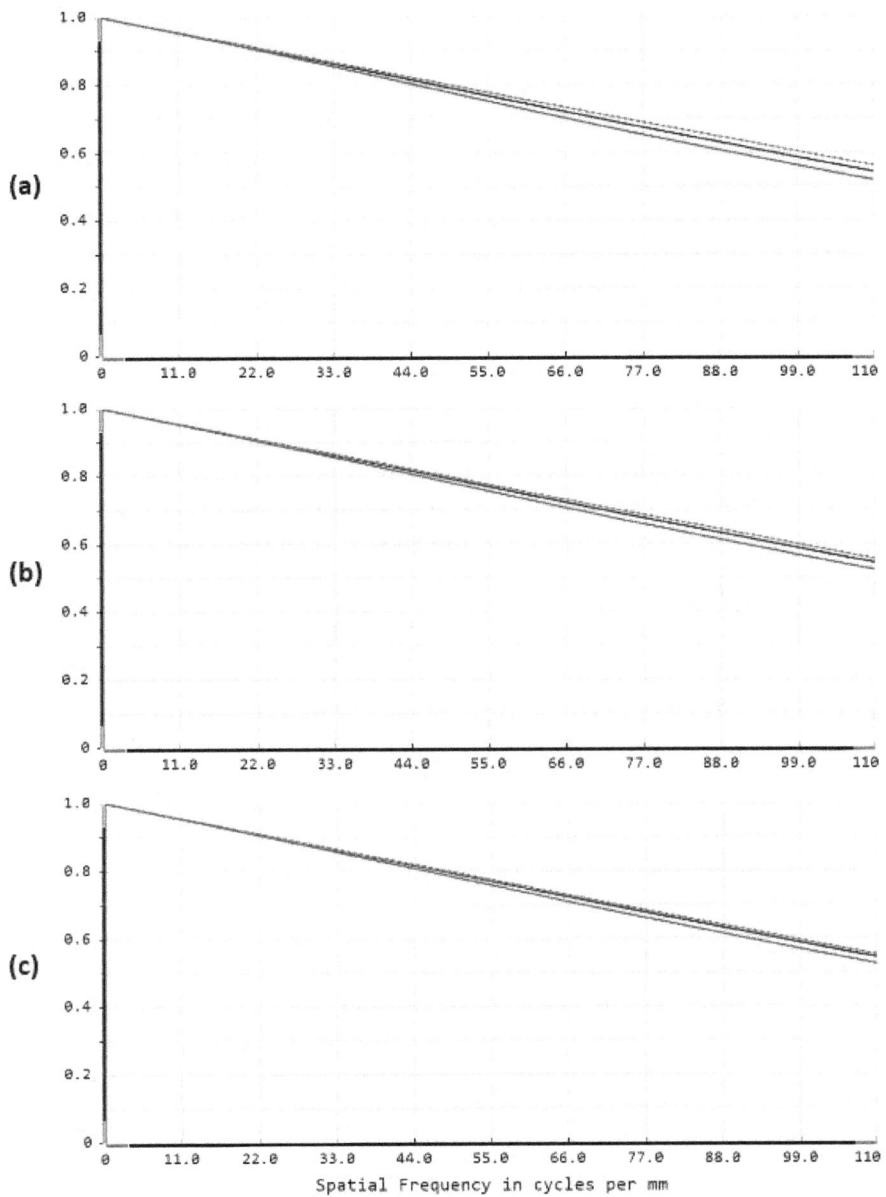

FIGURE 3.84 MTF for Fig. 3.83. (a) Configuration 1. (b) Configuration 2. (c) Configuration 3.

Now, if an available off-the-shelf liquid lens can be made to possess negative power, then, in the models that have been presented above, one isn't using the full focusing dynamic range of the liquid lens. The solution to this is to not set up the FFL to image at infinity. Instead, one may set the FFL to image at some finite object distance when the liquid lens's power is set to zero (i.e., when the liquid lens is essentially made into a window) and let this object distance be the middle or "nominal WD". Under such a condition, when the object is closer than the nominal WD, the liquid lens's power is tuned to be positive. When the object is farther from the nominal WD, the liquid lens's power is made to become negative. This enables the use of the full focusing dynamic range of the liquid lens. Additionally, such a setup is advantageous in terms of reducing the effect of temperature on the image's focal position. Fluid lenses in general are susceptible to focal length variations as a consequence of thermal loads, resulting in image defocus or "thermal defocus". Calibration minimizes such effects, but in those cases where calibration is an unwanted process, then setting up the FFL to image at finite conjugates minimizes thermal defocus [24].

3.2.7 Depth sensing with machine vision lenses

3.2.7.1 Triangulation with an imaging lens

Optical methods for distance measurement or "depth sensing" have generally been applied to non-contact dimensional measurements of 3D objects [25 – 27], and even to computer games [28]. In this and the three sections that follow, we focus on the application of imaging lenses to depth sensing. Let us take the example described by Berkovic and Shafir [29] where a laser shines a spot onto a distant screen at point P, and a thin lens images P onto P' at an image sensor (Fig. 3.85).

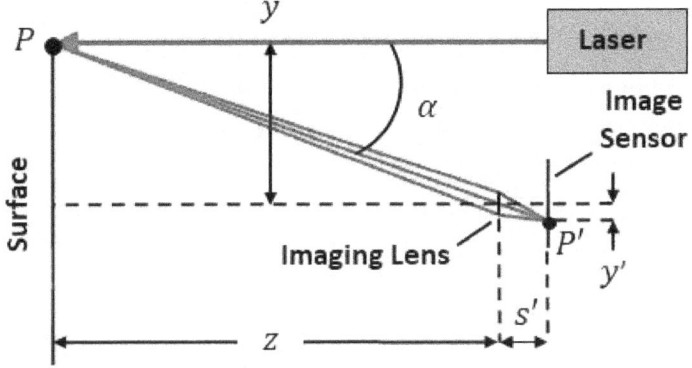

FIGURE 3.85 Distance measurement through triangulation and an imaging lens.

In Fig. 3.85, y and s' are known by design, so measurement of y' at the image sensor determines α. From this, the distance z is given by simple trigonometry:

$$z = y/\tan\alpha. \tag{3.55}$$

If image distortion D is present, then application of Eq. (2.130) to Eq. (3.55) and noting that $\tan\alpha = y'_p/s' = y'/[s'(1+D)]$ yields

$$z = [ys'(1+D)]/y'. \tag{3.56}$$

3.2.7.2 Impact of entrance and exit pupils on depth sensing

Eq. (3.56) is deceptively simple. It is only valid for an infinitesimally thin lens. A real lens (such as a multi-element system of lenses) has entrance and exit pupils, so the angle α is really the chief ray angle, which is measured from the vertex of the entrance pupil (Fig. 3.86). Consequently, determination of y' at the sensor (together with knowledge of s') does not yield α nor α'. Rather, it yields β', which is not an issue if the image of the laser spot is focused at the sensor. We recall from Sec. 3.2.2.2 that when the image is focused, the chief ray and b-ray intersect at the image plane (point P' in Fig. 3.86). Hence, Eq. (3.56) may be used to determine z, which is really the distance between the object and the front principal plane.

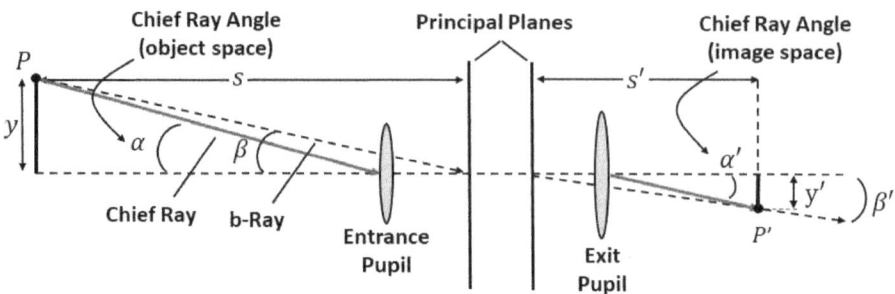

FIGURE 3.86 Multi-element representation of the lens in Fig. 3.85.

Now, consider a situation where the imaging lens in Fig. 3.85 is a fixed focus lens (FFL), so there is no adjustability for shifting it to focus an image. In this case, for any object distance that is not at the nominal distance (i.e., the distance at which the FFL attains best focus at the sensor), the image of the laser spot will be defocused, resulting in the chief ray and b-ray not intersecting at the image sensor. For example, when the object is farther from its nominal position, the laser spot image would be located s' from the rear principal plane, which would be in front of the image sensor (Fig. 3.87). At the sensor, the spot image is defocused, and its position P'' at the sensor is defined by the chief ray height, which is y''. In this case, Eq. (3.56) is not applicable, because this equation only holds when the image is focused (i.e., when the chief ray and b-ray intersect).

There are several solutions to the defocus problem illustrated in Fig. 3.87. Perhaps the first and most obvious solution is to consider provision of focus adjustment, either manual or autofocus. If instrument constraints somehow do not allow for focus adjustment, then another option would be to calibrate the lens-sensor system such that one determines the location and magnifications of the entrance and exit pupils, which would yield knowledge of α and α', and then to apply trigonometry to determine the object distance relative to the entrance pupil. But there is yet one other solution to this problem in the event that focus adjustment is not available: Use a lens with front-to-back symmetry about the stop. This is the subject of discussion in the next two sections.

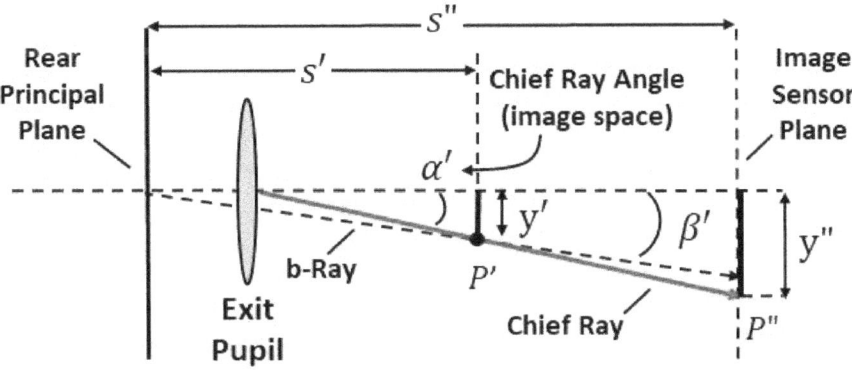

FIGURE 3.87 Case of defocused laser spot image at the image sensor.

3.2.7.3 Impact of lens symmetry on depth sensing

If a lens possesses front-to-back symmetry of its elements about the aperture stop, then its entrance and exit pupil locations are coincident with its front and rear principal planes, respectively. This is rather convenient, as it would allow the application of Eq. (3.56), where z would be the distance from the front principal plane (which is also the entrance pupil) to the object. One way to understand why the entrance and exit pupil planes coincide with the front and rear principal planes for a symmetric lens is to recall from Fig. 2.10 how principal plane locations are determined geometrically. The diameter of a parallel beam of light (e.g., from a point source at infinity) entering a lens must be equal to the diameter of the entrance pupil. If there is symmetry of elements about the aperture stop, then the diameter of the exit pupil is equal to the diameter of the entrance pupil. This means that a converging marginal ray that is emerging from the exit pupil back-intersects at the exit pupil itself, which makes the exit pupil the rear principal plane. Let us examine a practical example using Zemax OS in the following section.

3.2.7.4 Zemax OS example: Symmetric lens for depth sensing

Consider a symmetric lens* with 1 mm EFL operating monochromatically at 633 nm with prescription given in Fig. 3.88. The layout for this lens is shown in Fig. 3.89 with field height at 1.4 mm. This height therefore represents y in Fig. 3.85.

Lens Data

Update: All Windows

Surface 0 Properties | Configuration 1/1

	Surf:Type		Radius	Thickness	Material	Semi-Diameter
0	OBJECT	Standard	Infinity	5.000		1.400
1	(aper)	Standard	0.284	0.050	N-SSK5	0.150 U
2	(aper)	Standard	0.256	0.040		0.130 U
3	(aper)	Standard	2.077	0.050	N-BAF4	0.150 U
4	(aper)	Standard	0.190	0.041		0.130 U
5	(aper)	Standard	0.238	0.214	N-SK2HT	0.150 U
6	(aper)	Standard	-1.082	0.040		0.150 U
7	STOP	Standard	Infinity	0.040		0.070 U
8	(aper)	Standard	1.082	0.214	N-SK2HT	0.150 U
9	(aper)	Standard	-0.238	0.041		0.150 U
10	(aper)	Standard	-0.190	0.050	N-BAF4	0.130 U
11	(aper)	Standard	-2.077	0.040		0.150 U
12	(aper)	Standard	-0.256	0.050	N-SSK5	0.130 U
13	(aper)	Standard	-0.284	0.908		0.150 U
14	IMAGE	Standard	Infinity	-		0.322

FIGURE 3.88 Prescription for symmetric lens.

* Derived from US patent 4,490,019 (H. Shinohara, 1984). Scaled and re-optimized with Schott glasses.

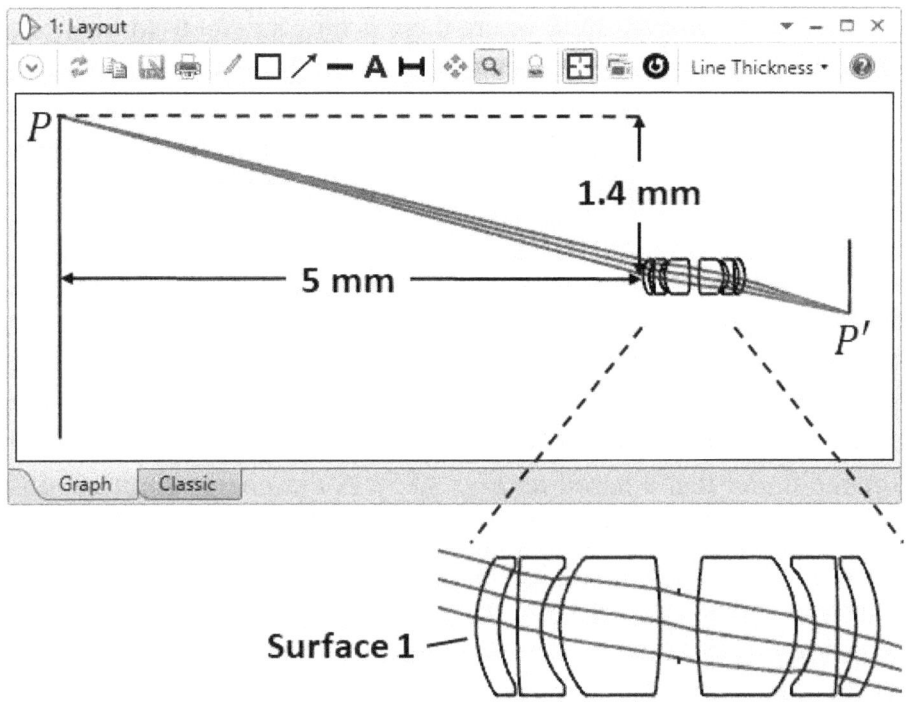

FIGURE 3.89 Layout of symmetric lens given by the prescription in Fig. 3.88.

Upon clicking on the "System Data" and "Cardinal Points" reports, one finds that the front principal plane and entrance pupil are both located 0.315613 mm to the left of surface 1. The same reports indicate that the rear principal plane and exit pupil are both located at 1.223916 mm to the left of surface 14 (the image plane). This is equal to 0.315613 mm to the left of surface 13. Thus, the entrance and exit pupils are coincident with the front and rear principal planes, as expected for this symmetrical lens. Since the thickness on surface 0 is 5 mm, the object is located 5 + 0.315613 = 5.315613 mm from the entrance pupil. Now, suppose we regard this system condition as the design condition, which makes 5.315613 mm the nominal object distance. This means that the design distance between the exit pupil and sensor is 1.223916 mm. Maintaining this distance, suppose the object were at an "unknown" distance of 20 mm from surface 1. The result is shown in Fig. 3.90, where there is now a defocused image of point P (of course, we are assuming that the laser's spot at P is an ideal point, which is just for a simple exercise). Our goal is to determine the "unknown" 20 mm distance. To do this, we'd first apply Eq. (3.56) to determine z, which is the distance between the object and entrance pupil. Then, subtracting 0.315613 mm from z should yield 20 mm. First, we need to measure y', which is the height of the defocused spot at the sensor (i.e., the distance from the center of the sensor to the center of the defocused spot image). An obvious approach would be to use the REAY operand to trace the chief ray through the system, but that's not reality (we're pretending to be using a range finder that's based on Fig. 3.85). A reasonable simulation is to use the Geometric Image Analysis feature. So, activate this feature and use the settings shown in Fig. 3. 91. The resulting output from the geometric image analysis settings shown in Fig. 3.91 should be as shown in Fig. 3.92. Note that the settings assume that the laser's spot on the target surface is a circle of diameter 100 microns. In reality, the beam would likely be from a laser diode, whose wavelength may not be 633 nm, and whose spot would likely have a Gaussian-like irradiance distribution with an elliptical shape (but you can get spots that are circular too, or you can make them circular using anamorphic systems [30]).

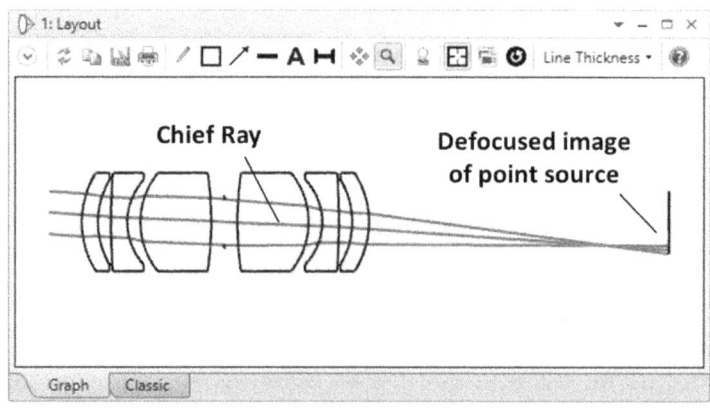

FIGURE 3.90 Layout of lens from Fig. 3.89 with object at 20 mm distance.

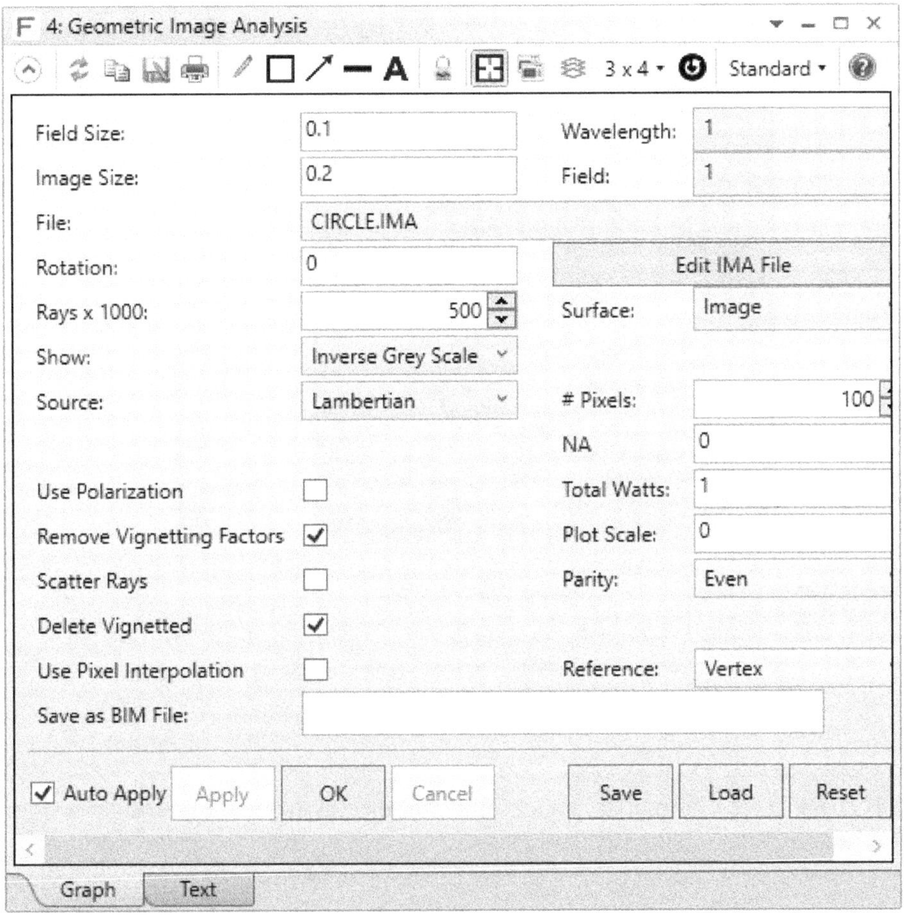

FIGURE 3.91 Geometric image analysis settings to simulate laser spot image.

Now, to obtain the height y' from the output shown in Fig. 3.92, I have found it to be more accurate if one chooses "Cross Y" for the "Show" option in the geometric image analysis setting (under the "Rays x 1000" option on the left side of the settings dialogue box). This results in a vertical scan of the plot as shown in Fig. 3.93 (make sure you choose the center column of pixels). From this, I obtain -0.084 mm for the position of the center of the spot.

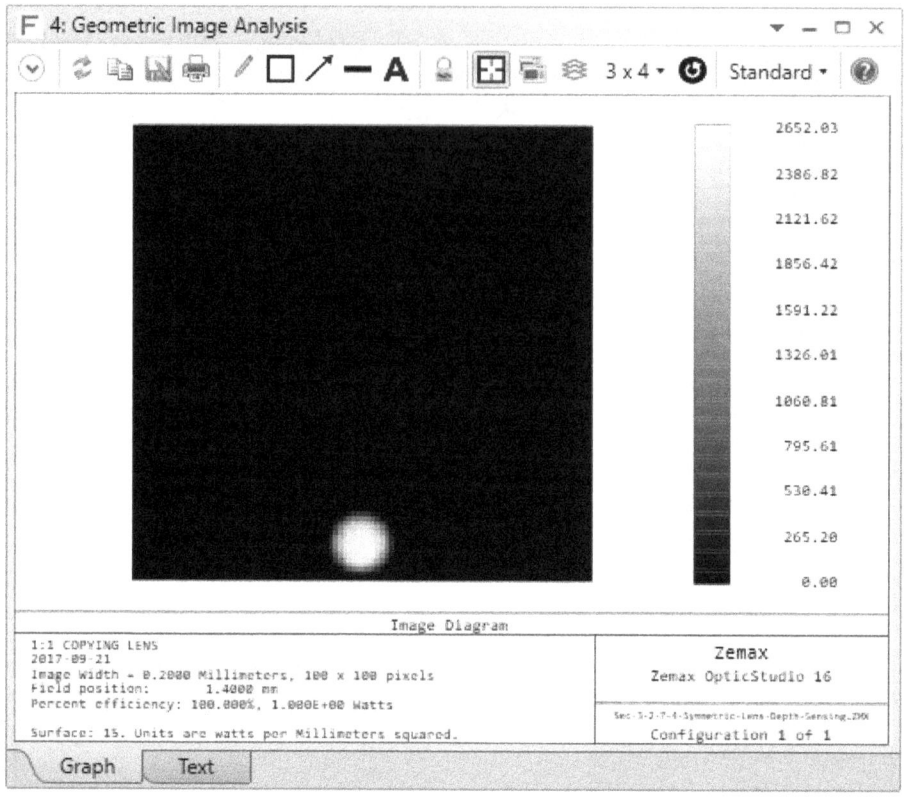

FIGURE 3.92 Geometric image analysis output from the settings in Fig. 3.91.

Now, we should account for image distortion, if any. We presume that image distortion has been determined through some calibration process. Reading this from the Field Curvature/Distortion plot in Zemax OS, the distortion at full field is about -0.0177%. This means $D = -0.000177$ in Eq. (3.56), which is negligible. So, taking $y' = 0.084$, $D = 0$, $s' = 1.223916$ and $y = 1.4$ in Eq. (3.56), we have $z = [1.4(1.223916)]/0.084 = 20.3986$ mm. This is the distance between the object and the entrance pupil. Subtracting 0.315613 mm from 20.3986 mm we obtain $z \approx 20.083$ mm. We have of course made some assumptions, such as ignoring the presence of a sensor cover glass. Also, a proper analysis should include uncertainty estimates and error bars. However, details aside, I think you get the idea.

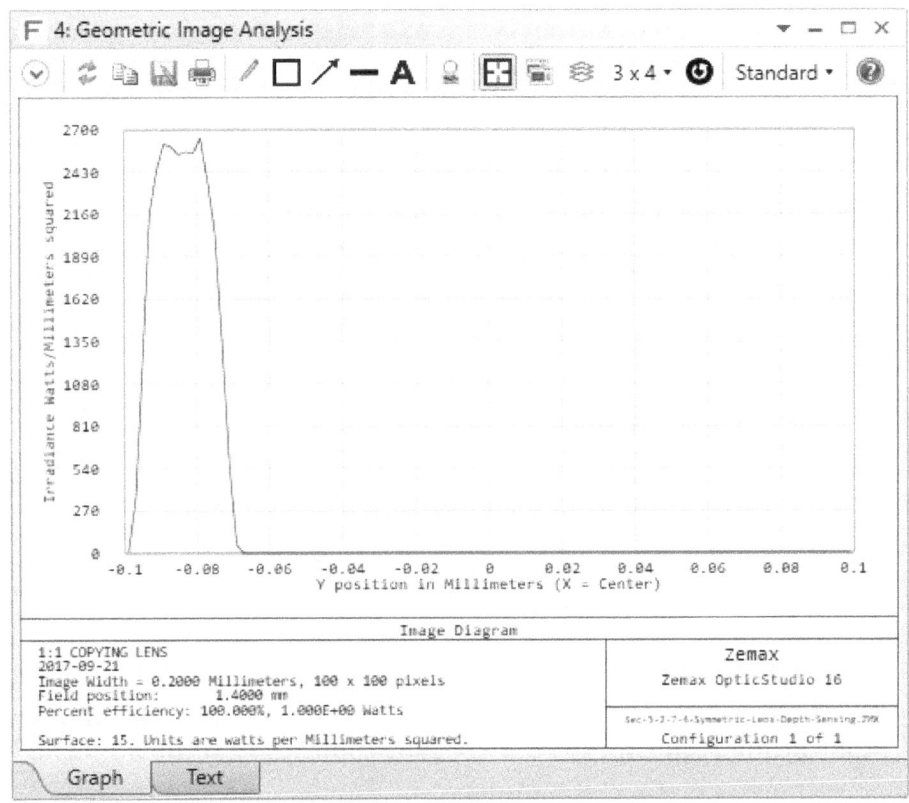

FIGURE 3.93 Column scan output for the geometric image analysis in Fig. 3.92.

If one attempts the exercise described above using, for example, the lens from Table 2.1, one will not get the same results, due to the asymmetry of elements about the stop. However, it is possible to perform the exercise in terms of the positions of the entrance and exit pupils, provided that one has determined their locations and relative magnifications (in order to relate α with α'). This seems reasonable, because perhaps most machine vision stock lenses are not symmetric about the stop. However, remember that one need not restrict one's choice of stock lenses to those listed specifically as "machine vision lenses" in a supplier's catalog. For instance, a stock relay lens designed for unit magnification would likely be a symmetric lens.

3.2.7.5 Depth determination by focus variation

In one patent [31], a method is described where lens focus variation yields depth information. This seems plausible. In Sec. 3.2.6, we applied Eq. (3.54) – which is derived from Eq. (3.53) – to compute the amount of lens shift required to yield a focused image for an object that has shifted an amount δ from some initial distance s. The inverse problem is therefore the estimation of the new object distance $s + \delta$ [using Eq. (3.54)] from knowledge of δ.

3.2.7.6 Extended depth of field imaging

We are ordinarily accustomed to the limits on depth of field imposed by the laws of geometrical and physical optics (see Sec. 2.1.6). However, certain techniques – collectively known as ***extended depth of field imaging*** – make it possible to push these limits, which has gained interest in modern applications [32]. Having a sufficiently extended depth of field could completely eliminate the need for focus adjustment, which is rather useful.

One example of a method for extending the depth of field is called "wave-front coding", first introduced by Dowski and Cathey [33]. At its core, wave-front coding may be understood in terms of Eq. (2.34), which expresses the final image irradiance distribution as a convolution of the ideal image with the point spread function (PSF) of the imaging system. This equation implies that a defocused image results from convolving the ideal image with a defocused PSF. It follows that if the PSF were insensitive to defocus, then the final image would be invariant with defocus. Wave-front coding is the act of modifying a lens system's PSF in such a way that the PSF is sufficiently invariant with defocus. This modification is performed by mounting an optical element within the path of rays in an imaging system that introduces a specially designed nonlinear phase function. In contrast to what we did when we apodized a lens in Sec. 2.2.2.3, this physical optical element (aka "phase mask") in wave-front coding introduces phase rather than amplitude, and it therefore does not block any light. The introduction of the known phase function therefore "codes" the wavefront entering the system. Because the phase mask's phase profile is known by design, the coded PSF is also a known function. The resulting image is given by the convolution of the coded PSF with the ideal image, which yields a distribution that's actually not recognizable as an image, albeit it is a distribution that's invariant with defocus. Thus, this image is regarded as an intermediate image, and the addition of the phase mask is regarded as a pre-processing step of imaging. In order to retrieve a recognizable image (which would serve as the final image), a post-processing digital deconvolution of the intermediate image is required, which would yield the final image. What makes the deconvolution possible is knowledge of the form of the coded PSF. Now, one might wonder, why go through all this trouble if, without the phase mask, one may simply de-convolve a "known" defocused PSF from the intermediate image formed by an ordinary lens? The reason that this is not done is because it is actually *not known* what a defocused PSF would look like for arbitrary ranges of object distances. This is precisely the condition that wave-front coding solves. Thus, the real value in wave-front coding is in not having to possess any knowledge of the axial displacement of the object about its normal distance to the lens. All that's needed is a focus-invariant intermediate image, whose de-convolved component yields the final image.

Actually, phase modification is not new to a "classical" lens designer. The wavefront aberration function is a phase modifier – the result of a lens introducing phase error to a converging spherical wave. Even if aberrations were not present, defocus may be considered a phase modifier [e.g., 34]. In fact, balancing spherical aberration with defocus is essentially a modification of a lens's total phase factor by adding the fourth powered spherical aberration phase factor with an appropriate negative magnitude of the quadratic defocus phase factor [e.g., 35]. Similarly, wave-front coding involves a special nonlinear phase factor that counteracts defocus. In the work by Dowski and Cathey [33], this nonlinear phase factor is a cubic function in rectangular pupil coordinates (i.e., a rectangular shaped pupil).

Another example of an extended depth of field system is the so-called **plenoptic camera** [36, 37], which enables refocusing of synthetic images that have been post-processed from the original images captured by a camera. To understand its core operating principle, we shall take an approach that follows an example provided by Adelson and Wang [36], which is depicted in Figs. 3.94a – 3.94c. The essential idea is that depth information about the object in a scene may be inferred from the spatial location of rays across pixels on the image sensor placed behind a microlens array.

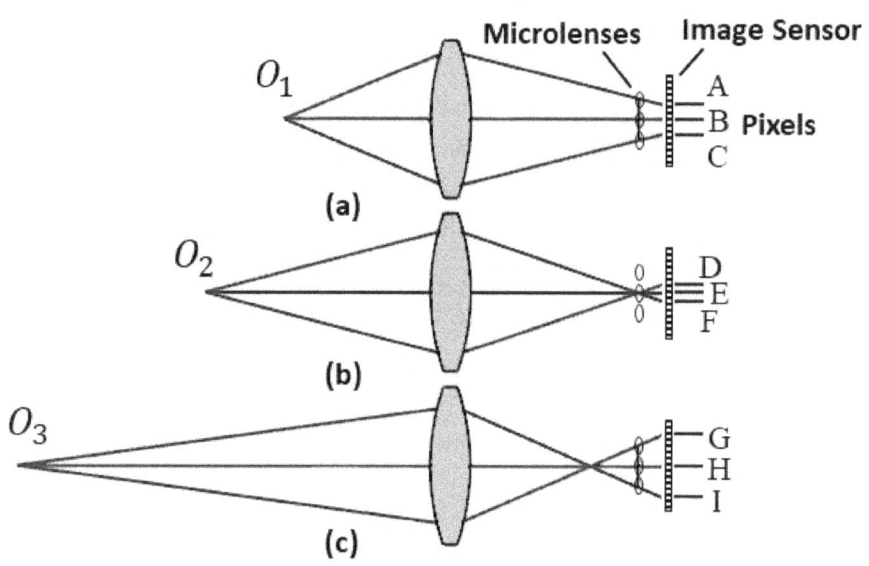

FIGURE 3.94 Schematic for illustrating the basic principle of the plenoptic camera.

Ng *et al.* [37] recognized that digital post-processing of the pixel signals may be used to generate synthetic images of the original digital image that may be refocused over a wide range, thus, extending the effective depth of field of the lens. To understand this, let's take a close look at Fig. 3.94b. In this figure, rays from a point source at O_2 are focused onto a microlens. These rays continue to strike pixels D, E, and F at the image sensor. The sum of the signals from pixels D, E, and F is the total signal from object O_2, which means that this integrated signal represents the total signal in an "effective pixel" (whose size is as large as the total area of pixels D, E, and F) that samples a sharply focused image of O_2. Pixels D, E, and F are therefore subpixels of the effective pixel that samples O_2. In Fig. 3.94a, object O_1 is closer to the lens, and its rays strike pixels A, B, and C. The sum total of the signals from pixels A, B, and C therefore represent the signal in an effective pixel that samples a sharply focused image of O_1. Similarly, in Fig. 3.94c, the sum total of the signals from pixels G, H, and I represent the signal in an effective pixel sampling a sharply focused image of O_3. By grouping subpixels into an array of effective pixels across a display, a synthetic image may be formed. Therefore, refocusing of these synthetic images would be a matter of displaying the appropriate array of effective pixels that have been appropriately grouped. Note also that, in this way, spatial resolution is traded-off with depth resolution.

3.2.8 The limits of miniaturization

There appears to be some trend in the miniaturization of machine vision lenses [38, 39], driven in part by social media and the use of mobile phones. Near the end of Sec. 3.2.2.6, I mentioned that miniaturization in some sense can partly be motivated by Eq. (3.10), which implies that reducing sensor size, increasing pixel count, and increasing aspect ratio increases the linear spatial sampling of the image field. Additionally, linear aberrations (e.g., not percent distortion) scale with size. In some cases, without scaling, aberrations may be overcome by digital processing methods [40]. However, scaling down a lens significantly reduces most aberrations. This is the reason why many of the Zemax OS examples in this book involve lenses at 1 mm EFL. It allows me to scale down the sizes of reference design forms from patent literature and re-optimize without significant effort.

But there are practical and fundamental limits to scaling. Some of the practical limits are, for example, flux collection from smaller pixels, and *increased* aberrations if one scales at constant "space-bandwidth product" (SBP) [41]. Moreover, considerations for practical manufacturing tolerances for large volume production of small scale optical elements become highly critical. For sub-millimeter glass lenses whose diameters are down to about 0.2 mm, production tolerances that have been published online currently range from about +/- 0.01 mm to +/- 0.02 mm for lens diameters and thicknesses, respectively. However, from experience, it is usually best to discuss all production matters closely with a supplier.

Concerning fundamental limits to scaling and miniaturization, consider the case of scaling with constant f-number. Recall from Eq. (2.24) that, at a fixed f-number, the Airy disk diameter does not scale with lens size. Rather, it scales with wavelength and f-number. Therefore, scaling down a lens at constant f-number yields a fixed Airy disk size (Fig. 3.95). Accordingly, by Eq. (2.45), the diffraction-limited MTF curve also does not scale with size. This brings about an interesting question: At a fixed f-number, as a lens reduces in size, at what point would Eq. (2.24) break down? To cite an extreme but plausible condition, surely Eq. (2.24) wouldn't apply to a lens made up of a few macro-molecules. In other words, at micron, sub-micron, and nanoscales, what, really, is a lens?

As Fig. 3.95 depicts, it seems unlikely that, at some lens diameter less than an undetermined value D_{min}, a lens would continue to behave as a lens. Would it continue to focus light to a spot? If there isn't a spot, then could there be an image? Recall that our classical view of image formation is in terms of the convolution integral given by Eq. (2.34). Therefore, if there were no spot (i.e., a point spread function), then could there be an image?

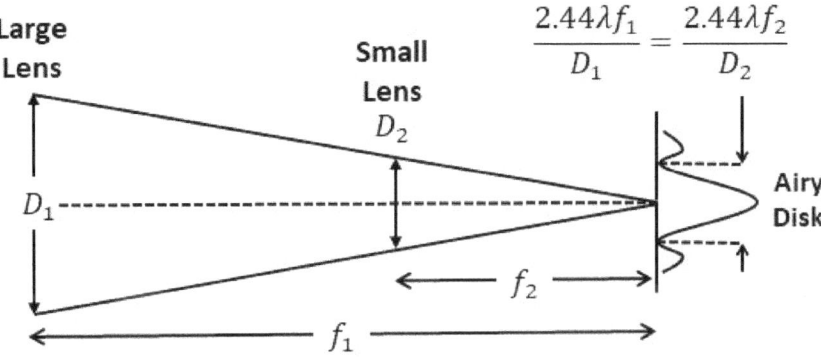

FIGURE 3.95 Miniaturization through scaling at constant f-number.

There are perhaps some reasonable answers. For instance, Y. Huo *et al.* [42] have shown that optical flux is focused by microlenses whose diameters are down to 2 microns, but their collection efficiency is reduced by diffraction effects. I have estimated (roughly) that the f-number of the microlens (in the geometrical optics sense) that Huo *et al.* investigated was somewhere between f/1 – f/3. This implies that microlenses at around 2 micron diameters and between f/1 – f/3 still behave like lenses in the sense that they focus light, but it remains unclear whether or not they form images with sufficient quality. At the extreme low f-number case of a ball microlens, M-S Kim *et al.* [43] show that the refraction limit is at a diameter of about 10 microns. They define the boundary between refraction and diffraction as the scale at which ray caustics are replaced by the observation of "photonic nanojets". The latter refers to a phenomenon where focused light by microspheres yield "focal lengths" of several wavelengths, propagating with a beam waist smaller than the "diffraction limit" [44]. By "diffraction limit" the authors mean the Airy disk spot radius (at the lowest possible f-number), which they express is about half the wavelength of light. That is, applying Eq. (2.24), if we take the lowest possible f-number to be f/0.5, then we have $r \approx 1.22\lambda(0.5) \approx \lambda/2$. In considering the above studies, it seems that, at such scales, the action of imaging might best be performed through some scanning means rather than with an image sensor.

3.3 Mobile phone imaging attachments

3.3.1 Lens attachments and the meaning of magnification

Figs. 3.96a – 3.96c depict a human eye looking at an object placed at distances equal to this eye's far point (FP), near point (NP), and some distance $s < NP$, respectively. Evidently, the image in the eye gets larger with reduced object distance.

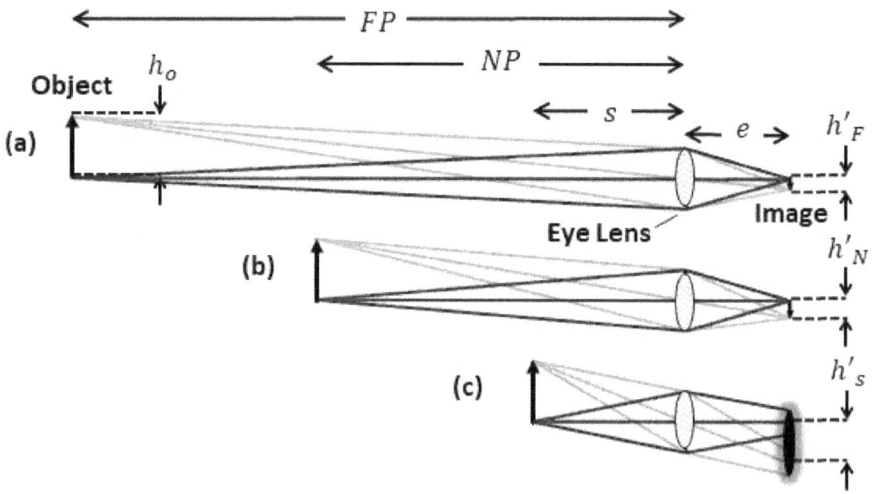

FIGURE 3.96 Human eye focusing. (a) Far Point. (b) Near Point. (c) Too close.

The far point distance FP for the eye in Fig. 3.96a is the farthest distance beyond which this eye cannot form a "sharp" image of the object onto the retina. Accordingly, the near point distance NP in Fig. 3.96b is the closest distance below which this eye cannot form a "sharp" image of the object onto the retina. What is a "sharp" image in the eye? For the human eye, it usually means being able to resolve two object points subtending about 1 arc minute from the vertex of the eye [45, 46]. The dimensions h'_F, h'_N and h'_s in Figs. 3.96a – 3.96c refer to the height of the image formed on the retina at the object distances FP, NP and at s respectively. The dimension e is the distance between this eye's lens and its retinal plane (we are neglecting effects from the refractive index of the medium between the eye lens and retina). In Fig. 3.96c, although the image is blur, it is still associated with a height h'_M defined by the angle the chief ray makes at the retinal plane. If this eye had been able to focus rays from an object at s sharply onto the retina, then the distance s would be this eye's near point, and the resulting relative magnification M between placing the object at the distance NP and at s would be

$$M = \frac{h'_s}{h'_N} = \frac{(eh_o)/s}{(eh_o)/NP} = \frac{NP}{s}. \tag{3.57}$$

Eq. (3.57) expresses the magnification of the image in the eye for an object at a distance s, relative to the image in the eye for an object at a distance NP. Accordingly, the relative magnification between placing the object at distance NP and at FP is $h'_N/h'_F = FP/NP$. And of course, the relative magnification between placing the object at distance s and at FP is $h'_s/h'_F = FP/s$. This all makes sense, because the closer the object, the bigger it looks. For this reason, without wearing any spectacle lenses, a person who is "more near-sighted" than you are (i.e., a person whose near point is less than your near point) would be able to see objects placed at closer distances to that person's eyes than you would. If the dimension e is the same for the both of you, then the relative magnification for viewing close-up objects between the both of you would be the ratio given by your near point distance divided by that person's near point distance. The eyes of a near-sighted person are, therefore, natural magnifying lenses.

For most of us, a ***magnifier*** is required to view any object placed at some distance $s < NP$. Such a magnifier would be a positive powered lens of some focal length f, placed directly in front of the eye in Fig. 3.96c. If the magnifier were made to form a virtual image of the object at some distance s' in front of the eye, then applying Eq. (2.1) for this lens we have

$$\frac{1}{f} = \frac{1}{s} + \frac{1}{-s'} \,. \tag{3.58}$$

Solving Eq. (3.58) for s and substituting the result into Eq. (3.57) yields

$$M = \frac{NP}{s} = NP\left(\frac{1}{f} + \frac{1}{s'}\right). \tag{3.59}$$

In Eq. (3.59), evidently, the distance s' must be such that $NP \leq s' \leq FP$ because the naked eye cannot focus sufficiently well for object distances less than NP and for object distances greater than FP. To vary s' between these two distances, one simply adjusts the distance between the eye and the object (i.e., by adjusting s), while at the same time making sure that the positive powered lens (i.e., the magnifier) is kept close to the eye. If, by chance, the eye is focused at its far point, then at some distance s, the magnifier may be made to form a virtual image of the object at FP so that $s' = FP$. Under this condition, by applying Eq. (3.59), the magnification of the image in the eye is $M = (NP/f) + (NP/FP)$. If the eye's far point is effectively infinity (i.e., if $FP \gg NP$), then $M = NP/f$, which is the well-known formula for a magnifier's magnifying power. Equivalently, placing the object at a distance f from the magnifier [i.e., letting $s = f$ in Eq. (3.58)] yields $s' = \infty \gg NP$, which yields $M = NP/f$. The expression $M = NP/f$ therefore assumes that the eye's far point is infinity. A somewhat greater magnification is achieved by shifting the magnifier closer to the object than f such that $s' = NP$. Under this condition, Eq. (3.59) yields $M = (NP/f) + 1$. This makes sense, because placing the virtual image of the object at the near point of the eye requires that $s < f$, which results in a larger chief ray angle, thereby creating a larger image.

This is what lens attachments are all about. A magnifier is a lens attachment for the human eye. A spectacle lens for the near-sighted person is a lens attachment for the near-sighted person, which produces a virtual image of a far object at some "comfortable" viewing distance in front of the near-sighted person's eyes. This "comfortable" distance could be, for example, somewhere between the patient's near and far points. On this note, it is evident that not everyone possesses the same near and far points. Therefore, by Eq. (3.59), not everyone experiences the same image magnification when looking through a magnifier. Hence, image magnification is a rather loosely defined semi-quantitative and relative quantity. It varies among individuals. It makes no difference whether or not NP is standardized in Eq. (3.59) (it's usually "standardized" at 250 mm). Physically, the size of an image formed inside the human eye varies from person to person.

But whatever the size of an image is inside anyone's eye, it can be made to increase further by using an additional attachment. This is what microscopes do. They scale M by a factor M_o, using another positive powered lens – the **microscope objective** – with focal length f_o to form a magnified image of the object at some plane a distance s in front of the magnifier (Fig. 3.97).

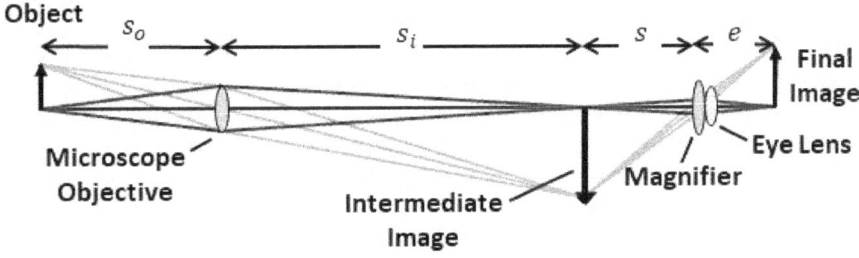

FIGURE 3.97 Optical principle of the microscope as an "eye attachment".

Applying Eq. (2.2), the objective lens's image magnification is $M_o = s_i/s_o$, where, quite reasonably, we would assume that $s_i > s_o$. If we let $L = s_i - f_o$, then simple geometry yields

$$M_o = \frac{L}{f_o}. \tag{3.60}$$

If f_e is the EFL of the magnifier, then applying Eq. (3.59) to Eq. (3.60), the final magnification M_m of a microscope-eye system may be expressed as

$$M_m = M_o M = \frac{L}{f_o} NP \left(\frac{1}{f_e} + \frac{1}{s'} \right). \tag{3.61}$$

In Eq. (3.61), we must again have the requirement that $NP \leq s' \leq FP$, and this time, s' may be varied by adjusting the distance between the magnifier and the intermediate image. In practice, the "magnifier" in Fig. (3.97) is really an **eyepiece**, and the eye must be placed at some finite distance away from the eyepiece. We'll examine these points in greater detail in Sec. 3.3.2.

Apparently, microscope objectives have a "standardized" L, which, according to one reference [47], $L = 160$ mm, and according to another [48], $L = 250$ mm. Modern microscope objectives are often "infinity corrected", which means that they have been optimized to yield an image at infinity – which also means that L is meaningless for such objectives unless a **tube lens** is used to form the intermediate image (which is what's done these days). The tube lens may therefore possess any "standard" focal length. It could be 160 mm, or it could be 250 mm, or virtually anything (it's usually 250 mm). Whatever magnitude L may be given, optical science tells us that one may regard a tube lens's EFL as the quantity L in Eq. (3.61).

The working principle of a telescope is a little more subtle. Theoretically, for a Keplerian telescope (whose objective lens and eyepiece are both positive powered), the operating principle is the reverse condition for the objective lens in Fig. 3.97, where we would now have it that $s_i < s_o$. But in this case, the final telescopic magnification of the image formed in the eye is given by a modified form of Eq. (3.61). In particular, we first note in Eq. (3.61) that $L/f_o = s_i/s_o$, and that, for a telescope, since $s_o \to \infty$, we have $s_i \to f_o$. Next, we note that NP appears in Eq. (3.61) because it originates from Eq.

(3.57), which expresses the magnification for the image on the retina for an object placed at s relative to an object placed at NP. For the telescope condition, we are placing an object at s relative to an object at s_o. Hence, we must replace NP in Eq. (3.61) by s_o. Doing this, and also replacing L/f_o by f_o/s_o, the final telescope-eye system magnification M_T is

$$M_T = f_o \left(\frac{1}{f_e} + \frac{1}{s'} \right). \tag{3.62}$$

In Eq. (3.62), we must again require that $NP \le s' \le FP$. In the event that a person's far point is such that $FP \gg f_e$, then $M_T = f_o/f_e$, which is the well-known formula for a telescope's magnification (thus, that formula assumes that $FP \gg f_e$).

So, an imaging attachment system for a mobile phone camera is simply one in which the eye in all of the above discussions is replaced by a mobile phone camera lens, except that the phone lens's EFL is fixed, and the dimension e in Figs. 3.96 and 3.97 is usually a focusing variable. But there is one additional complication: The entrance pupil of the phone's lens. In fact, this additional complication is not unique to the mobile phone. In the above discussions, we did not consider the impact of the exit pupil of both the microscope and telescope on the entrance pupil of the human eye. This is the subject of discussion in the next section.

3.3.2 Significance of entrance and exit pupils

The proper term for the magnifier in a microscope (or telescope) is "eyepiece". It differs from a simple magnifier in that it is optimized specifically to admit rays that fit through the diameter of the eye's entrance pupil (or simply "eye pupil" for short). In contrast, magnifiers are designed to admit rays over a larger "pupillary area" that allows an observer to shift the eye within some reasonable amount. When integrated with a microscope objective, an eyepiece forms an image of the objective lens at some distance ER from the eyepiece called the "eye relief". This image of the objective is therefore the exit pupil of the microscope, and the observer's eye pupil must be placed at the position of the eye relief to prevent vignetting (Fig. 3.98).

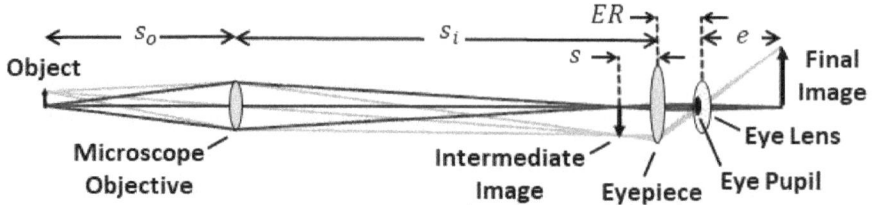

FIGURE 3.98 Schematic of a microscope with eyepiece and eye relief.

The eye relief of a microscope modifies Eq. (3.61) into

$$M_m = M_o M = \frac{L}{f_o} NP \left(\frac{1}{f_e} + \frac{1}{s' + ER} \right). \qquad (3.63)$$

Accordingly, for a telescope, we have

$$M_T = f_o \left(\frac{1}{f_e} + \frac{1}{s' + ER} \right). \qquad (3.64)$$

The rationale for the modification in Eqs. (3.63) and (3.64) is that one's eye is no longer s' away from the virtual image that's formed by the eyepiece. Rather, it is farther by an amount ER. As for a magnifier (which does not form any exit pupil because it is not used with an objective lens or any optic in front of it), there isn't any such eye relief, but we often design magnifiers *as if* there were an eye relief, which may be any reasonable distance behind the magnifier. As such, adding some distance ER to s' in Eq. (3.59) accounts for this "effective magnifier eye relief".

Replacing the eye in Fig. 3.98 by a mobile phone's camera lens does not change any of the conclusions made here. A phone lens's pupil must therefore be placed at the eye relief of a microscope or a telescope in order to avoid vignetting. In this way, the exit pupil of an imaging attachment must be at the entrance pupil of the mobile phone lens. Moreover, the exit pupil diameter must be at least equal to or larger than the entrance pupil of the phone lens. Let's call this "pupil matching". When the pupils of the imaging attachment and phone lens are matched, Eqs. (3.63) and (3.64) may be used to compute the magnification in a mobile phone camera provided that an appropriate near point is determined for the phone's lens (it is probably reasonable to assume that a phone lens's far point is approximately infinity).

Pupil matching prevents vignetting, but the final field of view (FOV) on the phone camera's sensor is ultimately limited by the FOV of the imaging attachment, which is usually determined by the size of a *field stop* in an eyepiece (Fig. 3.99).

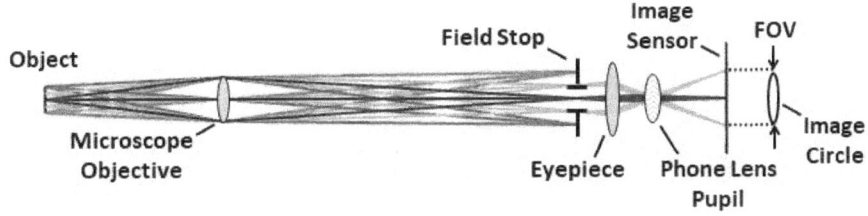

FIGURE 3.99 Image circle at the sensor resulting from an eyepiece's field stop.

The field stop is a physical aperture that is mounted at the plane of the intermediate image. Its primary function is to purposefully limit the spatial extent of the image field, covering those outer edges of the image that have not been optimized for sufficient quality. Additionally, a field stop provides a nice sharp window for the image (otherwise, one will see the edge of the objective lens inside the lens housing, along with everything else inside the lens barrel). Thus, the field stop is what you see whenever you peek through a telescope, microscope, or binoculars and find a nice sharp circle surrounding the image (and everything else outside the circle is pretty much pitch black). This means that an image circle forms onto your retina.

Accordingly, an image circle forms onto the sensor of a phone camera. This image circle is the image of the field stop. This would be the case for compound microscope attachments, as well as for Keplerian telescope attachments. Depending on an eyepiece's FOV (i.e., the size of the field stop), an image circle will either be inside or outside of the phone camera's sensor area. Take, for example, the images of the setting Sun shown in Figs. 3.100a (taken with just a mobile phone camera) and 3.100b (taken by placing the phone camera lens just behind one of the eyepieces of a 10 x 50 binocular). The field stop is clearly seen in Fig. 3.100b.

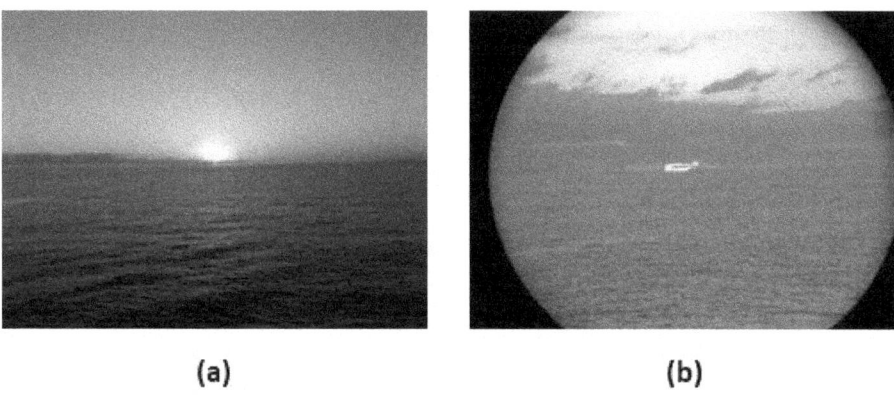

(a)   (b)

FIGURE 3.100 Photos of the Sun taken through my mobile phone camera. (a) Phone only. (b) Phone placed behind one of the eyepieces of a 10 x 50 binocular.

In some other cases, the image circle is not a sharp-looking window surrounding the image. This occurs for telescopes of the Galilean-type, whose eyepiece is a negative powered lens. In this case, the exit pupil is located inside the telescope and is therefore inaccessible by the phone's pupil (Fig. 3.101). Evidently, there is no eye relief for a Galilean telescope. Its exit pupil is a virtual image of the telescope objective, formed by the negatively powered eyepiece. This exit pupil appears as a virtual window floating in front of the eyepiece, and it is not conjugate with the image field. Therefore, when a phone lens looks through the eyepiece of a Galilean telescope, there will appear a defocused image of the exit pupil surrounding the image of distant objects on the sensor, if the diameter of this image circle is less than the sensor's diagonal length. To eliminate the presence of this image circle, one should increase the diameter of the exit pupil. Naturally, this means physically increasing the diameter of the telescope objective.

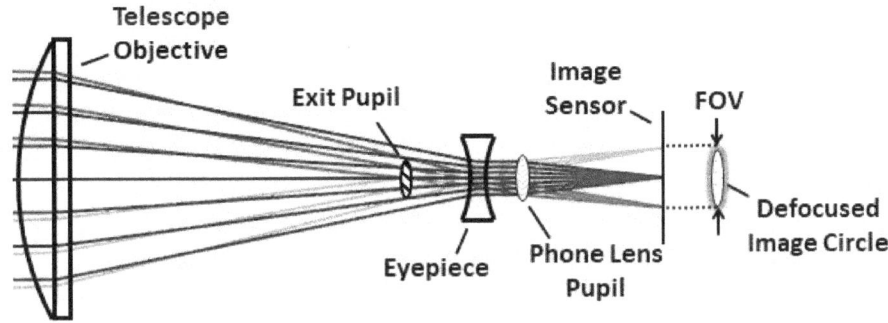

FIGURE 3.101 Galilean telescope attachment. The exit pupil is "inside" the telescope.

Telescope attachments are often called *afocal attachments*, the "afocal" referring to the condition of parallel exiting rays. The advantage of having a Galilean over a Keplerian telescope as an afocal attachment is the relatively shorter length of a Galilean telescope, for the same image magnification in both systems. On the other hand, because the exit pupil for a Galilean telescope is inside the telescope, one has to increase the size of the telescope objective, so it becomes somewhat a trade-off between length and width between the two types of telescope attachments. However, note that even a Keplerian telescope's objective lens diameter must be increased if its exit pupil size does not match the size of a phone lens's pupil. Moreover, note that having an exit pupil that's outside the imaging attachment does not necessarily make it easily accessible, because it also depends on the length of the eye relief. For high powered telescopes and compound microscopes, the eye relief is rather short. Moreover, the pupil of a mobile phone camera lens is usually not at the front plane of the phone. Rather, it is usually somewhere behind the phone lens's cover glass, or perhaps even behind the phone lens (Fig. 3.102).

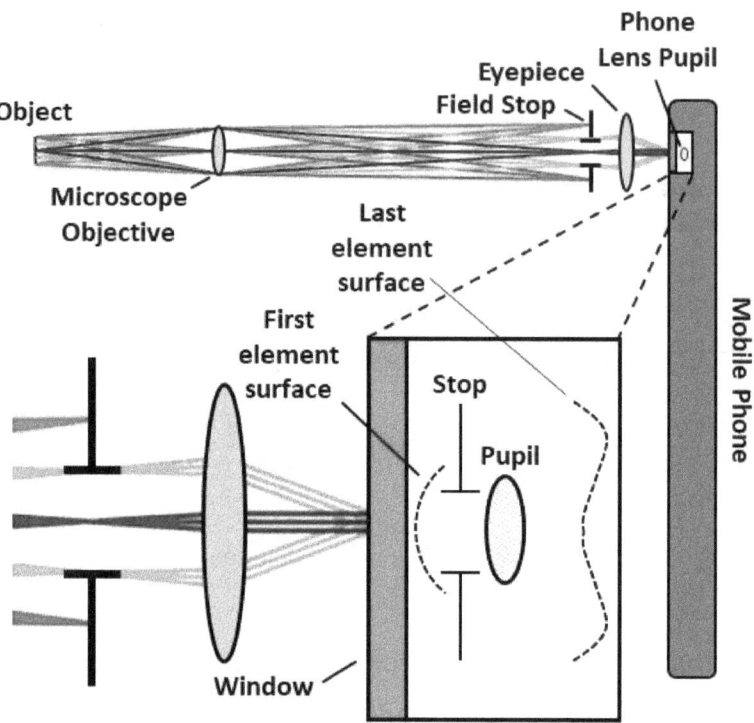

FIGURE 3.102   Illustration of the location of a phone lens's pupil.

3.3.3 The OOWA™ lens from Dynaoptics

OOWA™ [49] is the product trademark for a line of mobile phone imaging attachments from the company Dynaoptics [50], which have risen to popularity for their rather high performance. The key ingredient in an OOWA (pronounced "*oo – ah*") lens is the freeform surface. In theory, a freeform surface may take on any shape, may be described by virtually any polynomial, and may even be asymmetric about the optic axis if the image field isn't symmetric (for instance, when photographing a tilted scene). The surfaces of mobile phone lens elements (i.e., not the attachments) are often somewhere between aspheres and freeform, with the back elements being more of the latter in order to bend the chief ray appropriately whilst balancing aberrations [51]. The value of having freeform surfaces for the elements in phone attachments is in achieving high image quality at a short track length with minimal lens elements. Fig. 3.103a shows two of Dynaoptics' OOWA products: the 2.5x "75 mm zoom equivalent" attachment (bottom left), and the wide-angle attachment (top right). Fig. 3.103b shows OOWA lenses attached to mobile phones.

(a) (b)

FIGURE 3.103 (a) Two OOWA products. (b) OOWA attached onto mobile phones. (*Courtesy of Dynaoptics, with permission.*)

The 2.5x attachment is referred to as a "75 mm zoom equivalent" because it provides an effective "zoom ratio" of 75/30 = 2.5. The zoom ratio of any zoom lens is the ratio of the long EFL to the short

EFL. However, the OOWA 2.5x lens is more accurately an afocal of the Galilean type, which provides a magnification of 2.5x when attached to any phone lens that has been set to focus at infinity. Applying Eq. (3.64), we have

$$M_T = f_o \left(\frac{1}{f_e} + \frac{1}{-\infty} \right) = \frac{f_o}{f_e} = 2.5. \qquad (3.65)$$

The magnification is 2.5, but it is not known what f_o and f_e are. One may attempt to estimate them by noting that the length between the objective and eyepiece is equal to $f_o - f_e$, but note that these dimensions are taken relative to the principal planes of the lenses. Still, it is instructive to obtain rough estimates. For instance, for the OOWA 2.5x lens that I have with me, $f_o - f_e \approx 20$ mm. Combining this with the result in Eq. (3.65) we have $f_o = 20 M_T / (M_T - 1) = 20(2.5)/(2.5 - 1) \approx 33.33$ mm, which means $f_e = 33.33/2.5 \approx 13.33$ mm. Of course, this is the absolute value of the eyepiece's EFL. Now, if you're like me (a geek), there is really no other use for these data other than the satisfaction of having calculated them. But for this book, we can do a little more. We can estimate the position of the exit pupil. Applying Eq. (2.1) we have $1/(-f_e) = (1/20) + (1/s')$ where $f_e = 13.33$ and s' is the distance from the eyepiece to the exit pupil. Solving for s' yields $s' \approx -8$ mm, which means that the exit pupil is 8 mm in front of the eyepiece. Actually, I did this exercise to prove that the same value could have been obtained simply by dividing the distance between the objective and eyepiece by the magnification, which yields $20/2.5 = 8$ mm. Also, the size of the exit pupil is given by dividing the diameter of the objective lens (about 30 mm) by the magnification, which yields $30/2.5 = 12$ mm. Thus, the general rule is that whenever a scene is magnified M times by an afocal system, the exit pupil diameter is reduced M times. What else can we learn from these calculations? Well, I'll tell you what. First, let's estimate the OOWA 2.5x lens eyepiece's half angle FOV by taking the arc tangent of the ratio of the exit pupil's semi-diameter to its distance from the eyepiece, which yields $\tan^{-1}(6/8) \approx 37^0$. I estimate that my mobile phone camera has an effective half angle FOV of approximately 30^0, so this means that placing the OOWA 2.5x lens right against my mobile phone's lens should yield an image free from the appearance of an image circle, which is indeed the case as shown in the photograph of my bookshelf in Fig. 3.104c using the OOWA 2.5x lens.

(a)   (b)   (c)

FIGURE 3.104 Photos of my bookshelf. (a) Phone only. (b) The OOWA 2.5x lens about 10 cm from my phone. (c) OOWA 2.5x lens mounted (by hand) over my phone lens.

In theory, the OOWA wide angle attachment's operating principle would be the complete reverse of the 2.5x afocal: it is a reverse Galilean telescope (aka "opera glass") in which the objective lens is negative powered, and the eyepiece is positive powered. In this way, the negative front element refracts oblique angled rays towards the rear lens, yielding a wide-angle view. The value in the OOWA wide angle lens is in its low distortion without the application of any software correction. Without taking apart the OOWA lens, it is difficult to know how many elements are used for this design. Moreover, one would not know the shape of the freeform surfaces used on the elements unless one measures them with some profile measuring tool. However, even if one did measure the surface profile, and even if one did attempt to design a freeform lens into a phone imaging attachment, one may not know how to produce the lens. The ultimate value that Dynaoptics possesses is the knowledge of not just how to design such lenses, but also how to manufacture them at large volume and still maintain the quality in the images. Such knowledge is acquired through a combination of solid understanding of optical design theory and understanding the limits of manufacturing processes, production tolerances, and quality control. It is only through the above that one knows how to push the limits of designing and producing practical freeform lenses. In the meantime, why not simply enjoy the images they yield (Figs. 3.105a – 3.105b)?

(a)   (b)

FIGURE 3.105 Comparison of wide-angle images. (a) An OOWA competitor's lens. (b) The OOWA wide angle lens attachment. (*Courtesy of Dynaoptics, with permission*).

3.3.4 Spectrometer mobile phone attachments

3.3.4.1 Basic concepts for spectrometer attachments

Anyone who's ever looked through a transmissive diffraction grating towards a light source can relate to the experience of seeing diffracted orders of the source's spectra. This means that placing a diffraction grating over a mobile phone camera lens should produce the same effect. Therefore, neglecting any consideration to aberrations, *spectral resolution*, *angular dispersion*, *free spectral range*, and *spectral distortion*, seeing spectra using a mobile phone camera should be as simple as mounting a dispersing element over the phone's lens. In fact, it is, as illustrated by the spectra of a white LED desk lamp and white LED flashlight I photographed by holding a blazed transmission grating over my mobile phone's lens in Figs. 3.106a – 3.106d.

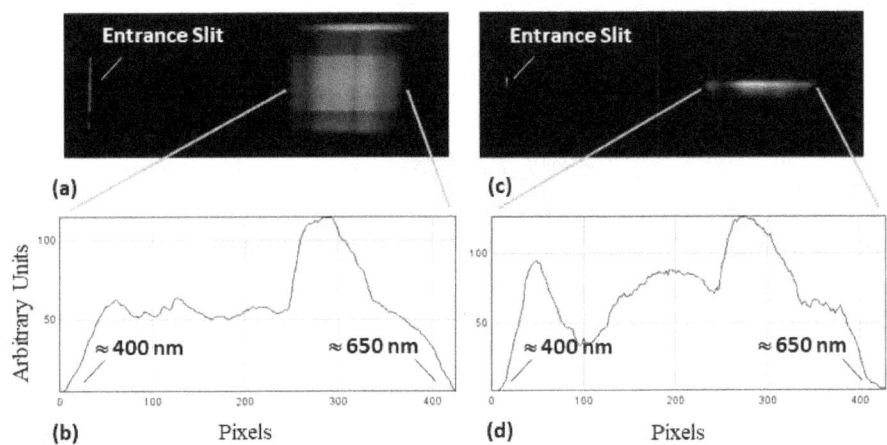

FIGURE 3.106 Spectra photographed through a blazed grating in front of my mobile phone lens. (a) – (b) White LED desk lamp. (c) – (d) White LED flashlight.

The "Entrance Slit" shown in Figs. 3.106a and 3.106c is cut out from cardboard (≈ 1 mm width) and taped over the LED desk lamp and flashlight. Additionally, I taped a pair of linear polarizers (crossed at about an 80-degree angle relative to each other) over my phone's lens in order to reduce the brightness of the image and avoid saturation. The LED desk lamp was placed about 5 ft. away from the phone, and the LED flashlight was about 1 ft. away. To obtain the spectrograms in Figs. 3.106b and 2.106d, I opened up the image files using ImageJ [52] (available free online). Fig. 3.107 shows my rather crude home-made spectrometer phone attachment. Indeed, I had to hold on to the grating using my fingers. This grating was an unwanted component salvaged from an old instrument.

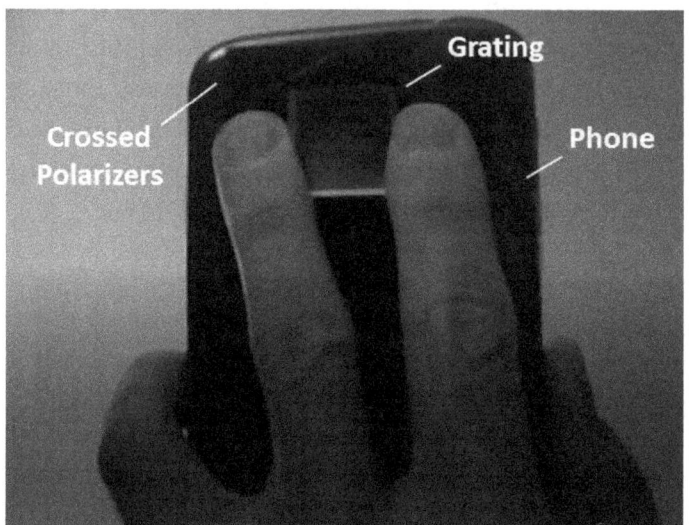

FIGURE 3.107   Home-made spectrometer phone attachment.

In recent years, scientific spectrometer and multi-spectral imaging phone attachments have gained wide interest [53 – 59]. In stark contrast to my simple home-made spectrometer phone attachment above, professional and scientific spectrometers provide quantitative data, which requires calibration. For example, initial images from sources with known spectra should be taken, and correspondences should be made between spectral wavelengths and sensor pixels. For my home-made attachment, I would have also needed to obtain photographs of known spectra with and without the crossed polarizers over the phone's lens in order to account for the spectral transmission of the polarizers. I also would have needed to obtain background images of the ambient environment and perhaps subtract their average from the spectrograms in Figs. 3.106 (I didn't have a cover over my phone and grating). Even the monochromatic and chromatic aberrations of the phone's lens should be characterized. Also, if there is an infrared cut-off filter in the camera, then this limits the spectral range of the system.

In the development of any imaging system, calibration philosophies and key instrument specifications are intimately related. In spectrometry, one should consider characteristics such as spectral resolution, angular dispersion, free spectral range, and spectral curvatures and distortions (just to name a few). Let us begin with **spectral resolution** R_s which, for a grating with N grooves (within the diameter of the incident light beam) that's diffracting into the m'th order, is given by

$$R_s = mN. \tag{3.66}$$

The spectral resolution given by Eq. (3.66) is often referred to as a grating's spectral **resolving power**. Evidently, for an incident beam of fixed diameter Na, where a is the pitch between two grooves, the resolving power is increased if the number of grooves per unit beam diameter is increased (i.e., if N increases and a decreases proportionately). Equivalently, one may maintain the groove pitch a while increasing N, which requires that the diameter of the incident beam be increased as well. This results in an increase in the width of the diffraction grating, which is consistent with the argument presented in Sec. 1.3.2 concerning the impact of the size of a dispersing element on spectral resolution. Another way to see this is to note that Eq. (3.66) may be derived from determining the ratio $\lambda/\Delta\lambda_{min}$ for a diffraction grating [60]. In this ratio, $\Delta\lambda_{min}$ is the minimum resolvable diffracted wavelength interval between λ and $\lambda + \Delta\lambda$. Now, consider the derivative in Eq. (1.1), which may be expressed as

$$\frac{d\phi}{d\omega} = \frac{d\phi}{d(2\pi c/\lambda)} = \frac{d\phi}{d\lambda}\left(\frac{-\lambda^2}{2\pi c}\right) \approx \frac{\Delta\phi}{\omega}\left(\frac{-\lambda}{\Delta\lambda}\right). \qquad (3.67)$$

As can be seen in Eq. (3.67), the ratio $\lambda/\Delta\lambda$ (which, again, is related to R_s) determines the spectral resolution. For a fixed beam diameter (and therefore, fixed grating size), the spectral **angular dispersion D_s** of a grating increases with reduced groove pitch a. To see this, consider Fig. 3.108, which depicts the geometry of rays diffracting through a grating at normal incidence:

$$a \sin\theta = m\lambda. \qquad (3.68)$$

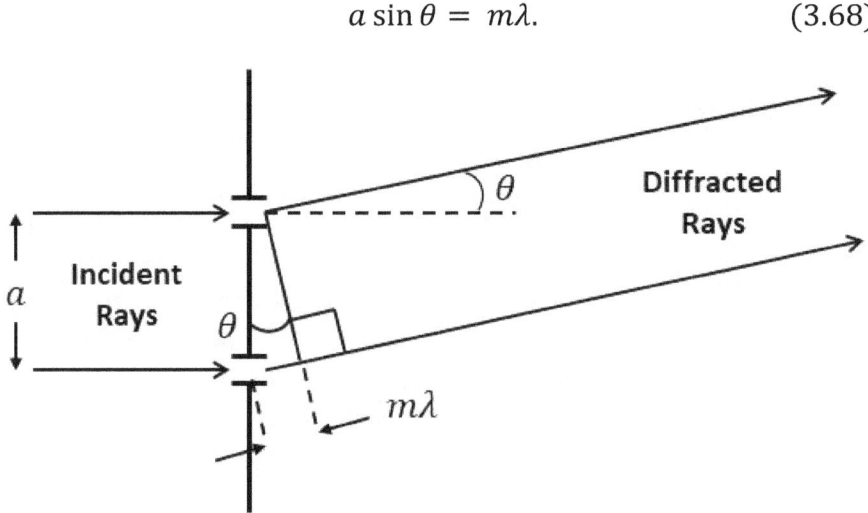

FIGURE 3.108   Ray geometry for diffraction of rays through a grating.

Differentiating Eq. (3.68) with respect to wavelength and solving for $d\theta/d\lambda$ yields

$$\frac{d\theta}{d\lambda} = \frac{m}{a\cos\theta} = D_s. \qquad (3.69)$$

Note from Eq. (3.69) that, the smaller the groove pitch a, the larger the angular dispersion. For this reason, gratings are often specified by the number of grooves per unit width [i.e., $N/(Na) = 1/a$] rather than the total number of grooves N.

The *free spectral range* F_s of a grating refers to the available non-overlapping wavelength range between two diffracted orders and is given [61, 62] by

$$F_s = \frac{\lambda_S}{m}.$$
(3.70)

In Eq. (3.70), the wavelength λ_S is the shortest wavelength in order $m + 1$. The free spectral range is evidently independent of grating design and is therefore, rather fundamental. For visible light at the first diffracted order, $F_s \approx 400$ nm, which explains why there is ample room between the red wavelength and the blue from the 2nd order spectrum in Figs. 3.106a and 3.106c.

Finally, let's consider *spectral curvatures and distortions*. These refer to visible curves ("smiles" or "frowns") and keystone effects of the projected spectrum onto a screen. Such distorted-looking spectra can occur, for example, when a grating tilts about an axis that is orthogonal to the grating's grooves (Figs. 3.109a – 3.109d). In Fig. 3.109a, I photographed some "slits" of sunlight that passed through my window blinds and landed on the surface of my metal filing cabinet. In Fig. 3.109b, I placed my blazed grating over my mobile phone's lens, ensuring that the grating was flat against the lens's window, and snapped a photo of the light "slits" on my cabinet. In Fig. 3.109c, I show how I tilted my grating, which resulted in the spectral distortions shown in Fig. 3.109d. These are a consequence of the fact that the optical path lengths for diffracted rays must satisfy Eq. (3.68) across the plane at which they diffract. Tilting a grating results in a sort of curved plane for satisfying these optical path differences. Spectral curvatures and distortions are important considerations [63 – 65] because a spectral line must align with a column of pixels that sample this wavelength. Doubling or tripling the number of columns of pixels can average the effect, but at the expense of spectral resolution. Incidentally, spectral resolution is also dependent on the entrance slit's width, as we will see in the following section with a Zemax OS example.

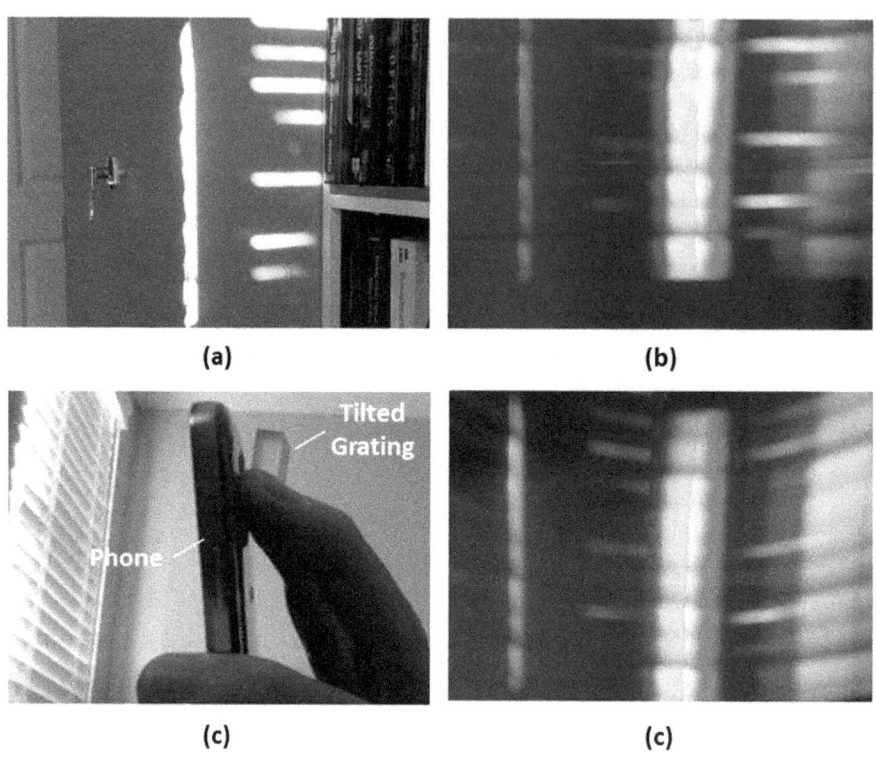

(a)

(b)

(c)

(c)

FIGURE 3.109 Photo of "slits" of light on my cabinet. (a) Phone only. (b) Grating flat against phone lens. (c) Tilting the grating. (d) Result of grating tilt on spectra.

3.3.4.2 Zemax OS example: Simple spectrometer phone attachment

First enter the prescriptions shown in Figs. 3.110 – 3.111 (units in mm) with field height 0, and wavelengths 400, 450, 500, 550, 600, 650, and 700 nm (primary wavelength = 550 nm). Although surface 17 represents an IR-cut filter (usually at 650 nm cut-off), we shall include wavelengths 650 and 700 nm for completeness.

	Surf:Type		Radius	Thickness	Material	Semi-Diameter	
0	OBJECT	Standard ▾	Infinity	42.970		0.000	
1	STOP (aper)	Standard ▾	Infinity	1.000		2.500	U
2	(aper)	Standard ▾	57.060	1.300	N-SF5	4.500	U
3	(aper)	Standard ▾	19.750	3.000	N-BK7	4.500	U
4	(aper)	Standard ▾	-27.830	10.000		4.500	U
5		Coordinate Break ▾		0.000	-	0.000	
6	(aper)	Standard ▾	Infinity	0.000	MIRROR	4.500	U
7		Coordinate Break ▾		-20.000	-	0.000	
8		Coordinate Break ▾		0.000	-	0.000	
9	(aper)	Standard ▾	Infinity	0.000	MIRROR	4.500	U
10		Coordinate Break ▾		20.000	-	0.000	
11		Coordinate Break ▾		0.000	-	0.000	
12		Diffraction Grating ▾	Infinity	5.000		12.500	U
13		Coordinate Break ▾		0.000	-	0.000	
14	(aper)	Standard ▾	Infinity	0.500	N-BK7	5.000	U
15	(aper)	Standard ▾	Infinity	2.000		5.000	U
16	(aper)	Paraxial ▾		3.858		1.000	U
17	(aper)	Standard ▾	Infinity	0.200	N-BK7	3.000	U
18	(aper)	Standard ▾	Infinity	0.512		3.000	U
19	IMAGE	Standard ▾	Infinity	-		0.411	

FIGURE 3.110 Prescription for the spectrometer phone attachment shown in Fig. 3.112. Note that the continuation for this prescription (which shows the coordinate break parameters) is shown in Fig. 3.111.

The IR-cut filter on surface 17 is based on specifications applied to an IR-cut filter for a miniature lens system discussed by Reshidko and Sasian [17]. Surface 16 is a paraxial thin lens model (PTLM) for a 2-mm diameter (f/2.25) miniature lens with EFL 4.5 mm. Surface 12 is a thin ideal diffraction grating surface model in Zemax OS, with 0.6 lines per micron, set to diffract at the +1st order. In reality, this grating could be, for example, part no. 49-580 from the supplier Edmund

Optics[*] with 600 grooves per mm. However, note that the actual part from this supplier has a thickness of 3 mm that must be accounted for. Also, note that the "blaze angle" for this part number does not refer to the angle of the diffracted 1st order from the grating. Rather, it refers to θ in Fig. 2.29 at some unspecified wavelength (which you may be able to request from the supplier).

	Surf:Type		Decenter X	Decenter Y	Tilt About X
0	OBJECT	Standard ▾			
1	STOP (aper)	Standard ▾			
2	(aper)	Standard ▾			
3	(aper)	Standard ▾			
4	(aper)	Standard ▾			
5		Coordinate Break ▾	0.000	0.000	25.000
6	(aper)	Standard ▾			
7		Coordinate Break ▾	0.000	0.000	25.000 P
8		Coordinate Break ▾	0.000	0.000	-15.366
9	(aper)	Standard ▾			
10		Coordinate Break ▾	0.000	0.000	-15.366 P
11		Coordinate Break ▾	0.000	0.000	-19.269
12		Diffraction Grating ▾	0.600	1.000	
13		Coordinate Break ▾	0.000	0.000	0.000
14	(aper)	Standard ▾			
15	(aper)	Standard ▾			
16	(aper)	Paraxial ▾	4.500	1	
17	(aper)	Standard ▾			
18	(aper)	Standard ▾			
19	IMAGE	Standard ▾			

Lens Data — Update: All Windows ▾ — Surface 13 Properties ‹ › — Configuration 1/1 ‹ ›

FIGURE 3.111 Continuation for the unseen portion of the prescription in Fig. 3.110.

---

[*] www.edmundoptics.com

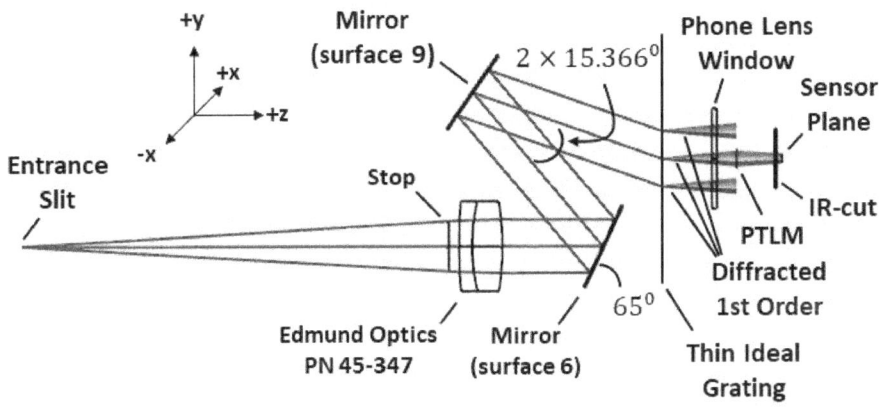

FIGURE 3.112 System layout for the prescriptions in Figs. 3.110 – 3.111.

Surfaces 2 and 3 together form an achromatic doublet model for part no. 45-347 from Edmund Optics. This doublet serves to collimate light from the entrance slit, whose width is along the x dimension. Note that in Zemax OS, the aperture setting should at this point be set to "float by stop size". The distance between the entrance slit and the stop is 42.97 mm, which is the back focal length (BFL) of the achromatic doublet lens (at some unspecified visible wavelength). Since an additional 1 mm air space is added between the stop and the doublet, the light from the slit is not fully collimated by the doublet. However, this has a minimal impact on the beam, which remains very much collimated overall. Ray aiming is unnecessary at this point, because the stop is in front of all optical surfaces. The angle for the first mirror shown in Fig. 3.112 is $90 - 25^0 = 65^0$. In the Zemax OS lens prescription of Fig. 3.111, an angle of 25 degrees is shown because this is measured relative to the plane normal to the optic axis. The second mirror (surface 9) tilts at 15.366 degrees relative to the rays that are normally incident on its surface, giving rise to reflected rays at 2×15.366^0 relative to the rays that are reflected off the first mirror. The thin ideal grating is tilted at 19.269^0 relative to the rays reflected off the second mirror because that is the diffracted 1st order beam angle at 550 nm wavelength [apply Eq. (3.68) for $a = 1/600, m = 1$ and $\lambda = 0.55 \times 10^{-3}$]. By allowing the incident rays to be at 19.269^0 relative to the grating's normal, we enable the 1st order rays to be diffracted normal to the grating's surface. In this way, we also enable the phone's lens to be normal to the grating. If we took a look at the footprint diagram at the image plane for this setup, it would be difficult to see the "spread" of the seven wavelength points across the plane, due to the rather small spot size of the focused rays. But if we defocused the image plane by +0.1 mm, one obtains the result shown in Fig. 3.113.

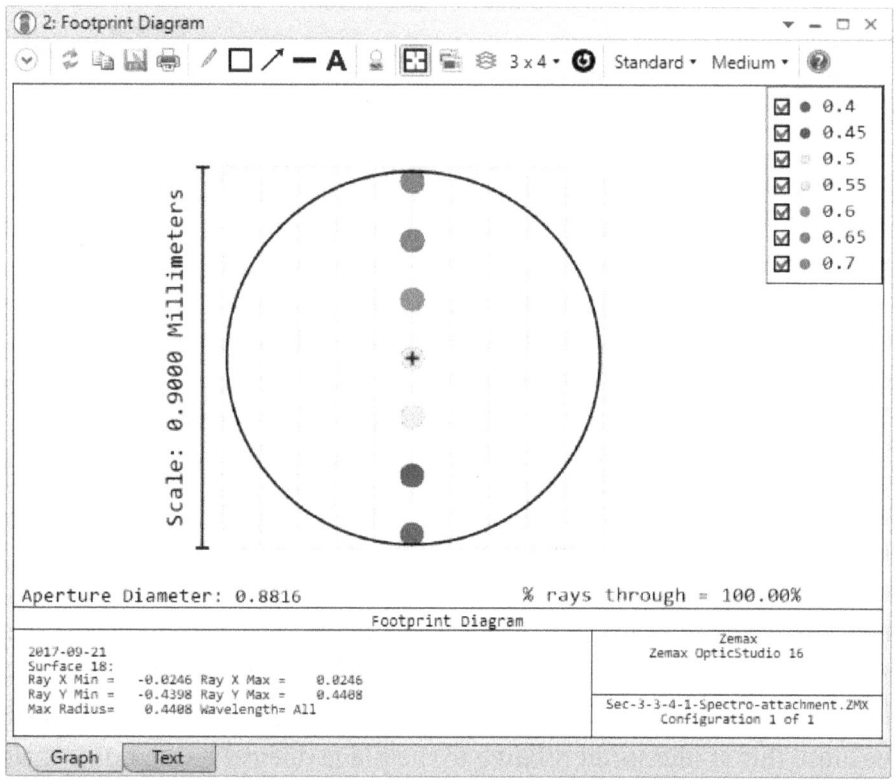

FIGURE 3.113 Print-screen from Zemax OS of the footprint diagram at the sensor plane in the model from Fig. 3.112 (defocused by +0.1 mm). Note that each spot has been color-coded to their respective spectral colors.

In Fig. 3.113, note that I have used the "Color Rays By Wavelength" option under the "Settings" menu of the footprint diagram. This lets Zemax OS approximate the visual rendering of colors corresponding to their respective wavelengths across the spectrum.

The lateral spread of focused spots across the bandwidth of 400 – 700 nm in Fig. 3.113 is about 940 microns across. If we assume that the pixels on a mobile phone's sensor is about 2 microns (a reasonable assumption), then this spread would be comprised of about 470 pixels, which is close to the number of pixels spread across the spectra shown in Fig. 3.106.

The result modeled in Fig. 3.113 is for a point source at the entrance slit (in fact, it is for a "point slit"). Let's take a look at the effect of having a finite entrance slit width on the spectral resolution for this simple spectrometer phone attachment model. To do this, first, if the image plane is still defocused by +0.1 mm, undo this defocus and ensure that the thickness on surface 18 is back to 0.512 mm. However, note that the 0.512 mm thickness on surface 18 is actually a purposefully slightly defocused thickness, yielding a "realistic" finite geometric spot size at the image plane that helps us examine the impact of a finite slit width on spectral resolution. Set Ray Aiming "on" (aimed at the paraxial entrance pupil). Set the PTLM (surface 15) as the aperture stop. Set field heights to y = 0, +0.1 mm, and -0.1 mm (i.e., all in the y dimension). This models a slit width of 0.2 mm. Remove all wavelengths, except for the primary wavelength (550 nm). Add one other wavelength at 560 nm (so you'd now have only two wavelengths, 550 nm and 560 nm). If we now look at the footprint diagram and select "Wavelength" for the "Color Rays By" setting under the footprint diagram settings menu, we obtain the plot shown in Fig. 3.114. In this figure, the image of the slit at 550 nm and 560 nm are clearly separated, which implies a *geometric* spectral resolution of about 10 nm for the current simple model of an imaging spectrometer phone attachment. In contrast to the spectral resolving power given by Eq. 3.66 (which assumes an infinitesimally thin entrance slit width), the geometric spectral resolution is a rather practical design consideration. The spectra observed on the image sensor is comprised of a spread of images of the entrance slit (this is also where aberrations take importance). Therefore, the thinner the entrance slit, the more resolvable are the individual wavelengths. In the limit that the entrance slit is infinitesimally thin (but wide enough to allow light to pass through), we may estimate the theoretical limit at which two wavelengths are resolvable using Eq. (3.66) and the fact that $R_s = \lambda/\Delta\lambda_{min}$:

$$R_s = mN = \lambda/\Delta\lambda_{min}. \tag{3.71}$$

In order to compare the theoretical limit of spectral resolution with the condition shown in Fig. 3.114 (i.e., at 550 nm), we would let $m = 1$ and $\lambda = 550$ in Eq. (3.71), but a decision must be made on what value to use for N. This is not the number of grooves per unit length. Rather, it is the number of grooves within the diameter of the incident beam that eventually forms the image of the spectrum of slits. In our simple spectrometer model, it is the PTLM's diameter (2 mm) that limits the rays that form the spectral image. Therefore, N in Eq. (3.71) must be the number of grooves across 2 mm. Since we have assumed a grating model with 600 grooves per mm, we have $N = 1200$ grooves per 2 mm. Therefore, using this value for N in Eq. (3.71), and for $m = 1$ and $\lambda = 550$ nm, we may solve for $\Delta\lambda_{min}$, yielding

$$\Delta\lambda_{min} = \lambda/N = 550 \text{ nm}/1200 \approx 0.5 \text{ nm}. \tag{3.72}$$

Pause for insight: $\rightarrow$ We have made an implicit fundamental assumption in our computation in Eq. (3.72) [and the application of Eq. (3.66)] that there is perfect spatial coherence across the grooves of the diffraction grating. Why? Because in order for a beam to be diffracted by N grooves, wavelets from every groove across the surface of a grating must add in amplitude and phase. This is only possible if wavelets from each groove are spatially coherent with one another. Fortunately, the spatial coherence across the width of a grating increases with decreasing entrance slit width [66].

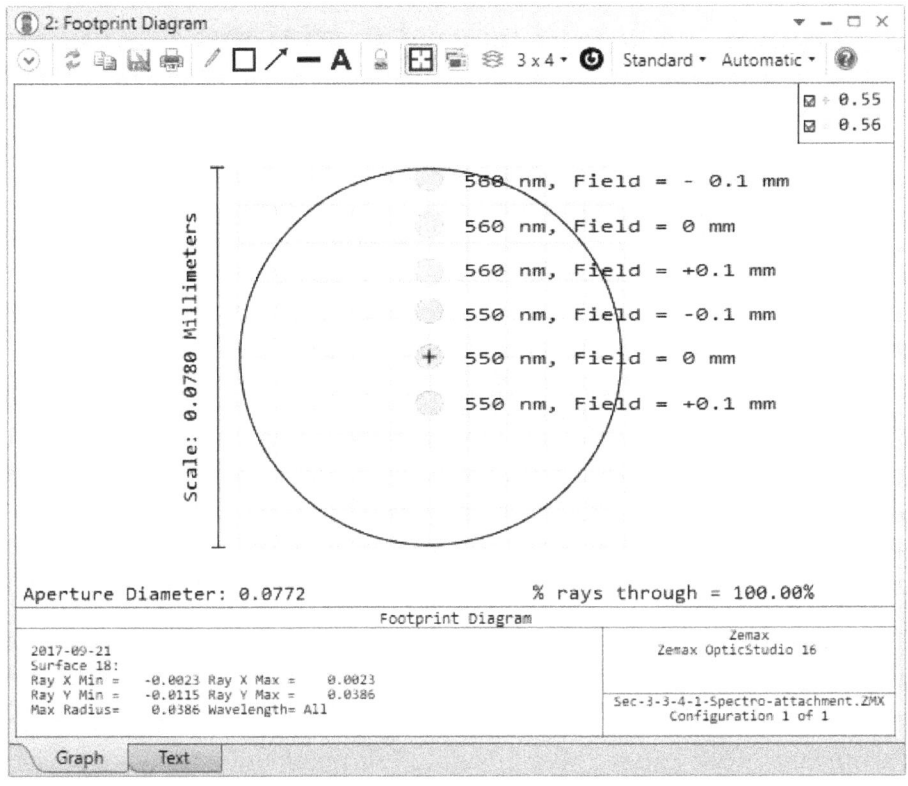

FIGURE 3.114 Footprint diagram at the sensor plane at 550 and 560 nm, slit width = 0.2 mm.

One last thing: → Those of you who know a little something about Fourier transform spectroscopy might wonder if there is some simple way to convert a mobile phone camera into a portable hand-held Fourier transform spectrometer. While it may not be too simple, in theory, it is possible. All that would be needed are the essential components of an interferometer whilst replacing the detector with a mobile phone camera. I haven't proved this. But in theory, it should be possible. However, beware of the impact of miniaturization on spectral resolution (see Sec. 1.3.2).

3.4 Google's Glass™ wearable computing device

Optical devices providing augmented reality have come into the spotlight in recent times [67 – 70]. Augmented reality (AR) refers to the modification of observed scenery in the human eye by the addition of digital imagery provided by external optical elements. One example is the "Glass™ wearable computing device" (aka "Glass") developed by the company Google [71]. Since its introduction, a number of practical applications for Glass have been explored. These have included, for example, head-worn information display systems for drivers of vehicles [72], portable imaging of rapid diagnostic test results in healthcare [73], and ophthalmoscopy [74], where the retinal image of a patient's eye is obtained through Glass's built-in camera, and the results are displayed to the examiner wearing Glass.

Fundus cameras (optical instruments that image the human eye's retina) are ordinarily rather complex systems [e.g., 75, 76]. Therefore, the successful application of Glass to fundus imaging by the researchers in Ref. 74 is a notable development for portable healthcare. Now, can we take this a step further? Note that, in Ref. 74, Glass is worn by the eye examiner, not the patient. For the purpose of remote point-of-care diagnostics, it would seem interesting to wonder if Glass may be used for self-examination – in particular – of the human eye fundus. In this section, we'll explore this as a practical exercise using Zemax OS. To begin, we'll examine the basic optical principles of Glass, based on Google's patents [77 – 79].

3.4.1 Basic imaging principle of the Glass™ eyepiece

In examining Refs. 77 and 79, the optical display device for Glass is actually referred to as "Lightweight Eyepiece for Head Mounted Display." For the purpose of our discussion, we'll simply call it "Glass". The basic ray path in Glass is depicted in Fig. 3.115. Note that this figure only illustrates the "ray path", not the optical principle. That is, Google's patent actually indicates that the two beamsplitters are polarizing beamsplitters, not just 50/50 beamsplitters. Moreover, there is a quarter wave plate between Beamsplitter 1 and the concave mirror. These all help to minimize loss in the transmittance of the rays in order to maintain a bright image of the LCoS (Liquid Crystal on Silicon) display onto the retina. LCoS displays are front illuminated, so the Google patents also indicate that there should be a source (say, an LED) in front of Beamsplitter 2 that illuminates the LCoS display.

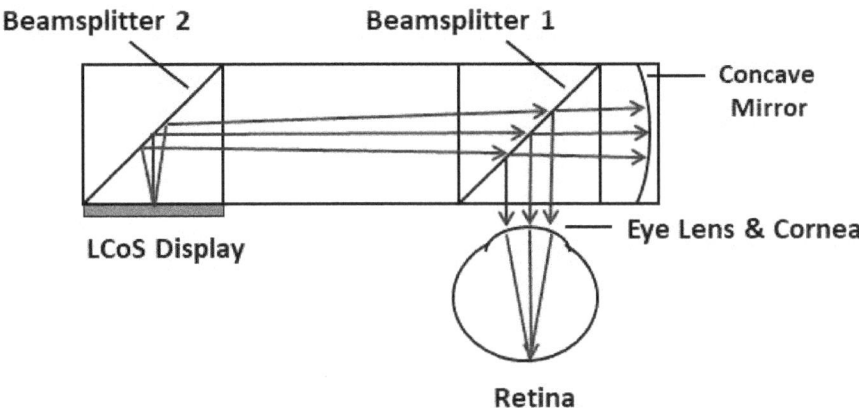

FIGURE 3.115 Trajectory of rays for displaying a digital image in Glass.

In the patents, there is also a "convex lens" cemented to the concave mirror. However, I believe that this "lens" simply acts as a substrate to which the concave mirror's reflective coating is applied. Hence, this "lens" is effectively just a slab of glass, with no actual optical power other than the introduction of some spherical aberration (deliberate or not). In any event, for our simple exercise, polarizing components and "convex lenses" aren't essential. The basic point is that rays from the LCoS display are collimated by the concave mirror, which then reflects into the eye from Beamsplitter 1. Thus, the eye looks at "infinity", where it sees a clear image of the LCoS display.

3.4.2 Zemax OS example: Glass™ as a fundus imager

Near the beginning of Sec. 2.1.1, I mentioned that image and object points are conjugates in an imaging system. This means that, while an image point is an image of the object point, reversing their positions makes the object point an image of the image point (Fig. 3.116). Hence, if we were to replace the LCoS display in Glass with a sensor, then an image of the retina may be formed onto the sensor provided that some form of illumination is provided onto the retina (Fig. 3.117). However, for our exercise, we will make many simplifications and modifications to our optical model of Glass. In particular, the two polarizing beamsplitters shall be replaced by 50/50 beamsplitters, and the quarter wave plate shall be eliminated. Also, the concave mirror shall be made simply a front surface concave mirror rather than the coated surface of a convex lens (I guess it could equivalently be considered the coated surface of a concave lens).

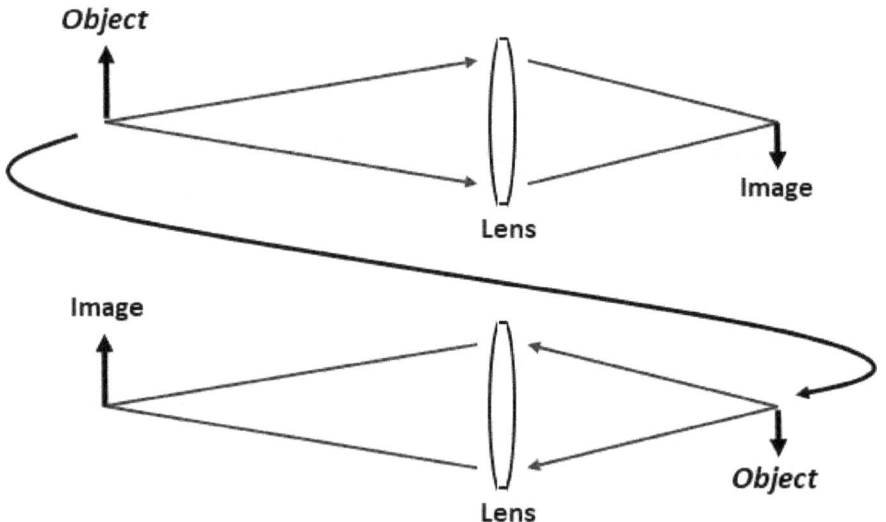

FIGURE 3.116 Conjugates in an imaging system may switch places.

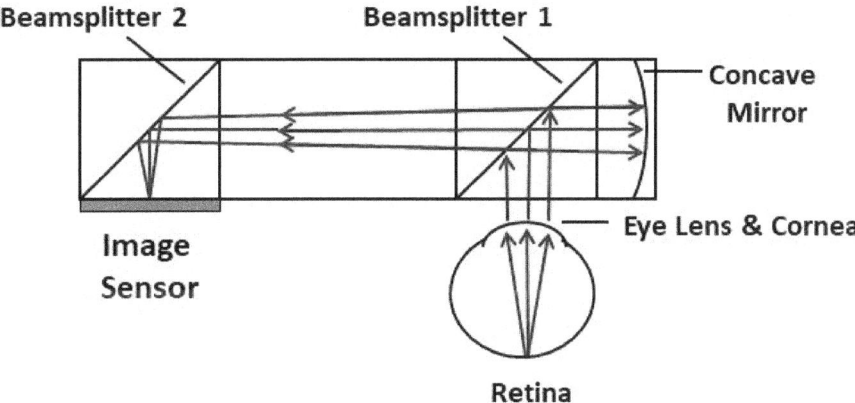

FIGURE 3.117 LCoS display substituted with an image sensor (compare with Fig. 3.115) in a model for "Glass".

Since the image sensor takes the place of the LCoS display, we would not need an illuminator in front of Beamsplitter 2. However, we need a source to shine light into the eye. There are two plausible ways, the first being less plausible. Nonetheless, it is instructive to discuss the first method without performing a full simulation. In the first method, consider the concept shown in Fig. 3.118, where an occulting disk placed in front of a condenser causes a "ring" of illuminated light at the retina. The essential idea for the concept illustrated in Fig. 3.118 is to avoid having any light from the point source to strike the image sensor from reflecting off the first beamsplitter. This is the reason for mounting the occulting disk behind the condenser. However, as a consequence, the central portion of the retina [i.e., the actual region of interest (ROI) that we want to image onto the sensor] is not directly illuminated. The hope then is that the illuminated surrounding area (i.e., the "ring") diffusely reflects and scatters within the "sphere" of the eye, such as happens in an *integrating sphere* [80]. In this way, diffuse scattered light would eventually strike the ROI, whose surface scatter would emerge from the pupil and be directed by the concave mirror onto the sensor. This seems plausible, but it is rather difficult to simulate the many millions of rays that would be needed to scatter onto the ROI, where we would then require to scatter rays back through the eye lens and cornea.

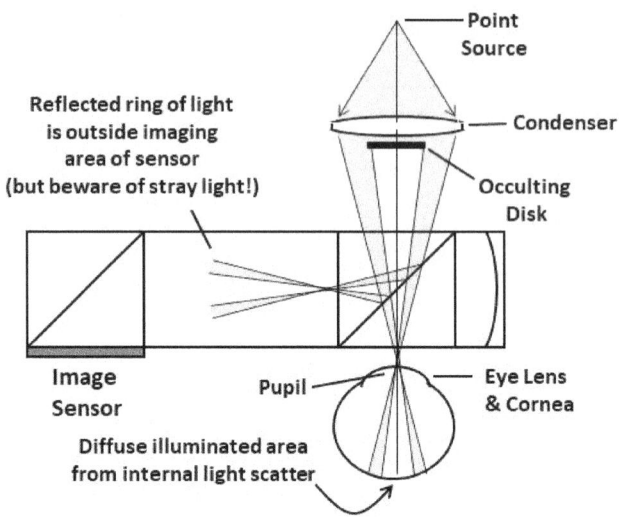

FIGURE 3.118 "Ring illumination" at the retina from an occulting disk and point source.

Because of the reasons above, we shall explore a second more plausible illumination method, which is to place a third beamsplitter between the eye lens and Beamsplitter 1 where light from below would reflect into the eye (Figs. 3.119a and 3.119b). Note in this concept that a "Ring Source" is used, which is imaged by the condenser onto the cornea. The reason for this is to prevent specular back-reflections of the illumination from the cornea onto the image sensor. This is the practice in all fundus cameras. However, it should be noted that one usually needs to have additional optics and an appropriate secondary aperture to ensure that this back-specular reflected ring of light is blocked. In our simulation, we shall purposely avoid this back-reflection by placing near 100% transmission coatings on the cornea (and in fact, on all of the reflective interfaces that are at the eye lens, cornea, and pupil).

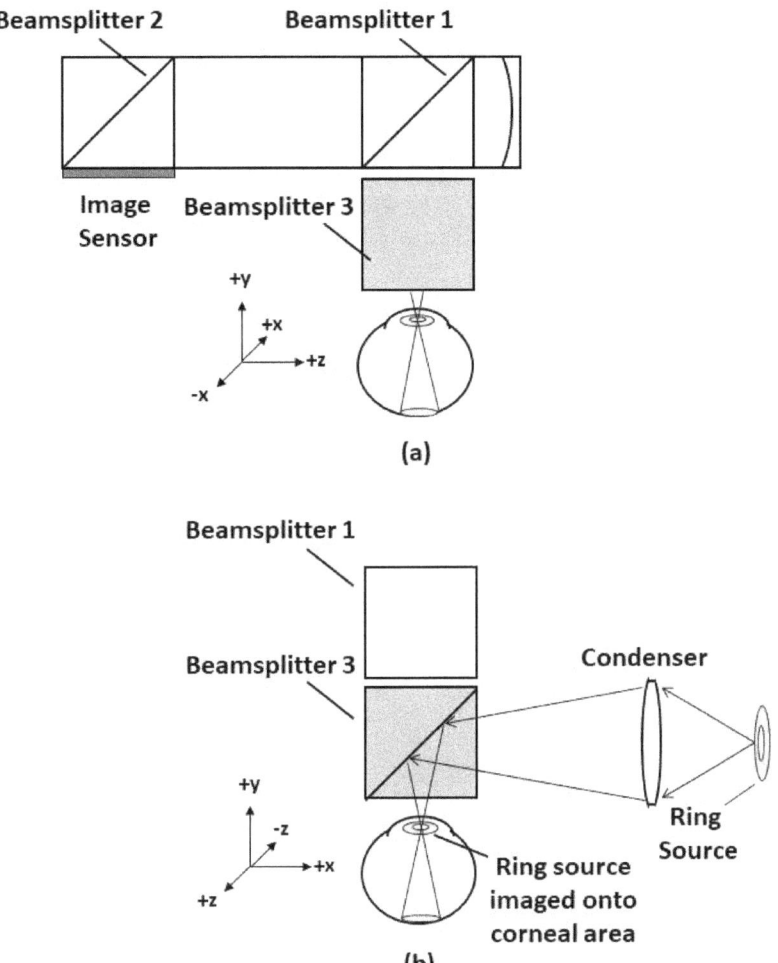

FIGURE 3.119 Retinal illumination. (a) Top view. (b) Rotated about the y-axis.

So, to simulate all that's been said, first ensure that your Zemax OS file is in the non-sequential (NSC) mode. Then enter the prescription and data provided in Figs. 3.120a – 3.120d into the NSC component editor, whose output 3D layout should appear as shown in Figs. 3.121a – 3.121c. The eye model we are applying is based roughly on that provided by Walker [81], where Vd = 55 for all media in the eye.

	Object Type	Comment	Ref Object	Inside Of	X Position	Y Position	Z Position
1	Null Object ▾	BS Refs	0	0	0.000	-5.000	30.000
2	Standard Lens ▾	Condenser	1	0	5.000	0.000	10.000
3	Source Ellipse ▾	Ring Source	2	0	0.000	0.000	-10.000
4	Cylinder Pipe ▾	Light House	2	0	0.000	0.000	-10.000
5	Standard Lens ▾	Cornea	0	0	0.000	-21.000	30.000
6	Standard Lens ▾	Aquous	5	0	0.000	0.000	0.600
7	Annulus ▾	Pupil	6	5	0.000	0.000	3.000
8	Standard Lens ▾	Eye Lens	6	0	0.000	0.000	3.000
9	Standard Lens ▾	Vitreous	8	0	0.000	0.000	4.000
10	Rectangle ▾	ROI	9	9	0.000	0.000	16.860
11	Rectangular Volume ▾	BS1 Cube	1	0	0.000	0.000	2.500
12	Rectangular Volume ▾	BS3 Cube	1	0	0.000	0.000	7.500
13	Rectangular Volume ▾	BS3	1	12	-0.150	0.000	10.000
14	Rectangular Volume ▾	BS1	1	11	0.000	0.000	5.000
15	Standard Surface ▾	CC Mirror	11	0	0.000	6.000	2.500
16	Rectangular Volume ▾	BS2 Cube	0	0	0.000	-10.000	-10.000
17	Rectangle ▾	BS2	16	16	0.000	0.000	2.500
18	Rectangular Pipe ▾	COVER	16	0	0.000	-2.500	2.500
19	Detector Rectangle ▾	SENSOR	0	0	0.000	-13.000	-7.500

(a)

	Object Type	Tilt About X	Tilt About Y	Tilt About Z	Material	Par 1(unused)
1	Null Object ▾	90.000	0.000	0.000	-	
2	Standard Lens ▾	0.000	-90.000	0.000	BK7	3.000
3	Source Ellipse ▾	0.000	0.000	0.000	-	0
4	Cylinder Pipe ▾	0.000	0.000	0.000	ABSORB	2.600
5	Standard Lens ▾	90.000	0.000	0.000	1.38,55.0	7.800
6	Standard Lens ▾	0.000	0.000	0.000	1.34,55.0	6.400
7	Annulus ▾	0.000	0.000	0.000	ABSORB	4.000
8	Standard Lens ▾	0.000	0.000	0.000	1.40,55.0	10.100
9	Standard Lens ▾	0.000	0.000	0.000	1.34,55.0	-6.100
10	Rectangle ▾	0.000	0.000	0.000		0.200
11	Rectangular Volume ▾	0.000	0.000	0.000	BK7	2.500
12	Rectangular Volume ▾	0.000	0.000	0.000	BK7	2.500
13	Rectangular Volume ▾	0.000	45.000	0.000	BK7	3.400
14	Rectangular Volume ▾	-45.000	0.000	0.000	BK7	3.400
15	Standard Surface ▾	-90.000	0.000	0.000	MIRROR	-80.500
16	Rectangular Volume ▾	0.000	0.000	0.000		2.500
17	Rectangle ▾	45.000	0.000	0.000	MIRROR	3.400
18	Rectangular Pipe ▾	90.000	0.000	0.000	ABSORB	2.500
19	Detector Rectangle ▾	90.000	0.000	0.000	ABSORB	2.000

(b)

FIGURE 3.120   (a) NSC editor for Glass fundus model. (b) Continuation.

	Object Type	Par 2(unuse	Par 3(unu	Par 4(unu	Par 5(unu	Par 6(unuse	Par 7(unused
1	Null Object ▾						
2	Standard Lens ▾	0.000	2.500	2.500	2.300	0.000	0.000
3	Source Ellipse ▾	1000000	1.000	0	2	2.000	2.000
4	Cylinder Pipe ▾	12.300	2.600				
5	Standard Lens ▾	0.000	4.000	4.000	0.600	6.400	0.000
6	Standard Lens ▾	0.000	4.000	4.000	3.000	10.100	0.000
7	Annulus ▾	4.000	1.000	1.000			
8	Standard Lens ▾	0.000	4.000	4.000	4.000	-6.100	0.000
9	Standard Lens ▾	0.000	4.000	4.000	16.920	-10.000	0.000
10	Rectangle ▾	0.600					
11	Rectangular Volume ▾	2.500	5.000	2.500	2.500	0.000	0.000
12	Rectangular Volume ▾	2.500	5.000	2.500	2.500	0.000	0.000
13	Rectangular Volume ▾	3.400	0.100	3.400	3.400	0.000	0.000
14	Rectangular Volume ▾	3.400	0.100	3.400	3.400	0.000	0.000
15	Standard Surface ▾	0.000	3.500	0.000			
16	Rectangular Volume ▾	2.500	5.000	2.500	2.500	0.000	0.000
17	Rectangle ▾	3.400					
18	Rectangular Pipe ▾	2.500	0.500	2.500	2.500	0.000	0.000
19	Detector Rectangle ▾	2.000	100	100	0	0	0

(c)

	Object Type	Par 8(unu	Par 9(unu	Par 10(unu	Par 11(unu	Par 12(unus	Par 13(unus
1	Null Object ▾						
2	Standard Lens ▾	2.500	2.500				
3	Source Ellipse ▾	0.000	40.000	0.000	0.000	0.000	0.000
4	Cylinder Pipe ▾						
5	Standard Lens ▾	4.000	4.000				
6	Standard Lens ▾	4.000	4.000				
7	Annulus ▾						
8	Standard Lens ▾	4.000	4.000				
9	Standard Lens ▾	10.000	10.000				
10	Rectangle ▾						
11	Rectangular Volume ▾	0.000	0.000				
12	Rectangular Volume ▾	0.000	0.000				
13	Rectangular Volume ▾	0.000	0.000				
14	Rectangular Volume ▾	0.000	0.000				
15	Standard Surface ▾						
16	Rectangular Volume ▾	0.000	0.000				
17	Rectangle ▾						
18	Rectangular Pipe ▾	0.000	0.000				
19	Detector Rectangle ▾	0	0.000	0	0	-90.000	90.000

(d)

FIGURE 3.120 (c) Parameters 2 – 7. (d) Parameters 8 – 13.

Note that only object 3 and 19 possess data beyond parameter 13. In particular, object 3 has the value "1" for both parameters 14 and 15, and object 19 has the values -90 and 90 for parameters 14 and 15, respectively. The following is an explanation for what each object (nos. 1 – 19) is in the NSC component editor along with their most essential object property settings:

Object 1: "BS Refs". This is a "Null Object" which simply serves as a reference coordinate position for placing Beamsplitters 1 and 3 (i.e., objects 11 – 14) into the model.

Object 2: "Condenser". This lens images the Ring Source onto the cornea. Ensure that its front and back surfaces are "coated" with an ideal transmission of 0.99999999, and that its side surface (i.e., its edge surface) is absorbing. This is done by entering values into the "Coat/Scatter" tab in this object's properties dialogue box. Use "I.99999999" for the ideal front and back surface coatings. For the side surface, select "Absorbing" under the "Face is" pull-down menu.

Object 3: "Ring Source". No special property settings required.

Object 4: "Light House". This is a cylindrical hollow absorbing pipe used as a housing for the condenser and Ring Source. No special property settings required.

Object 5: "Cornea". This is the model of the eye's cornea [81]. Ensure that it has the same surface coating settings for Object 2.

Object 6: "Aquous". This is the model of the aqueous volume of fluid between the cornea and eye lens [81]. Ensure that it has the same surface coating properties as Object 2.

Object 7: "Pupil." This is modeled as an absorbing annulus. No special property settings required.

Object 8: "Eye Lens". This is the model of the eye's lens [81]. Ensure the same surface coating settings as Object 2.

Object 9: "Vitreous". This is the model for the fluid between the eye's lens and the retina [81]. Ensure that the front surface coating has 0.99999999 transmission (i.e., same as Object 2), but ensure that the side and back surfaces are absorbing (we are using Object 10 as the region of interest for imaging at the retina).

Object 10: "ROI". This is our region of interest (ROI). It is a 0.4 mm x 1.2 mm rectangular surface at the center of the retina that back-scatters light almost directionally through the pupil when rays from the Ring Source are incident on it. Ensure that its Coat/Scatter properties are as follows: Use "Gaussian" for the Scatter Model. Set the "Sigma" value to 0.1. Set Scatter Fraction to 1. Set Number of Rays to 5. Select "Reflective" under the "Face is" pull-down menu.

Object 11: "BS1 Cube". This is a solid glass cube made out of BK7 material that models the medium for Beamsplitter 1. The idea is to simply insert a thin rectangular volume (Object 14) into this cube, and let it serve as the reflective 50/50 coating for the beamsplitter (it's a simplistic way to get our model going without having to model the full complexities of an actual beamsplitter cube). Ensure that all surfaces have a coating of 0.99999999 transmission.

Object 12: "BS3 Cube". As with Object 11, this is a cube for Beamsplitter 3, made out of BK7 material, and has the same surface coating properties as Object 11.

Object 13: "BS3". This is a thin solid rectangular volume that serves as the 50/50 beam splitting interface for Object 12, and it is inserted into Object 12. Its front surface should have 0.99999999 transmission, and its back surface should have 0.5 transmission. Its side surfaces (i.e., the four edges) should be absorptive (set this property as is done for Object 2).

Object 14: "BS1". As with Object 13, this is the model for the 50/50 beam splitting interface for Object 11. Its surface coating properties follow that of Object 13.

Object 15: "CC Mirror". This is the concave mirror that images the center of the retina onto the image sensor (Object 19). No special property settings required.

Object 16: "BS2 Cube". This serves as the cube for Beamsplitter 2, but unlike Objects 11 and 12, this cube has no special property settings, and its walls are completely transmissive. In the model, it only serves to allow us to visualize the placement of Beamsplitter 2, which for our purposes, is a 100% reflective surface. Hence, Beamsplitter 2 is not a beam splitter at all in our model. It is simply a mirror.

Object 17: "BS2". As mentioned above, this is a 100% reflective rectangular mirror surface. No special property settings required.

Object 18: "COVER". This is a "housing" for the image sensor, which is simply a rectangular pipe whose four side walls are absorptive to prevent any stray light to strike the sensor other than rays that reflect off BS2. No special property settings required.

Object 19: "SENSOR". This is the image sensor model. It is simply a 4 mm x 4 mm square detector with 100 x 100 square pixels.

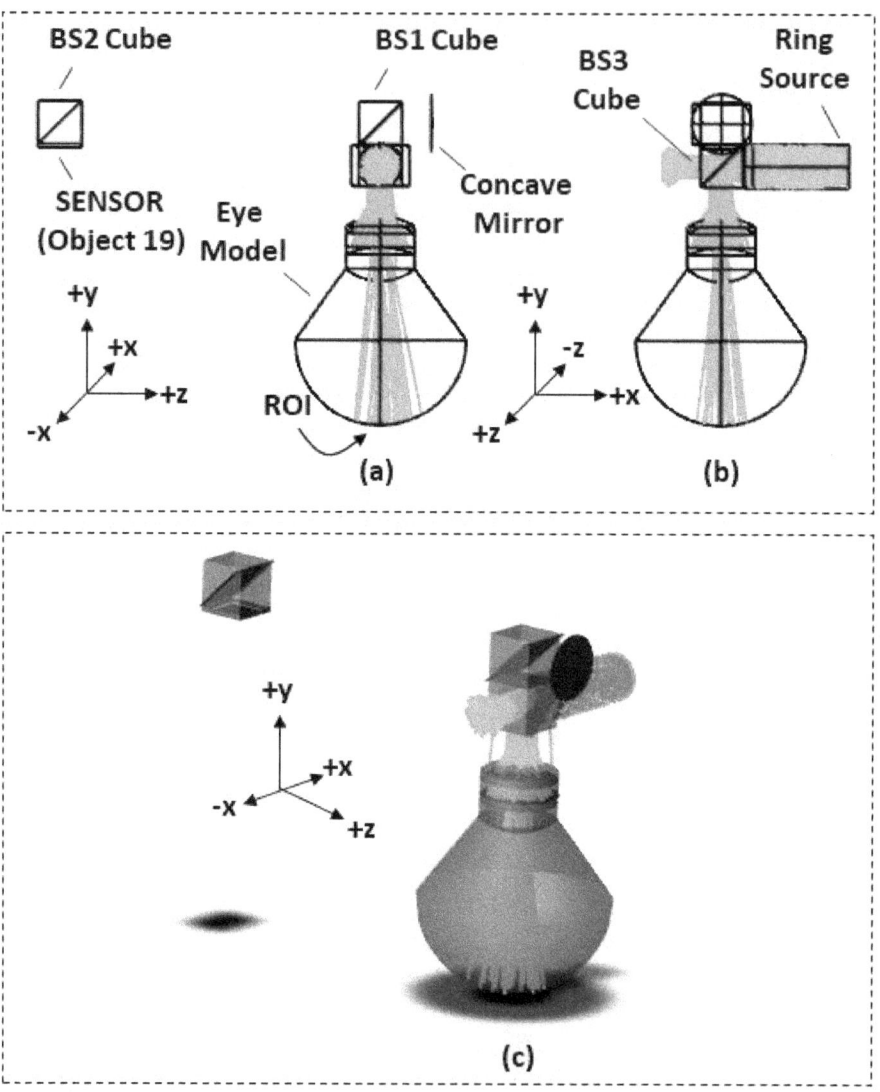

FIGURE 3.121 Glass fundus imager model with 1000 source rays. (a) Top view. (b) Rotated about y-axis. (c) 3D view.

Note that all of the dimensions in our model are mere guesses, plucked – in some cases – from nowhere, for a simple simulation. In Figs. 3.121a – 3.121c, 1000 source rays are displayed from the Ring Source, but no scattering properties have been set for the rectangular ROI (Object 10). A ray trace of 5,000,000 rays from the Ring Source results in the output distribution at the sensor (Object 19) shown in Fig. 3.122 at inverse grey scale, and smoothing = 2. Ensure that the following four boxes have been "ticked" in the "Ray Trace Control" menu: 1. Use Polarization. 2. Split NSC Rays. 3. Ignore Errors. 4. Scatter NSC Rays). A rectangular ROI image is clearly visible. Thus, our model works. Note the size of the ROI image on the sensor, whose vertical length is about 3 mm, so the magnification is $3/1.2 \approx 2.5$x.

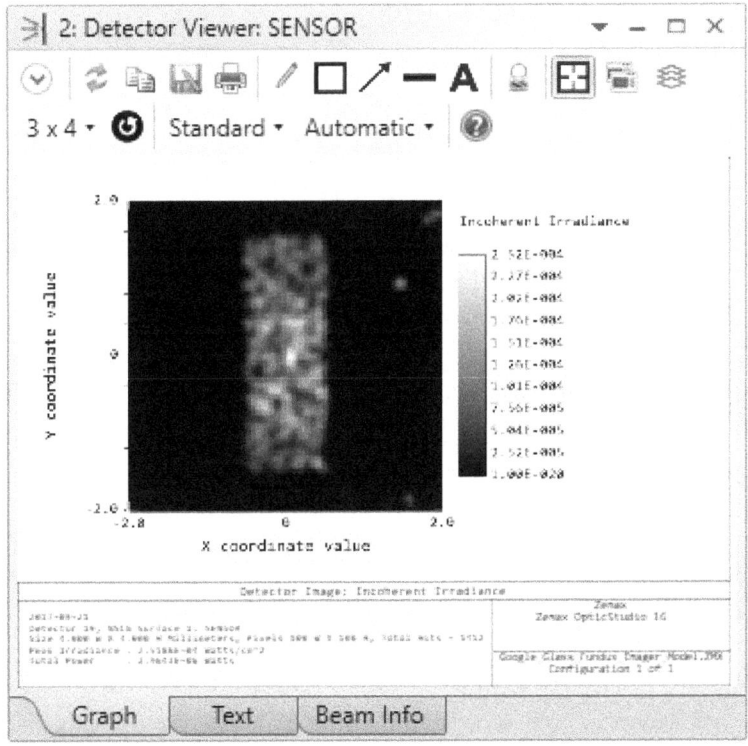

FIGURE 3.122 Sensor (Object 19) output from tracing 5×10^6 rays.

3.4.3 Use of adaptive optics

Since our Glass fundus imager model mimics Google's Glass eyepiece, the imaging component is merely a concave mirror. This is a single image-forming optical surface, so we would not expect the image quality to be as good as what might be achieved using a multi-element imaging lens. So, suppose we were to replace this concave mirror by a high quality multi-element imaging lens. Then we would also have to turn Beamsplitter 1 around such that it reflects the retinal back-scattered light towards the +z direction (Fig. 3.123). The lens would then form an image onto a sensor placed at the lens's focal plane, and "Beamsplitter 2" would be for the illumination.

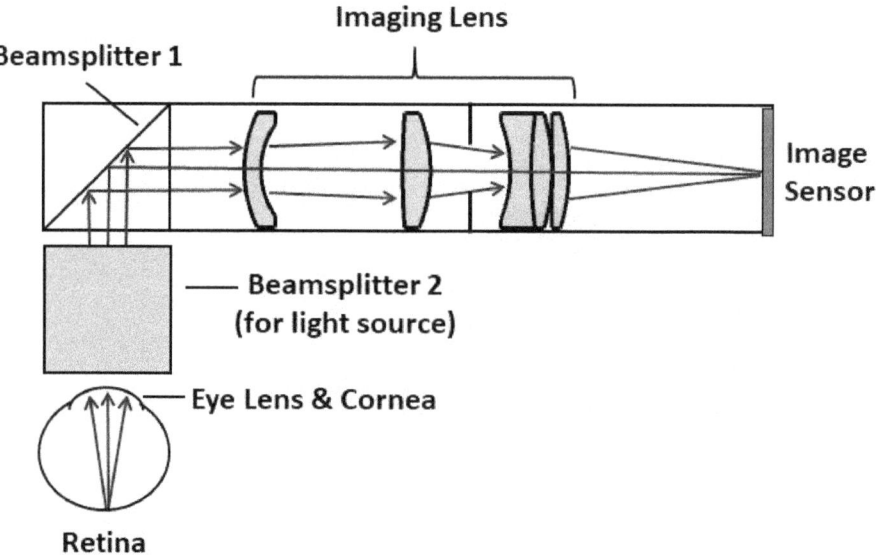

FIGURE 3.123 Modified Glass fundus imager using an imaging lens.

Now, we would normally expect a multi-element lens to form high quality images. However, we would soon find that the image quality is limited by the aberrations of the human eye's image forming components [e.g., 82], such as the cornea and eye lens. Fortunately, it has been shown that adaptive optics systems may be used to correct the eye's aberrations in order to obtain high quality images of the retina [for an overview, see, e.g., 83]. Therefore, it seems viable to apply the same technique to Glass-based fundus camera (Fig. 3.124). In this concept, we envision using Beamsplitter 3 for directing the illumination into the eye. I'd invite readers to try modelling a miniature adaptive optics system integrated with Glass. The results could be rather interesting.

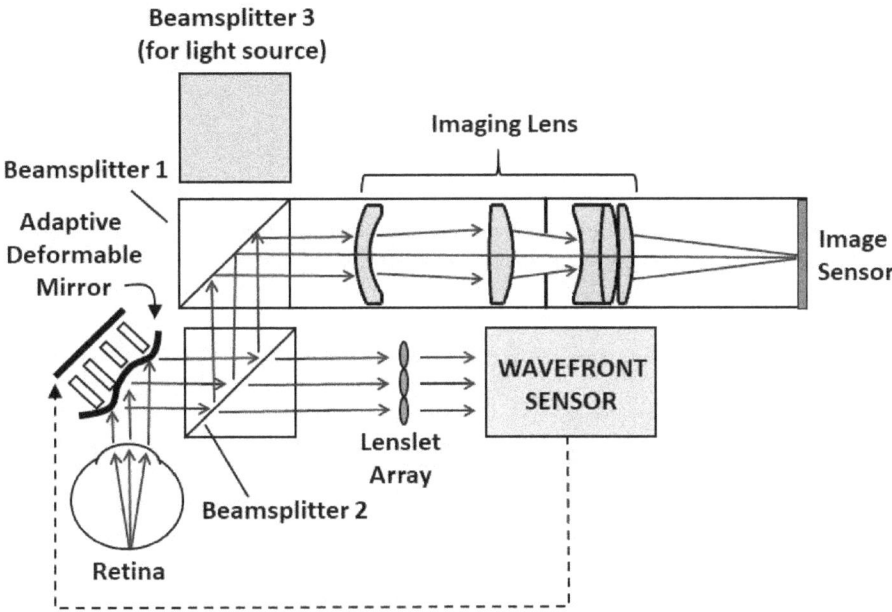

FIGURE 3.124 Theoretical concept sketch of Glass modified with an imaging lens, integrated with an adaptive optics system to correct the eye's aberrations for producing high quality images of the fundus.

3.5 Stereo imaging systems

The re-emergence of 3D cinema in the past decade or so – driven, perhaps, by the widespread application of electronic imaging and projection technologies – has renewed public interest in 3D imaging (though, ironically, I still prefer watching movies in 2D!). The basic principle is well-known: for humans, visible depth perception of a 3D scene requires two views, one from each eye[*] [e.g., 84]. It follows that if two photographs of a 3D scene (each taken from left and right eye positions) are each presented to respective left and right eyes of an observer, a similar depth perception of the original scene is achieved for the observer. This is of course how 3D movies are made and presented to the audience. During filming, two scenes are captured by separate video cameras. At the cinema, the same two scenes are projected and superimposed at the screen., each scene possessing separate light characteristics (e.g., each may be orthogonally polarized relative to one another.) The audience is made to wear visual accessories to separate the superimposed images towards respective left and right eyes. Such accessories are the so-called "3D glasses". If the light from each of the two projected scenes are polarized, then the 3D glasses are polarizers. If the two scenes are color-interlaced, then the 3D glasses are so-called "anaglyphs". And so on.

3.5.1 Stereoscopy vs. Autostereoscopy vs. Holography

The preceding description of depth perception from stereo images requires the use of a visual accessory, aka "3D glasses". You may have heard of *autostereoscopic displays*: 3D displays that do not require the observer to wear such an accessory. But this does not change the fundamental requirement that depth perception requires two different views for each eye. In the case of autostereoscopic displays, it is the display that "wears" the accessory. That is, an autostereoscopic display must possess the characteristic that it presents perspective views of a 3D scene to each human eye. This is usually done by physically projecting (or "allowing") separate beams toward each eye [e.g., 85, 86]. Elements at the display that separate the view of left and right eye scenes may be considered the "3D accessories" (e.g., lenticular arrays). This is what I mean by the display "wearing" the accessory instead of the observer.

---

[*] It should be noted that "depth cues" need not always be obtained from parallax. Even shadows and shapes can provide depth information for a single eye.

Now, autostereoscopic displays are not holographic*, but holograms are, indeed, autostereoscopic in the sense that you don't need to wear any visual accessory to perceive depth from looking towards a hologram. Holograms aren't even images at all. A hologram is a diffraction pattern of the 3D scene. To create a hologram, we let waves that are scattered (actually, diffracted) by an object† coherently interfere with a reference wave at the surface of a recording medium (such as film or an image sensor). *The recorded interference pattern on the recoding medium is the hologram*. To create a *holographic image* of the object, we shine light back onto the hologram (i.e., the medium that recorded the interference pattern), which diffracts the incident beam into Fourier components that synthesize (i.e., interfere) to essentially reproduce waves from the original object. Since a hologram reproduces waves from the original object, one perceives depth from looking at the holographic image of the object. However, the same "depth rule" applies: you still need to use both eyes to look at a holographic image produced by a hologram in order to perceive depth. Looking at a holographic image with a single eye does not produce the same depth effect (though the image would indeed still possess certain depth cues such as shadows, corners...etc.)

Confused? Figs. 3.125a – 3.125d might be helpful. In Fig. 3.125a, an observer looks at the actual 3D object, and features (i.e., points P and Q) from the object are imaged by the eyes onto specific locations across the observer's retina. The key point here is that it is not the presence of the object that matters for viewing a 3D scene of the object. Rather, *it is the presence of the image of the features of the object across the retinal plane that are important*. In order to re-create the perception of depth of the original object, one need only to *re-create the same irradiance pattern on the retina*. This is achieved in Fig. 3.125b when separate left and right eye photographs of the original object are placed at the correct location in front of each eye.

---

* Some studies suggest that there are subtle depth information differences between stereograms and holograms [87].

† Recall from the beginning of Sec. 2.2 that features on an object are like diffraction gratings.

The two photographs aren't the original object, but they reproduce the positions of image points P' and Q' in the left-eye, and P" and Q" in the right-eye (i.e., they reproduce the same irradiance patterns on the retina as in Fig. 3.125a). In Fig. 3.125c, a lenticular array bends rays from left and right eye interlaced columns of the scene towards respective left and right eyes of the observer, resulting in the reproduction of the irradiance pattern on the retina from the original object. In this case, the observer need not wear 3D accessories such as polarizing filters. As mentioned, it is the autostereoscopic display that wears this accessory, which, in this case, is the lenticular array. Finally, in Fig. 3.125d, a hologram reproduces wavefronts with amplitude and phase that resemble the waves from the original object. Thus, when the observer's eyes focus these wavefronts onto the retina, the same irradiance pattern is again reproduced.[*]

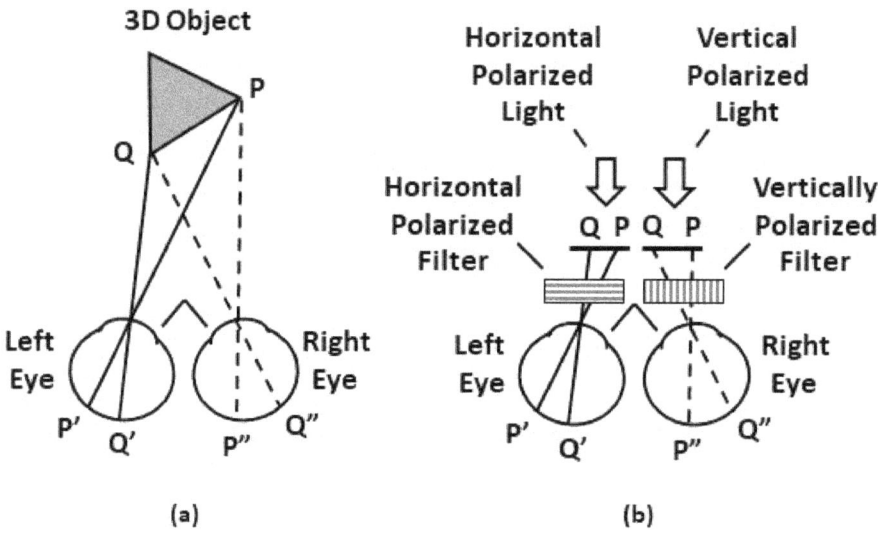

FIGURE 3.125 Four different conditions that the human brain and eyes could perceive depth from a 3D object. (a) Look at the object with the naked eye. (b) Look at two perspective photographs (aka "stereograms") of the object. (*Figs. 3.125c and 3.125d continue on the next page*).

---

[*] By the way, *holographic stereograms* can also produce the same effect [e.g., 88].

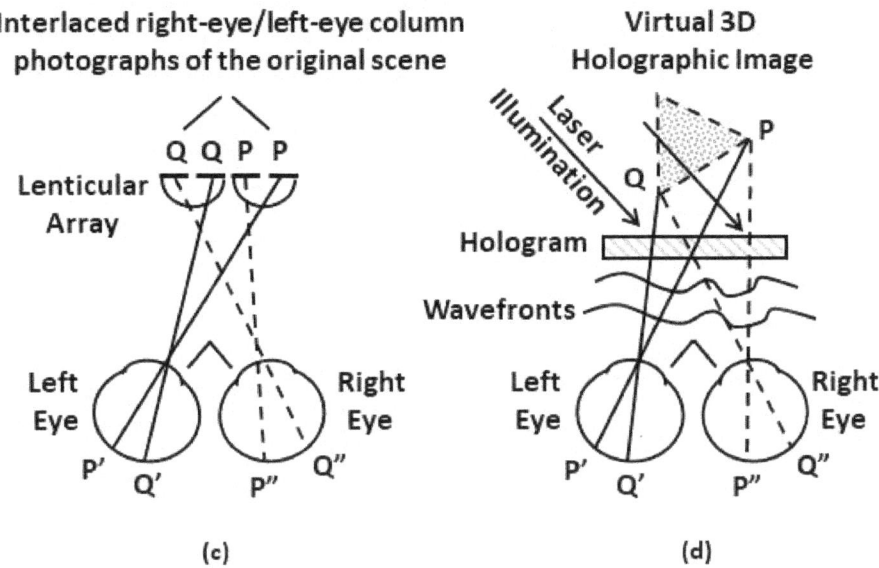

Interlaced right-eye/left-eye column photographs of the original scene

Virtual 3D Holographic Image

(c)   (d)

FIGURE 3.125 *Continued.* (c) Project interlaced columns of the two perspective photographs towards the eye. (d) A hologram reproduces wavefronts from the original object, and the eyes focus them onto the retina.

Now note that, in Figs. 3.125b and 3.125c, stereo photographs are taken either by using two cameras separated by some distance apart, or by a single camera taken at two different positions. Have you heard of achieving the same thing by using a single camera at a single position? This is indeed, possible, and it is the subject of discussion in the following section.

3.5.2 Stereo imaging with a single stationary lens

In their 1992 paper on the plenoptic camera [36] (which we discussed in Sec. 3.2.7.6), Adelson and Yang describe the basic principle of single lens stereoscopy. In fact, the title of their paper is "Single Lens Stereo Using a Plenoptic Camera". Oddly, in the years 2012 – 2013, two United States Patent Applications were published [89, 90] describing precisely the same fundamental concept, but they do not cite the work of Adelson and Yang. At any rate, the basic concept of single lens stereo is rather fascinating: shift the aperture stop position rather than shift the lens. So, the basic premise of single lens stereo imaging is that two images of a 3D object formed through shifted pupil locations within the same lens would appear different. Does it work? Let's examine its plausibility with a simple Zemax OS simulation.

3.4.2.1 Zemax OS example: Single lens stereoscopy

Let us use Zemax OS to examine the image of two points across a 3D scene, through a single lens at two different aperture stop positions. To that end, consider again the lens given by Table 2.1 (Sec. 2.2.2.3). Use a monochromatic wavelength at 588 nm. Set ray aiming "on" (aimed at the paraxial entrance pupil). Set the aperture to "float by stop size". Set a single field point at object height = 0.5 mm. Modify the prescription by inserting coordinate breaks at surfaces 6 and 8 as shown in Fig. 3.126. Note also that the distance from the last element's back vertex to the image plane is 1.359 mm. Then set up operands into the multi-configuration editor as shown in Fig. 3.127. The first operand in the multi-configuration editor decenters the aperture stop by 0.1 mm in the +y direction, and 0.1 mm in the −y direction. This is done by setting Parameter 2 on surface 6 (i.e., the coordinate break) to +0.1 and -0.1. Be sure to set Parameter 2 on surface 8 to the opposite numbers by placing a "pickup" solve such that, when Parameter 2 on surface 6 is +0.1 mm, then Parameter 2 on surface 8 is -0.1 mm, and so on. The idea in the second and third operands in the multi-configuration is the following: We are simulating two object points across a tilted object plane such that the first point is at a distance of 4.8 mm from surface 1, at a field height of 0.5 mm above the optic axis, and the second point is at a distance of 2 mm from surface 1, at a field height of -0.5 mm below the optic axis. The resulting layout (with all configurations superimposed) is shown in Fig. 3.128. Note how the shifted stop positions serve to change the locations of the image points. For example, rays from object point Q strike different locations Q' and Q" at the image plane at stop positions y = -0.1 mm and y = +0.1 mm respectively (the image points P' and P" for the two stop locations are also shifted but not as significantly as for Q' and Q"). This seems to indicate that single lens stereo imaging works.

In practice, one would image the two object points P and Q when the stop is at, say, the -0.1 mm position, followed by imaging these points at the +0.1 mm stop position. Two images of the scene are then presented to the observer as a stereo pair (i.e., a stereogram), and one may decide how the separate views are observed (i.e., through polarizers or anaglyphs). In the US patent applications by Mitchell and Shinkoda [89, 90], an LCD panel serves as an electronic shutter for the aperture stop, and an image sensor is mounted at the image plane. By synchronizing image capture (and display) with the electronic switching of left and right shutter positions on the LCD, stereo image pairs may be presented to an observer in real-time.

Lens Data — □ ×

Update: All Windows ▾   Surface 4 Properties ‹ ›   Configuration 1/4 ‹ ›

		Surf:Type	Radius	Thickness	Material	Semi-Diameter
0	OBJECT	Standard ▾	Infinity	4.800		0.500
1		Standard ▾	Infinity	0.200		0.177
2	(aper)	Standard ▾	0.584	0.071	N-LAK9	0.250 U
3	(aper)	Standard ▾	0.355	0.608		0.250 U
4	(aper)	Standard ▾	3.409	0.120	N-BAF52	0.250 U
5	(aper)	Standard ▾	-0.668	0.165		0.250 U
6		Coordinate Break ▾		0.000	-	0.000
7	STOP	Standard ▾	Infinity	0.181		0.050 U
8		Coordinate Break ▾		0.000	-	0.000
9	(aper)	Standard ▾	-0.559	0.091	N-SF11	0.210 U
10	(aper)	Standard ▾	1.794	0.089	N-LAF2	0.250 U
11	(aper)	Standard ▾	-0.947	2.392E-03		0.250 U
12	(aper)	Standard ▾	-10.515	0.072	N-SSK5	0.250 U
13	(aper)	Standard ▾	-0.871	1.359		0.250 U
14	IMAGE	Standard ▾	Infinity	-		0.300 U

FIGURE 3.126 Prescription for single lens stereo.

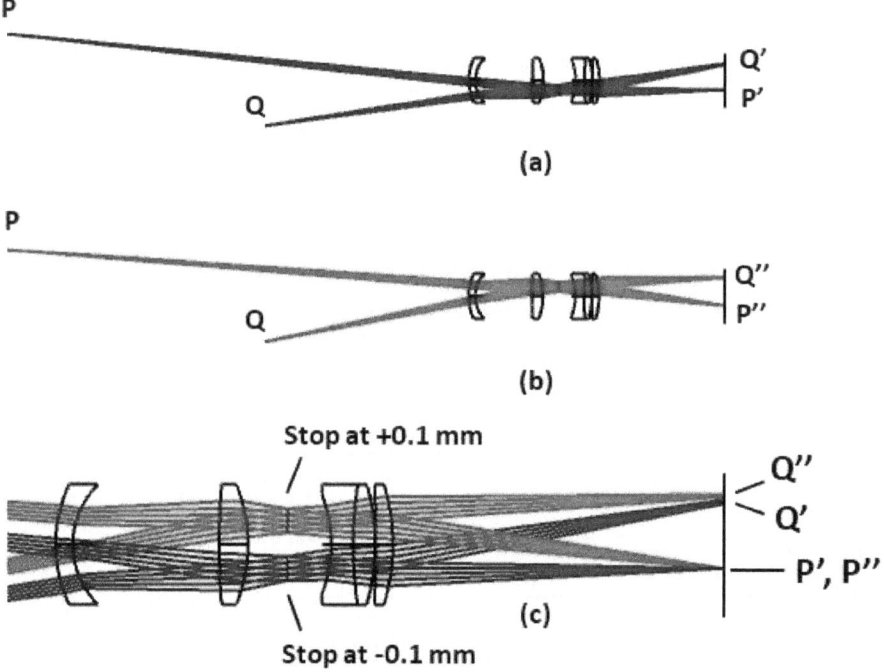

FIGURE 3.127 Multi-configuration settings for lens given by Fig. 3.126.

FIGURE 3.128 Layout of single lens stereo given by Figs. 3.126 and 3.127, imaging two object points P and Q at different distances (i.e., a 3D scene). (a) Stop at y = -0.1 mm position. (b) Stop at y = +0.1 mm position. (c) Superimposed layouts of (a) and (b).

3.5.3 Autostereoscopic 3D displays in mobile phones

3D displays for mobile phones have been studied and in development in the past decade or so [91, 92]. Today, mobile phones with autostereoscopic displays are commercially available [e.g., 93, 94]. And, apparently, those who have seen them have expressed that the image quality is quite good [95]. It also seems conceivable that the approach to producing autostereoscopic 3D on a mobile phone would take advantage of some aspect of the usual "classic" lenticular arrangement of interlaced columns of left and right-eye views across the display (i.e., Fig. 3.125c). That is, in all such types of autostereoscopic displays, a column of *display pixels* must consist of two columns of *stereo display subpixels*, one for each eye, such that a stereo pair of display pixels always represents a single 2D display pixel. Consequently, there is usually a reduction in the resolution of autostereoscopic displays by half of the normal 2D display resolution.

It would certainly be advantageous to somehow be able to maintain the full resolution of a mobile phone's display when it is presenting autostereoscopic imagery to the observer. Additionally, it would also be useful if a display could switch between 3D and 2D views. In fact, this is possible [96, 97]. The operating principle is simple and elegant: use an electronically switchable lenticular micro liquid crystal (LC) lens array whose focal length may vary between infinity and some finite value. Moreover, by essentially switching the micro LC lens's lateral position across the display, the light from interlaced left/right column pixels may be steered towards any of the two eyes, at any time. This means that time-multiplexing the left-right column pixel displays synchronously with the beam steering direction for these pixels enables each eye to view respective columns interchangeably, resulting in viewing the full resolution of the display (Figs. 3.129a and 2.129b). Evidently, viewing a normal 2D scene on the display would be a simple matter of switching off the beam steering or "lensing" characteristic of the micro LC lenses.

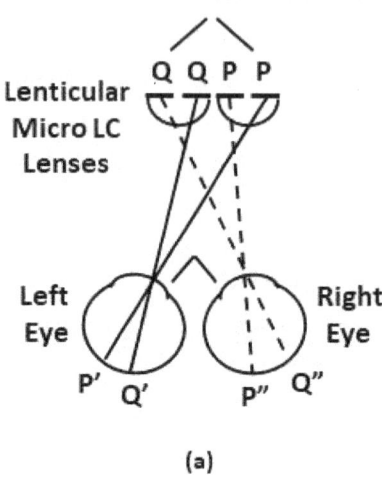

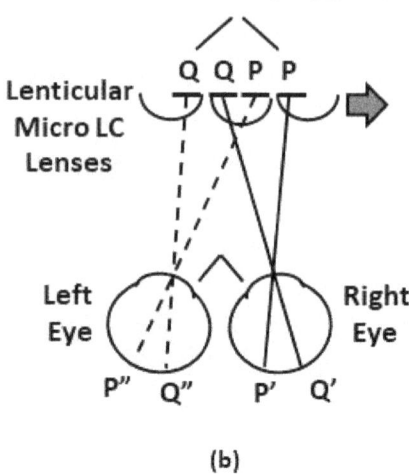

(a)   (b)

FIGURE 3.129 Time-multiplexing left-right eye views using lenticular micro LC lenses for autostereoscopic mobile phone display as described by the authors of Refs. 96 and 97. (a) Left column pixels (with left eye scenery) are directed to the left eye, while right column pixels (with right eye scenery) are directed to the right eye. (b) Left column pixels (with right eye scenery) are directed to the right eye, while right column pixels (with left eye scenery) are directed to the left eye.

3.4.3.1 Zemax OS example: Time-multiplexed autostereoscopic 3D display optical system for mobile phone

As an exercise, it would be rather interesting and instructive to simulate the action of time-multiplexing left and right eye views of a switchable lenticular array, and to see how left-right eye column display pixels are projected towards each eye. Let's do this in Zemax OS non-sequential (NSC) mode. Set up the NSC Component Editor with the objects (1 – 36) shown in Fig. 3.130, and set a monochromatic wavelength of 550 nm in the wavelength dialogue box. In Fig. 3.130, the "X Position" column is unused (ignore it), and object parameters beyond the "Z-position" are not displayed, but shall be explained next.

	Object Type	Comment	Ref Ob	Inside	X P	Y Position	Z Position
1	Null Object ▾		0	0	0.0...	0.000	0.000
2	Source Rectangle ▾	Right Eye Pixs1	1	0	0.0...	0.259	-0.150
3	Source Rectangle ▾	Right Eye Pixs2	1	0	0.0...	0.198 P	-0.150
4	Source Rectangle ▾	Right Eye Pixs3	1	0	0.0...	0.137 P	-0.150
5	Source Rectangle ▾	Right Eye Pixs4	1	0	0.0...	0.076 P	-0.150
6	Source Rectangle ▾	Right Eye Pixs5	1	0	0.0...	0.015 P	-0.150
7	Source Rectangle ▾	Right Eye Pixs6	1	0	0.0...	-0.046 P	-0.150
8	Source Rectangle ▾	Right Eye Pixs7	1	0	0.0...	-0.107 P	-0.150
9	Source Rectangle ▾	Right Eye Pixs8	1	0	0.0...	-0.168 P	-0.150
10	Source Rectangle ▾	Right Eye Pixs9	1	0	0.0...	-0.229 P	-0.150
11	Source Rectangle ▾	Right Eye Pixs10	1	0	0.0...	-0.290 P	-0.150
12	Null Object ▾	0+/-0.0305	1	0	0.0...	0.000	0.000
13	Paraxial Lens ▾	Lenticular 0	12	0	0.0...	0.336	0.000
14	Paraxial Lens ▾	Lenticular 1	12	0	0.0...	0.275 P	0.000
15	Paraxial Lens ▾	Lenticular 2	12	0	0.0...	0.214 P	0.000
16	Paraxial Lens ▾	Lenticualr 3	12	0	0.0...	0.153 P	0.000
17	Paraxial Lens ▾	Lenticular 4	12	0	0.0...	0.092 P	0.000
18	Paraxial Lens ▾	Lenticular 5	12	0	0.0...	0.031 P	0.000
19	Paraxial Lens ▾	Lenticualr 6	12	0	0.0...	-0.030 P	0.000
20	Paraxial Lens ▾	Lenticular 7	12	0	0.0...	-0.091 P	0.000
21	Paraxial Lens ▾	Lenticular 8	12	0	0.0...	-0.152 P	0.000
22	Paraxial Lens ▾	Lenticular 9	12	0	0.0...	-0.213 P	0.000
23	Paraxial Lens ▾	Lenticular 10	12	0	0.0...	-0.274 P	0.000
24	Paraxial Lens ▾	Lenticular 11	12	0	0.0...	-0.335 P	0.000
25	Standard Lens ▾	Cornea (LEFT)	0	0	0.0...	-30.000	304.800
26	Standard Lens ▾	Aqueous (LEFT)	25	0	0.0...	0.000	0.600
27	Annulus ▾	Pupil (LEFT)	26	26	0.0...	0.000	3.000
28	Standard Lens ▾	Eye Lens (LEFT)	26	0	0.0...	0.000	3.000
29	Standard Lens ▾	Vitreous (LEFT)	28	0	0.0...	0.000	4.000
30	Detector Rectangle ▾	Retina Det (LEFT)	29	29	0.0...	0.000	16.910
31	Standard Lens ▾	Cornea (RIGHT)	0	0	0.0...	30.000 P	304.800 P
32	Standard Lens ▾	Aqueous (RIGHT)	31	0	0.0...	0.000	0.600
33	Annulus ▾	Pupil (RIGHT)	32	32	0.0...	0.000	3.000
34	Standard Lens ▾	Eye Lens (RIGHT)	32	0	0.0...	0.000	3.000
35	Standard Lens ▾	Vitreous (RIGHT)	34	0	0.0...	0.000	4.000
36	Detector Rectangle ▾	Retina Det (RIGHT)	35	35	0.0...	0.000	16.910

FIGURE 3.130 Zemax OS NSC Component Editor setup for mobile phone autostereoscopic display simulation. The X Position is unused.

Object 1: This is a null object that serves simply as a reference position (in fact, the global origin) for Objects 2 – 11.

Objects 2 – 11: These are rectangular sources that represent 10 columns of "right-eye" display pixels at the center of a typical mobile phone display. Each column is separated by a blank column of pixels representing the "left eye" column of pixels. Hence, there are 10 line-pairs, each pair consisting of a bright column of pixels and a dark column of pixels. The bright columns are Objects 2 – 11, while the blank columns are simply not included as objects into the NSC Component Editor (thus, a blank column-space separates each object for objects 2 – 11). Each of the 10 bright columns (i.e., Objects 2 – 11) has a full width of 0.0305 mm, and a full height of 0.61 mm. The full-width dimension for these pixels are to model the width of the display pixels used by the authors of Ref. 97. Objects 2 – 11 together form a square display comprised of 10 bright columns and 10 dark columns, whose total dimension is 0.61 mm x 0.61 mm. At the state of the NSC Component Editor shown, the light from the 10 bright columns would be directed towards the right eye (Fig. 3.131). To display the layout in Fig. 3.131, I have used 100 Layout Rays per object for objects 2 – 11. For these objects, I have set 200,000 Analysis Rays per object that is meant for the simulation that shall be described later. Set the Color # to any color you wish (I have used yellow/green). Set the "Cosine Exponent" to "0" for collimated rays (this is by far the simplest model for our exercise). As for the "Pickups" for the Y-position of objects 2 – 11, you may simply ignore them and just use the values provided in Fig. 3.130. But if you'd like to know, those pickups are to place the source rectangles at proper locations across the display plane. Object 3's Y-position "picks up" from object 2 with a scale factor of "1", and an offset of -0.061 mm. Object 4's Y-position "picks up" from object 3 with a scale factor of "1", and the same offset value of -0.061 mm. And so on.

Object 12: This is a reference coordinate for objects 13 – 24 which serves to control the lenticular array (objects 13 – 24) to simulate "switching" of the position for the lenticular micro LC lenses through entering an appropriate value for the Y-position. A value of "0" retains the lenslets' current position (Fig. 3.132a), which directs bright column pixels towards the right eye. A value of +0.0305 mm shifts the lenslets up, directing the bright column pixels towards the left eye (Fig. 3.132b).

Objects 13 – 24: These are paraxial thin lens models (PTLMs) for 12 cylindrical lenslets. Together, they model the lenticular micro LC lens array that serve to direct the bright column pixels (objects 2 – 11) towards either the right eye or the left eye, depending on the value entered for the Y-position for object 12. Set the X-Half-Widths for each lens to 0.61 mm. Set the Y-Half-Widths to 0.03 mm. Leave "0" for the X-Focal Length. Set the Y-Focal Length to 0.15 mm. The pickups are as follows: Object 14 picks up from object 13 with a scale factor of "1", and an offset of -0.061 mm. Object 15 picks up from object 14 with the same scale factor and offset values, and so on.

Objects 25 – 30: These objects together form a model of the average human eye (the left eye), using the same properties that were applied to the eye model used in Sec. 3.4.2 for the Google "Glass" simulation. However, in the present model, I have made a number of modifications. The pupil (object 27) is now *inside of* the Aqueous (object 26), whereas in Sec. 3.4.2, it was *inside of* the Cornea. This actually does not make any difference to the simulations. However, subsequent to the Glass simulation performed in Sec. 3.4.2, I had felt that it made more sense to place the pupil inside the Aqueous. The present pupil also has twice the diameter of the one used for the Glass simulation. In other words, for the present pupil model, set the Minimum X-Half-Width and Minimum Y-Half-Width to 2 mm, thus, making the diameter of the pupil to be 4 mm. The Cornea in the present model has a model index of 1.377. The Aqueous has a model index of 1.336. The Eye Lens (object 28) has a model index of 1.411. The Vitreous (object 29) has a model index of 1.337. All of these indexes follow those provided by Walker [81]. Note that in Sec. 3.4.2., the indexes were rounded to the second decimal place. Here, they are precise to the third decimal place. All Abbe Vd values also follow the model by Walker [81], which is to set them to Vd = 55. Set the "Tilt About X" value for the Cornea (object 25) to +5.6 degrees. Set Radius 2 for the Aqueous (object 26) to +6.078 mm. This lets the eye lens focus on the surface of the objects 2 – 11. Since Radius 2 of object 26 coincides with Radius 1 of the eye lens (object 28), set Radius 1 on object 28 to -6.078 mm (note the negative sign on this radius). For the left eye's retinal detector (object 30), set the "Tilt about Z" value to - 90 degrees. Set "ABSORB" for this detector. Set the Half-Widths for the X and Y directions each to 0.03 mm. Set 100 pixels for both the X and Y pixel numbers.

Objects 31 – 36: Together, these serve to model the right eye. Due to symmetry, the Y-position and Z-position pickups for object 31 are simply the negative values from object 25. Accordingly, set the "Tilt About X" parameter for object 31 to -5.61 degrees (note the minus sign). For all other parameters, use symmetry from the left eye.

Note the gaps between pixels in Figs. 3.132a and 3.132b. Those gaps represent areas where display pixels would be present for the left eye. As shown in Fig. 3.133, entering a value of Y = +0.0305 mm for object 12 directs the rays from the bright column pixels towards the left eye. One may also enter a value in the opposite direction (i.e., Y = -0.0305 mm) for object 12, which yields the same result. This is the reason for having 12 lenslets for objects 13 – 24. When the lenslets shift in the +Y direction, there needs to be another lenslet at the bottom of the array to cover the bottom pixel. Similarly, when the lenslets shift in the -Y direction, the extra lenslet at the top covers the top pixel. And so on.

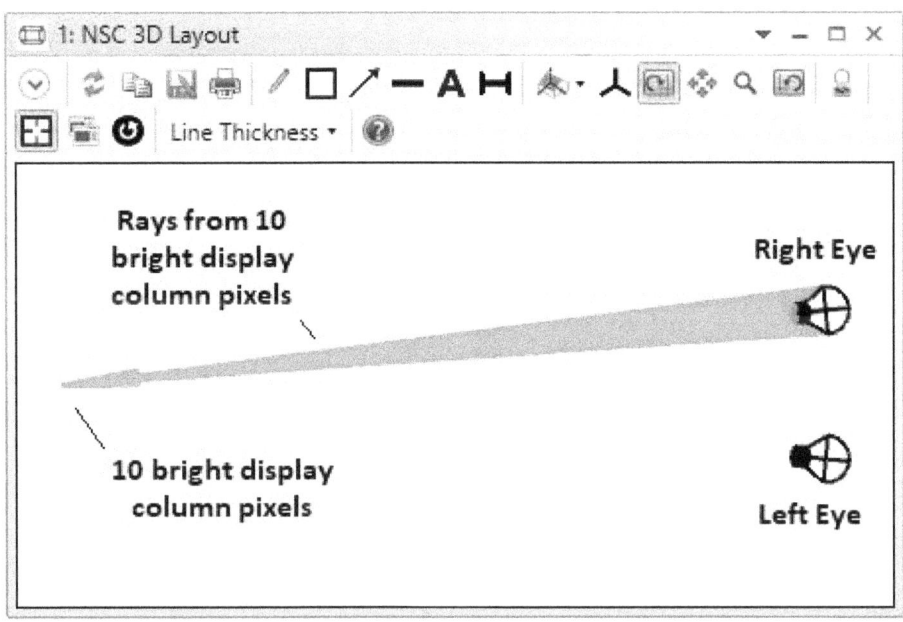

FIGURE 3.131 Zemax OS 3D Layout for the NSC Component Editor prescription given by Fig. 3.130, with 100 Layout Rays traced.

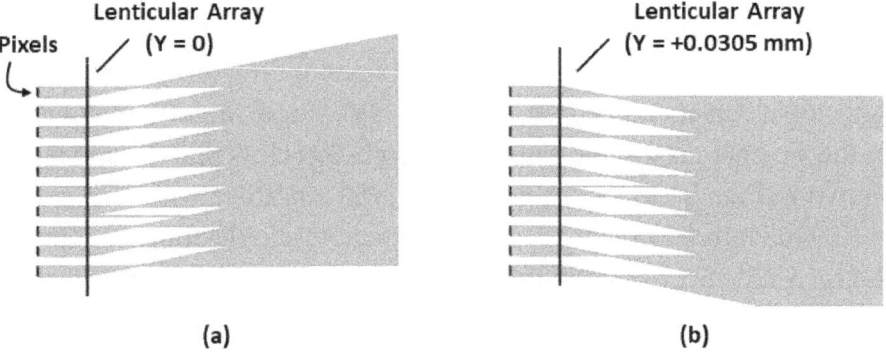

FIGURE 3.132 Zoomed-in view of the 10 bright display column pixels in Fig. 3.131. (a) Y = 0 for object 12. (b) Y = +0.0305 for object 12.

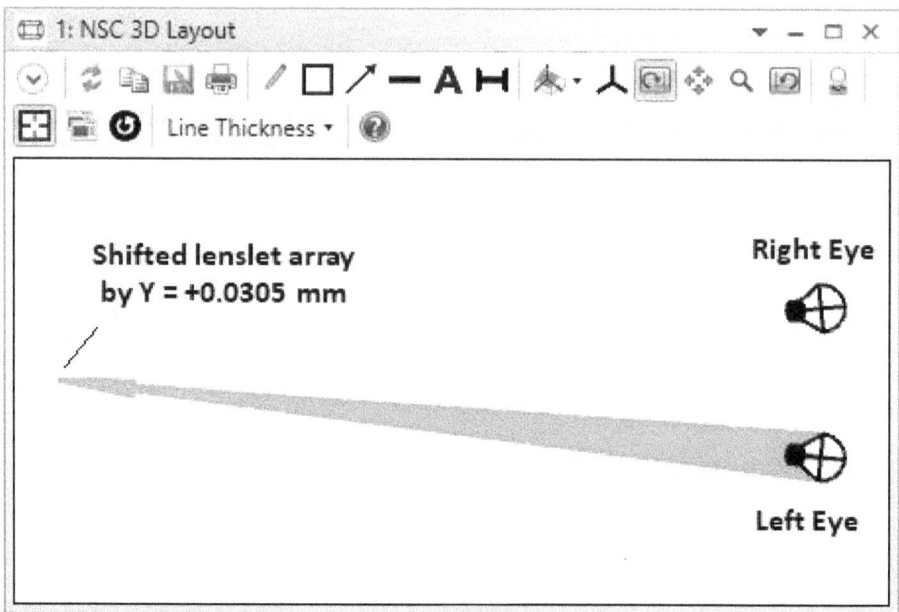

FIGURE 3.133 Zemax OS 3D Layout with Y-position = 0.0305 mm for object 12 in the prescription given by Fig. 3.130.

Now, let's take a look at what the image of the 10 display column pixels look like in each eye. If you haven't already done so, enter 200,000 Analysis Rays for objects 2 – 11. Perform a NSC ray trace with Y-position = 0 for object 12, and open up the left and right eye detectors (objects 30 and 36, respectively). For a smoothing value of 2, the inverse grey scale output for both detectors should appear as shown in Fig. 3.134. Next, perform a NSC ray trace with Y-position = +0.0305 mm for object 12, and the detectors should appear as shown in Fig. 3.135. So, here's an idea: In theory, if left-right eye perspective images are sliced and somehow made into interlaced rectangular sources in our model, then perspective images could in principle be formed onto the retina detectors. Now, if you're using a LCD monitor for your personal computer, then its display would already be polarized. If you could rotate the polarization of one of the detector images, then viewing these images with appropriate polarizers over your eyes could potentially yield simulation results in 3D. In theory, that is. By the way, an aspect of this approach was done in Zemax sequential mode for the analysis of biocular imaging systems [98].

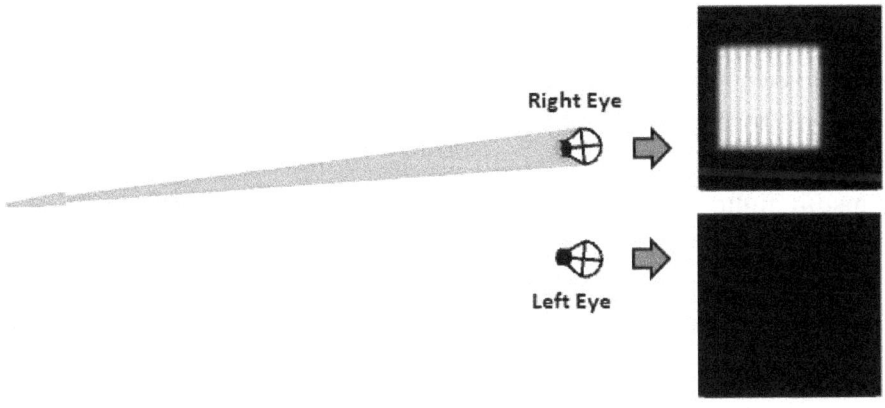

FIGURE 3.134 Retina detector outputs for Y = 0 on object 12.

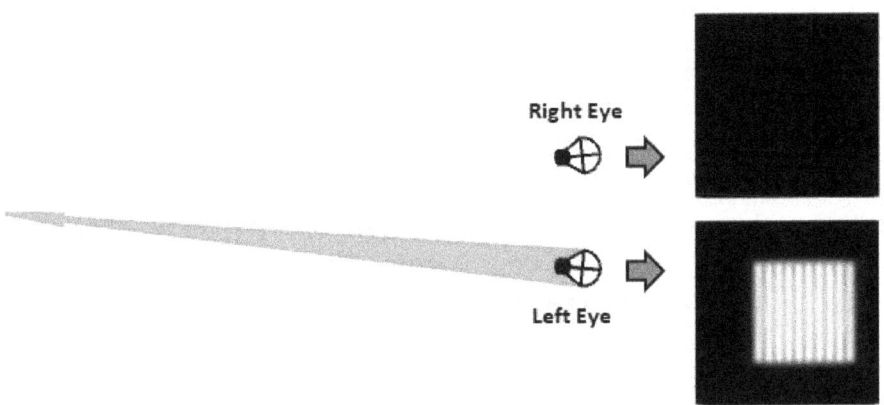

FIGURE 3.135 Retina detector outputs for Y = +0.0305 on object 12.

3.6 Imaging using the Grating Light Valve™

All modern digital displays are ***spatial light modulators***. This includes LCD screens, LED screens, and micro-electromechanical system (MEMs) devices such as Texas Instruments' DLP® chip – often known as the Digital Micromirror Device (DMD). LCD-based displays control (i.e., spatially modulate) the on/off switching of display pixels through the use of polarization to either transmit or not transmit light. LEDs clearly modulate pixels by literally switching diodes on and off. Texas Instruments' DLP® chip's pixels are micron-sized mirrors that reflect light either towards or away from imaging optics. And then there is the Grating Light Valve™ (GLV), which is a MEMs spatial light modulator invented during the early 1990s by David M. Bloom, Francisco S. A. Sandejas, and Olav Solgaard while they were at Stanford University [99]. It was subsequently productized by Silicon Light Machines, Inc. [100], which is today owned by SCREEN [101].

3.6.1 Operating principle of the GLV

The GLV is a rather unique device. Physically, it looks like a shiny metallic one-dimensional (1D) linear array of about 1088 pixels, each pixel measuring about 25 microns in length, and anywhere between 200 to 300 microns in width [100]. To use the GLV as a spatial light modulator, you typically illuminate the entire array of pixels with a coherent laser beam that has been shaped – perhaps by anamorphic optics – into a line. The "reflected" light from the pixels is fed into an imaging system that projects the image of the entire illuminated linear array of GLV pixels onto a screen. Naturally, a complete two-dimensional (2D) scene is produced when the imaging system scans the image of the GLV pixel array across the screen, using, perhaps, combinations of electro-mechanically controlled scan mirrors.

Perhaps the most fascinating aspect of a GLV-based optical imaging system is the manner in which GLV pixels switch on and off. As Fig. 3.136a depicts, a single addressable GLV pixel consists of 6 thin strips of silicon nitride "ribbons", each coated with a thin layer of aluminum [102]. Each ribbon's width is anywhere between 2 – 4 microns, and their lengths are between 100 – 300 microns. The pitch between ribbons is about 4 microns [103]. Under normal conditions, the ribbons act like thin reflective mirrors, but collectively possess the characteristic of a reflective diffraction grating (Fig. 3.136b). Therefore, there would be constructive interference in the 0th order forward direction, which results in a central bright patch of light in the far-field. When every other ribbon is made to deflect by a quarter of the incident light's wavelength (Fig. 3.136c), a total optical path difference of a half wavelength is imparted on the light that is reflected by those deflected ribbons, relative to the light from the un-deflected ribbons. This results in destructive interference in the forward 0th order direction. Thus, optical interference is the mechanism by which GLV pixels switch on and off.

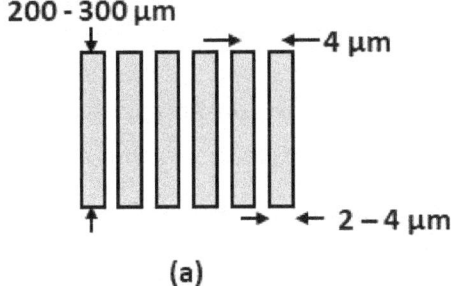

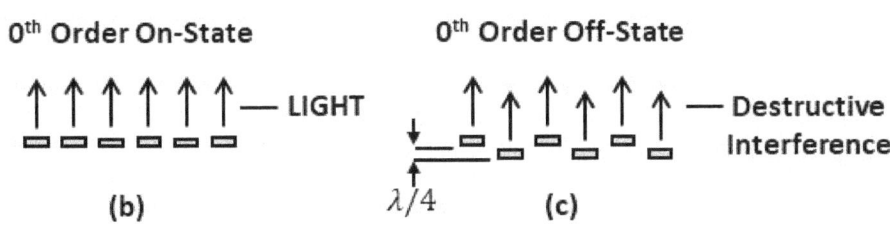

FIGURE 3.136 A single GLV pixel. (a) Top view. (b) Side view, normal state. (c) Side view, ribbons deflected by a quarter wavelength.

Let us take a closer look at the optical interference effect from GLV ribbons by deriving a simple expression for the far-field diffraction distribution. To do this, it is not necessary to consider diffraction from all 6 ribbons. Take, for example, two ribbons that act as a reflective double-slit pair (Fig. 3.137).

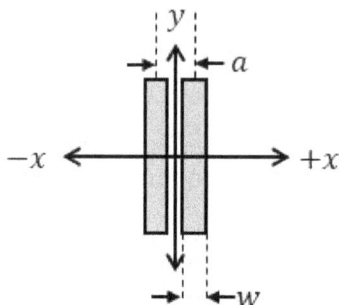

FIGURE 3.137 Coordinate axes for deriving the far-field distribution of diffracted light from two GLV ribbons.

Applying the methods from Sec. 2.2.2 and the coordinate axes shown in Fig. 3.137 for diffraction from two GLV ribbons, the one-dimensional far-field amplitude distribution along the η axis on an observation plane may be expressed as

$$A(\eta) \propto \int_{-\frac{a}{2}-\frac{w}{2}}^{-\frac{a}{2}+\frac{w}{2}} e^{-i\frac{k\eta x}{z}} dx + \int_{\frac{a}{2}-\frac{w}{2}}^{\frac{a}{2}+\frac{w}{2}} e^{-i\frac{k\eta u}{z}} dx. \qquad (3.73)$$

In Eq. (3.73), as usual, we shall take the definition that $k = (2\pi)/\lambda$. If we account for the provision of a relative phase shift ϕ that may be imparted on the left ribbon in Fig. 3.137, then Eq. (3.73) may be expressed as

$$A(\eta) \propto \int_{-\frac{a}{2}-\frac{w}{2}}^{-\frac{a}{2}+\frac{w}{2}} e^{-i\frac{k\eta x}{z}+i\phi} dx + \int_{\frac{a}{2}-\frac{w}{2}}^{\frac{a}{2}+\frac{w}{2}} e^{-i\frac{k\eta x}{z}} dx. \qquad (3.74)$$

Performing the integration in Eq. (3.74) and multiplying the result by its complex conjugate yields the following expression for the far-field irradiance distribution at a screen a distance z from the ribbons:

$$E(\eta) \propto |A(\eta)|^2 \propto \text{sinc}^2\left(\frac{k\eta w}{2z}\right)\left[1 + \cos\left(\frac{k\eta a}{z} + \phi\right)\right]. \quad (3.75)$$

If the left ribbon in Fig. 3.137 deflects by an amount d, then its reflected light acquires a relative phase shift of $\phi = 2kd = (4\pi/\lambda)d$. Inserting this into Eq. (3.75) and taking $k = (2\pi)/\lambda$ we have

$$E(\eta) \propto \text{sinc}^2\left(\frac{\pi\eta w}{\lambda z}\right)\left[1 + \cos\left(\frac{2\pi\eta a}{\lambda z} + \frac{4\pi d}{\lambda}\right)\right]. \quad (3.76)$$

Suppose $w = 0.003$ mm, $a = 0.004$ mm, $\lambda = 0.000588$ mm, and $z = 5$ mm. Using these values, a plot of Eq. (3.76) for $d = 0$ and $d = \lambda/4$ is shown in Fig. 3.138.

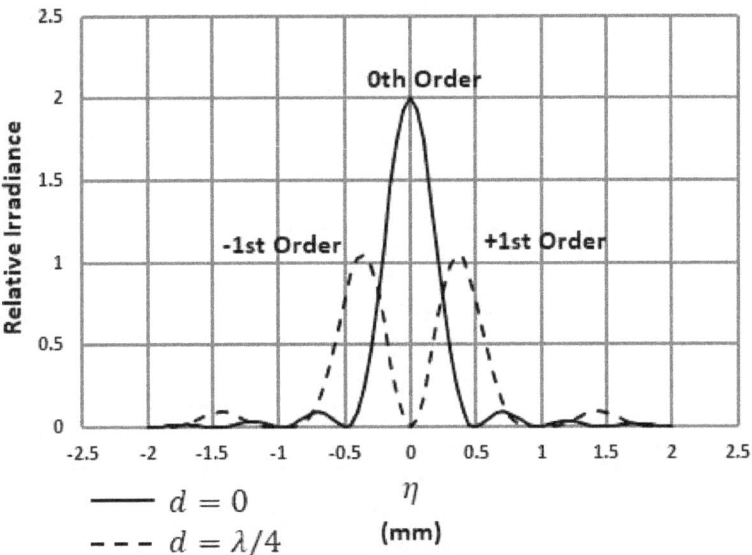

FIGURE 3.138 Irradiance distribution [Eq. (3.76)] at 5 mm distance from two GLV ribbons for $d = 0$ (solid curve), and $d = \lambda/4$ (dashed curve).

For the plot shown in Fig. 3.138, we have, of course, assumed that a distance of 5 mm is sufficient to place the observation plane at the far-field. Is this a reasonable assumption? According to the so-called "antenna designer's formula" [104], the far-field Fraunhofer diffraction approximation is satisfied if the distance z between a diffracting aperture of diameter D and the observation plane is

$$z > \frac{2D^2}{\lambda}. \tag{3.77}$$

For the plot in Fig. 3.138, if we let $D = a = 0.004$ mm and $\lambda = 0.000588$ mm, then applying the criterion given by Eq. (3.77) we have $z > 2(0.004^2)/0.000588 = 0.054$ mm. Since 5 mm > 0.054 mm, we're safe in our assumption.

From examining the plot in Fig. 3.138, we know how the 0th order and +/- 1st order irradiance distributions change with ribbon deflection. But how does this enable making a GLV pixel appear and disappear when an imaging system projects this pixel onto a screen? The answer lies in using an imaging system with a layout that generally places the aperture stop between a front and rear lens group [106, 107] as illustrated in Figs. 3.139a and 3.139b. In these figures, because the GLV is located at the front focal plane of the front lens group, the amplitude distribution at this lens group's back focal plane is precisely given by the far-field Fraunhofer diffraction pattern if no vignetting is present. The plane of the stop is therefore the Fourier transform plane or "Fourier plane". In Fig. 3.139a, by placing the aperture stop at the center of the Fourier plane, the +/- 1st order distributions are blocked for the pixel whose odd (or even) numbered ribbons are deflected. In this way, a GLV pixel whose odd or even ribbons are deflected by $\lambda/4$ would appear dark. This mode of imaging the GLV is called "0th order imaging". If, on the other hand, a double aperture is placed at the +/- 1st order locations on the Fourier plane (Fig. 3.139b), then a GLV pixel whose odd or even ribbons are deflected by $\lambda/4$ would appear *bright*. This mode of imaging the GLV is known as "1ˢᵗ order imaging". In either mode, deflecting odd or even ribbons by magnitudes between 0 and $\lambda/4$ produces grey levels. For the systems shown in Figs. 3.139a and 3.139b, the rear lens group performs a Fourier transform of the distribution that is at the Fourier plane, which results in the image of GLV pixels at the image plane. The magnification of the image would depend on the choice of the focal lengths f_1 and f_2. Again, we are assuming that vignetting by the lens groups is not present (i.e., we are assuming that the lens groups have infinite diameters).

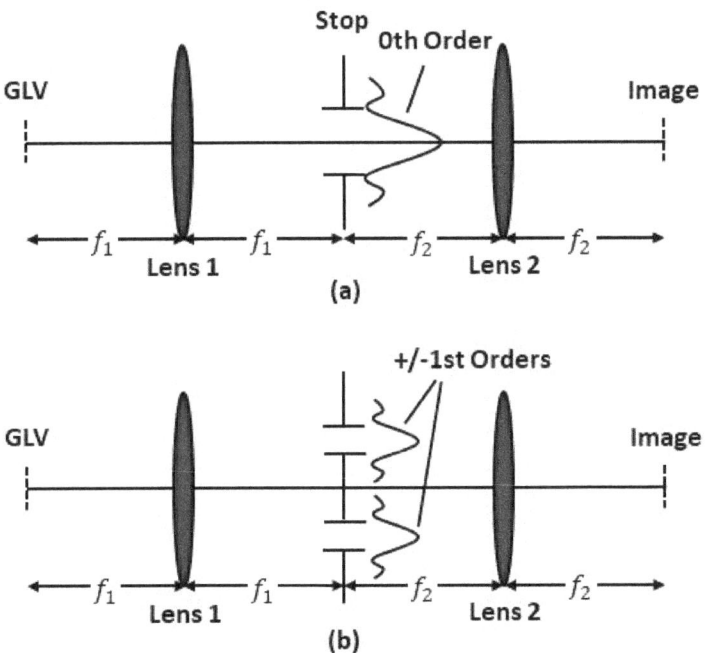

FIGURE 3.139 Optical configurations for GLV imaging. (a) 0ᵗʰ order imaging. (b) 1ˢᵗ order imaging. The aperture stop is at the Fourier plane of Lens 1, and the image is at the Fourier plane of Lens 2.

Pause for insight: → How dark can a GLV pixel really be in its "off" state? In examining Fig. 3.138, there is much "spill-over" of light from the +/- 1ˢᵗ orders into the 0ᵗʰ order area. The same can be said about light from the 0ᵗʰ order spilling over to the +/- 1ˢᵗ order areas. Therefore, one would expect that the apertures in Fig. 3.139 would have to be quite small in diameter. Actually, this need not be the case. Recall that the plot in Fig. 3.138 applies to diffraction from just a pair of GLV ribbons. What happens when more ribbons are included into the diffraction integral in Eq. (3.74)? We need only recall the results discussed elsewhere [105] on multiple slit diffraction. In particular, it is known that, as more slits are included, the principal maxima (i.e., the 0th order and +/- 1st order distributions) become narrower. Thus, one will find that the use of 6 ribbons is sufficient to reduce the width of either the 0th or +/- 1st orders within acceptable magnitudes without having to "squeeze" the apertures down to impractical diameters.

But there seems to be still a problem with the above discussion. In particular, why stop the interference at 6 ribbons? In reality, when all GLV pixels are in their normal state (i.e., when no ribbons are being deflected), then light from all ribbons across 1088 GLV pixels would interfere, would they not? Hence, at the Fourier plane, one would expect to see an extremely narrow 0th order distribution. Now, this is true. Indeed, they would interfere altogether, and, indeed this would result in a highly narrow 0th order distribution (assuming, as usual, zero vignetting by the lenses). However, this reasoning assumes that each GLV ribbon sums in amplitude and phase. As was mentioned at the "pause for insight" discussion near the end of Sec. 3.3.4.2, we have implicitly assumed that there is complete spatial coherence across diffracting GLV ribbons. In general, the illumination across ribbons and pixels on the GLV chip may not necessarily possess complete spatial coherence (though if the GLV were illuminated by a highly coherent line of laser light across all 1088 pixels, then indeed there would be excellent spatial coherence across 1088 pixels). In fact, complete spatial coherence even across all 6 GLV ribbons is not really necessary in order for a single GLV pixel to be switched on and off. In prior work [108], I had found that a complex degree of coherence magnitude of about $|\gamma| = 0.6$ across a dimension of about 6 microns is sufficient to switch off a GLV pixel to within acceptable dark levels.

Therefore, one usually only requires partial spatial coherence for the illumination across GLV ribbons. In fact, it would be theoretically possible to control the amount of interference between ribbons and pixels if one were to control the degree of spatial coherence of the light source [109].

Pause for another insight: $\rightarrow$ In Eq. (3.74), we sum the amplitudes from two ribbons because we have assumed complete spatial coherence for the light that bounces off the two ribbons. But if we can sum amplitudes from ribbons, then we can also sum amplitudes from pixels. Why not? What happens when pixels interfere with each other? Let us take, for example, the 0th order imaging of GLV pixels in an optical system layout as illustrated in Fig. 3.139a. Under the condition of complete spatial coherence for the illumination across the entire array of pixels on a GLV chip, if $A_1(\eta)$ is the amplitude distribution at the Fourier plane from pixel 1, and $A_2(\eta)$ is the amplitude distribution at the Fourier plane from pixel 2, then the total amplitude distribution from these two pixels at the Fourier plane may be expressed as

$$A(\eta) = A_1(\eta) + A_2(\eta). \tag{3.78}$$

The total irradiance at the Fourier plane from these two pixels is

$$E(\eta) \propto |A_1(\eta) + A_2(\eta)|^2$$

$$\propto |A_1(\eta)|^2 + |A_2(\eta)|^2 + 2\text{Re}[A_1(\eta)A_2(\eta)^*]. \tag{3.79}$$

In Eq. (3.79), $A_2(\eta)^*$ is the complex conjugate of $A_2(\eta)$. The presence of the third term (aka "cross-term") in Eq. (3.79) represents coherent interference at the Fourier plane between the two pixels. How does one account for this effect at the image? At the image plane, if there is any overlap between the amplitude distributions of two pixel images, then they would interfere. Therefore, if $A_1(x')$ is the amplitude distribution at the image plane from pixel 1, and $A_2(x')$ is the amplitude distribution at the image plane from pixel 2, then the total amplitude distribution from these two pixels at the image plane may be expressed as

$$E(x') \propto |A_1(x') + A_2(x')|^2$$

$$\propto |A_1(x')|^2 + |A_2(x')|^2 + 2\text{Re}[A_1(x')A_2(x')^*]. \quad (3.80)$$

It has been demonstrated in the lab that such pixel-to-pixel coherent interference occurs [108]. Under partially coherent illumination, it has also been shown that the cross-terms in Eqs. (3.79) and (3.80) are negligible [108].

Prerequisite for the next section: → From the discussion above, we find that there are actually two ways to understand the operational principle of the GLV. The most common approach is to consider what happens at the Fourier plane when ribbons are relaxed and deflected. I call this the "Fourier plane formulation" of GLV physics. Eqs. (3.73) – (3.76) are Fourier plane formulations of GLV physics (for two ribbons). The less common formulation is what I call the "image plane formulation" of GLV physics, which follows the spirit and motivation for Eq. (3.80). That is, just as pixel-to-pixel interference at the image would occur if two GLV pixel distributions overlap, we expect the images of ribbons at the image plane to overlap as well. In fact, they must. If they don't, then no interference would occur between the light from un-deflected and deflected ribbons, and in this case, on/off switching of GLV pixels would not occur.

What causes ribbon image distributions to overlap at the image plane? And under what conditions would they not overlap? We recall from the Abbe theory of image formation that an illuminated object diffracts light into spatial frequencies. In the case of the GLV, all of the light incident across the Fourier plane represent those spatial frequencies. Take, for example, the 0th order imaging of GLV pixels as in Fig. 3.139a. If the aperture at the Fourier plane blocks off all the diffracted orders of light except for the 0th order, then higher spatial frequencies would be missing, which prevents the rear lens group (i.e., Lens 2) from synthesizing them to re-produce sharp images of the GLV ribbons. Consequently, the irradiance and amplitude distribution of ribbon images would spread out, overlapping one another, which enables ribbon-to-ribbon interference to occur. If one then opens up the aperture in Fig. 3.139a to include the 0th order and +/- 1st order light (which, simply by their amplitude magnitudes, contain much high frequency spatial components), then the ribbon image distributions would be sharpened such that no light overlap could occur between ribbon images. Consequently, ribbon-to-ribbon interference can no longer take place, and GLV pixels would always remain bright regardless of the occurrence of ribbon deflection. This is consistent with the Fourier plane formulation of GLV physics.

Interference at the image plane is the way I like to think about the operational principle of the GLV. It provides a rather physically appealing formulation, and it was the theoretical foundation that drove much of the work that I had been involved in, using the GLV [108, 109]. In my prior work [108], I did not refer to this formulation as the "image plane formulation". Instead, I credited Ernst Abbe for his amazing and inspiring insight on the theory of image formation, which motivated me to think about GLV physics in terms of relating Fourier components at the Fourier plane to ribbon interference at the image plane. In the following section, we shall apply this principle to simulate the on/off switching of GLV pixels using Zemax OS.

3.6.2 Zemax OS example: On/off switching of GLV pixels in a simple GLV-based imaging system

In Zemax OS non-sequential (NSC) mode, a detector object can be made to record coherent interference simply by choosing the "Coherent Irradiance" option under the "Show Data" option for the detector settings. Additionally, according to the OpticStudio "Help Pdf" manual (page 540 for version 16 of OpticStudio), it states that "It is important to understand that OpticStudio considers ALL sources to be coherent with respect to one another, and for the phase of the ray to be zero at the starting coordinate of the ray, wherever it may be." This is actually very useful for us, because we can then create 6 individual rectangular sources to serve as GLV ribbons. By creating these "ribbon sources", we eliminate the need to model and assemble a coherent source of illumination, and we also eliminate the need for modeling 6 reflective ribbons.

Now, how shall we model diffraction in Zemax OS NSC mode? Well, we really don't need to. In our simulation, we shall apply the image plane formulation of GLV operation to our model for 0th order imaging of GLV pixels. In particular, we will model an optical imaging system whose aberrations are sufficiently large such that pairs of ribbon images are not resolvable. In this way, ribbon image distributions overlap, and when they do, they undergo interference at the detector. And since Zemax OS's detector objects can record interference, we would be able to observe the on/off switching of GLV pixels when every other ribbon source in our model is imparted a phase shift. This phase shift is created by physically shifting the z-position of every alternate ribbon source by $\lambda/2$ (half wavelength because these "ribbons" in our model are actual sources, not mirrors). In our model, we shall use $\lambda = 588$ nm for the sources.

So, open up Zemax OS in NSC mode and set a monochromatic wavelength at 0.588 microns in the Wavelengths menu under System Explorer. Next, set up the NSC Component Editor with the objects shown in Fig. 3.140. Object parameters beyond (i.e., to the right of) "Z Position" are not displayed, but shall be explained. But first, note in Fig. 3.140 that the appearance of the columns "Ins" (Inside Of) and "YP" (Y Position) have been purposely "squeezed". This is because they aren't being used in our model, and you may ignore them. Note also that there are 13 ribbons (objects 3 – 15) in our model, which approximately represents two GLV pixels. Now, let's go through the details and parameters for each object in Fig. 3.140.

	Object Type	Comment	Ref Object	Ins	X Position	Y P	Z Position	
1	Null Object ▾	Rib Gap	0	0	6.000E-04	0.0..	0.000	
2	Null Object ▾	Rib Ref	0	0	-0.024	0.0..	0.000	
3	Source Rectangle ▾	Rib1	2	0	-3.600E-03 P	0.0..	0.000	
4	Source Rectangle ▾	Rib2	3	0	4.600E-03 P	0.0..	0.000	P
5	Source Rectangle ▾	Rib3	4	0	4.600E-03 P	0.0..	0.000	P
6	Source Rectangle ▾	Rib4	5	0	4.600E-03 P	0.0..	0.000	P
7	Source Rectangle ▾	Rib5	6	0	4.600E-03 P	0.0..	0.000	P
8	Source Rectangle ▾	Rib6	7	0	4.600E-03 P	0.0..	0.000	P
9	Source Rectangle ▾	Rib7 (mid)	8	0	4.600E-03 P	0.0..	0.000	P
10	Source Rectangle ▾	Rib8	9	0	4.600E-03 P	0.0..	0.000	P
11	Source Rectangle ▾	Rib9	10	0	4.600E-03 P	0.0..	0.000	P
12	Source Rectangle ▾	Rib10	11	0	4.600E-03 P	0.0..	0.000	P
13	Source Rectangle ▾	Rib11	12	0	4.600E-03 P	0.0..	0.000	P
14	Source Rectangle ▾	Rib12	13	0	4.600E-03 P	0.0..	0.000	P
15	Source Rectangle ▾	Rib13	14	0	4.600E-03 P	0.0..	0.000	P
16	Null Object ▾	Surf 1	0	0	0.000	0.0..	24.000	
17	Standard Lens ▾	Surf 2/3	16	0	0.000	0.0..	1.000	
18	Standard Lens ▾	Surf 4/5	16	0	0.000	0.0..	4.399	
19	Annulus ▾	STOP	16	0	0.000	0.0..	5.823	
20	Standard Lens ▾	Surf 7/8	16	0	0.000	0.0..	6.730	
21	Standard Lens ▾	Surf 8/9	16	0	0.000	0.0..	7.184	
22	Standard Lens ▾	Surf 10/11	16	0	0.000	0.0..	7.641	
23	Detector Rectangle ▾	IMAGE	22	0	0.000	0.0..	7.128	

FIGURE 3.140 Print-screen of the Zemax OS NSC Component Editor with objects that model the on/off switching of GLV pixel images.

Object 1: This is a null object for letting us enter a value into the "X Position" column for the gap spacing between ribbons. Note that a value of 0.6 microns (0.0006 mm) has been entered.

Object 2: This is a null object that sets the X Position location for the farthest ribbon in the -X direction. The full width of each ribbon in our model is 4 microns, and if there is zero gap between ribbons, then the position of the center of the farthest ribbon in the -X direction is -

6 ribbons x 0.004 mm = -0.024 mm. Hence, a position of -0.024 mm has been entered for this position.

Objects 3 – 15: These are rectangular sources, each representing a ribbon. Each has a full width of 0.004 mm in the X-dimension, and a full length of 0.04 mm in the Y-dimension. Set "100" for the Cosine Exponent parameter for each ribbon source (we are modeling a condition where reflected light from each ribbon is highly directional, due to the use of some laser illumination). Use 1,000,000 Analysis Rays for each object. Leave the rest of the parameters at their default values. Under the X Position column, the pickup Solve on object 3 "picks up" from object 1 with a scale factor of -6 (note the minus sign). This is so that, when the gap between ribbons is nonzero, the location of the center of the first ribbon (which in our model is the farthest ribbon in the minus X direction) is -6g from the center of object 2, where g is the gap between ribbons (currently, at 0.0006 mm as shown in Fig. 3.140). Object 4 picks up from object 3 with a scale factor of 1, and an offset of 0.004 mm. Object 5 picks up from object 4 with a scale factor of 1, and an offset of 0.004 mm. And so on. Note that each subsequent object for objects 4 – 15 are referenced to the object above it. Under the Z Position column, object 4 picks up from object 3 with a scale factor of -1 (note the minus sign). Object 4 picks up from object 3 with a scale factor of -1. Object 5 picks up from object 4 with a scale factor of -1. And so on. This is to model the deflection of every other ribbon, starting with a deflection of $\lambda/2$ for object 3, as we will see later when we simulate the off state of a double pixel.

Objects 16 – 22: These together model a multi-element imaging lens with EFL = 5 mm. They were obtained by scaling the lens given by the prescription in Table 2.1 (Sec. 2.2.2.3) to an EFL of 5 mm. For reference and convenience, the prescription for the current lens that has been made into objects 16 – 22 is shown in Fig. 3.141. Note in Fig. 3.141 that the "Object" is 24 mm from Surface 1. This 24 mm dimension is the Z Position for object 16, which is why object 16 is named "Surf 1" in the NSC Component Editor. Thus, each GLV ribbon in Fig. 3.140 in their normal state is 24 mm from object 16. Note that the Maximum X and Y Half Widths of object 19 (the lens's aperture stop) is set at 1.8 mm, respectively. Ensure that the coating on all "Side Faces" (i.e., lens edges) is "Absorbing". Ensure also that all element front and back surfaces have the ideal transmission coating of I.99999999, which makes those surfaces into near 100% transmitting surfaces with virtually no Fresnel refection losses.

◢	Surf:Type		Radius	Thickness	Material	Semi-Diameter
0	OBJECT	Standard ▾	Infinity	24.000		4.000E-03
1		Standard ▾	Infinity	1.000		0.611
2	(aper)	Standard ▾	2.919	0.357	N-LAK9	1.250 U
3	(aper)	Standard ▾	1.777	3.042		1.250 U
4	(aper)	Standard ▾	17.047	0.600	N-BAF52	1.250 U
5	(aper)	Standard ▾	-3.338	0.824		1.250 U
6	STOP	Standard ▾	Infinity	0.907		0.900 U
7	(aper)	Standard ▾	-2.797	0.453	N-SF11	1.050 U
8	(aper)	Standard ▾	8.970	0.445	N-LAF2	1.250 U
9	(aper)	Standard ▾	-4.737	0.012		1.250 U
10	(aper)	Standard ▾	-52.573	0.358	N-SSK5	1.250 U
11	(aper)	Standard ▾	-4.356	6.770		1.250 U
12	IMAGE	Standard ▾	Infinity	-		1.500 U

FIGURE 3.141 Prescription in Zemax OS sequential model for the lens modelled by objects 16 – 22 in Fig. 3.140.

Object 23: This is a square detector object with X and Y half widths of 0.020 mm. Thus, it is a 0.04 mm x 0.04 mm square detector, placed at the image plane (i.e., Surface 12 in Fig. 3.141). Set the number of x and y pixels to 200, respectively. Now, the following is important: Place your mouse cursor on object 23 and open the "Object Properties" menu. Scroll all the way to the right side of the menu and ensure that the "Normalize Coherent Power" box is NOT ticked (Fig. 3.142). If this box is ticked, then Zemax OS will redistribute the flux that was "lost" through coherent destructive interference across the image plane. Now, in reality, this is the correct physical phenomenon if there is no aperture stop to block off diffracted light (and if all lenses do not vignette rays). But in our model, we want to simply mimic the action of destructive interference, so we do not want destructively interfered flux to be redistributed across the image plane. We are assuming that any flux lost through destructive interference is redistributed to +/- 1st orders at the Fourier plane and is blocked by the aperture stop.

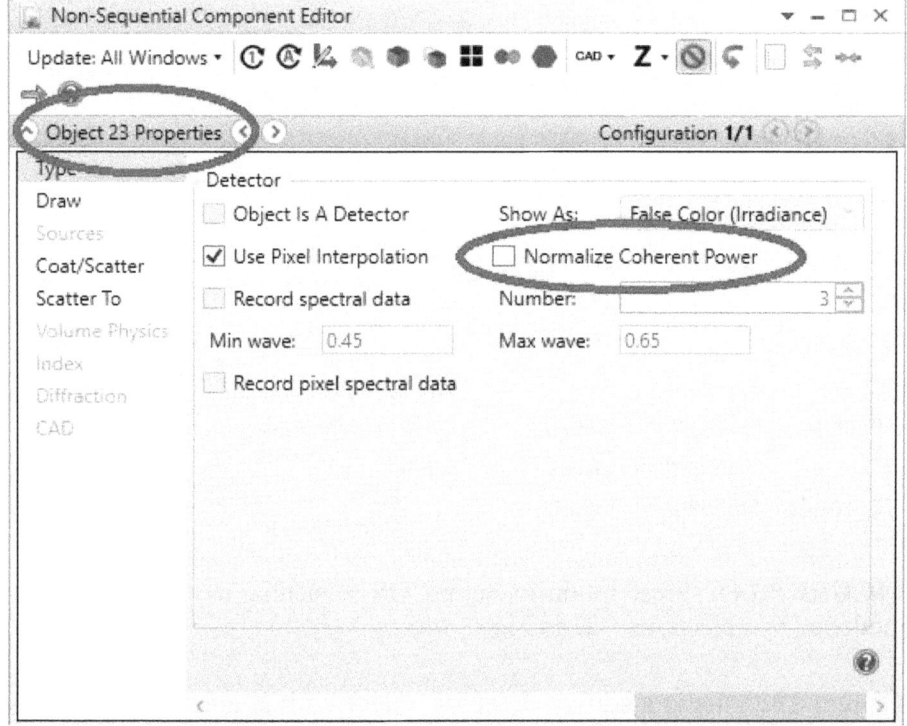

FIGURE 3.142 Ensure that the "Normalize Coherent Power" box is NOT ticked for the "Object 23 Properties" settings.

Let's do a sanity check. If you're all properly set up, the NSC Shaded 3D model for objects 16 – 22 should look as shown in Fig. 3.143 if the x, y, and z coordinates in this 3D layout are set to 0.908566, -

11.7884, and 14,8496, respectively. And setting Layout Rays for object 9 to "200" results in the layout shown in Fig. 3.144.

FIGURE 3.143 Zemax OS NSC 3D layout of the lens system (objects 16 – 22) given by Fig. 3.140.

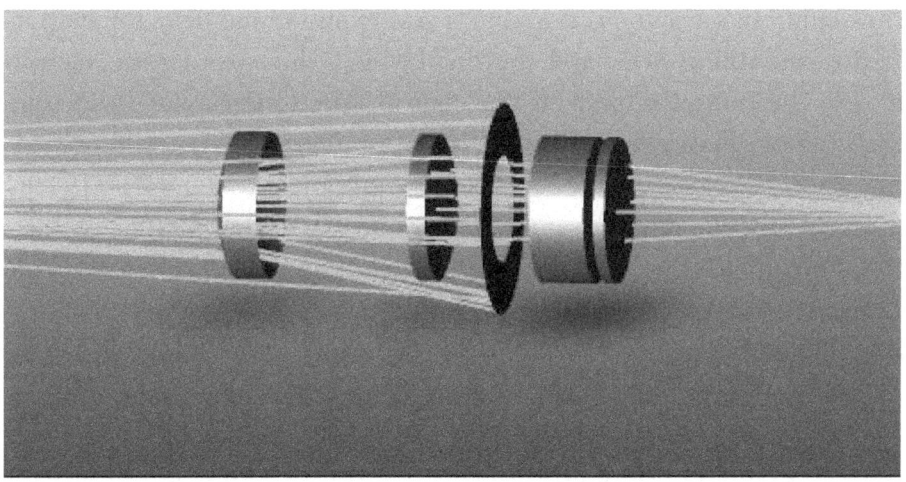

FIGURE 3.144 Zemax OS NSC Shaded 3D Model of the lens system with 200 Layout Rays traced from the central ribbon source (object 9).

If one now zooms into object 9 and set x, y, z to 0, 0, and +90 degrees for the 3D layout, one should see the ribbon sources as shown in Figs. 3.145a and 3.145b.

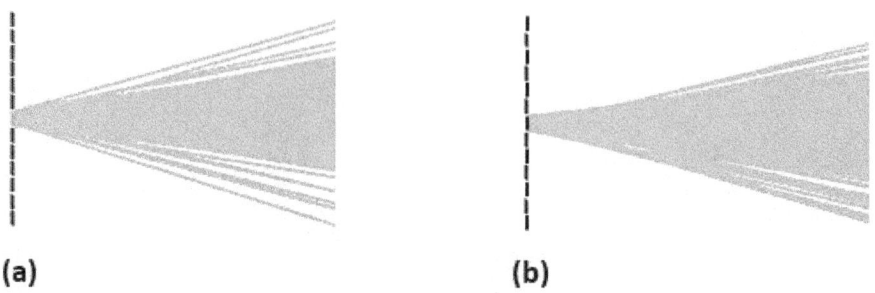

(a)   **(b)**

FIGURE 3.145 Zoomed-in view of the ribbon sources in Zemax OS NSC 3D Layout, with object 9's Layout Rays set at 200, and x, y, z = 0, 0, +90 in the layout settings for "Rotation". (a) No ribbon deflection. (b) Every other ribbon *source* deflected by $\lambda/2$ (i.e., Z Position on object 3 = 2.94E-04).

Now, perform a ray trace for all ribbon sources (each at 1,000,000 rays, and click on "Ignore Errors") and examine the detector output. If the smoothing is 2, and the display is "Show As Cross Section Row", the output would be the central one-dimensional irradiance distribution as shown in Fig. 3.146. Note that individual ribbons are not resolved. Now, set the "Z Position" on object 3 to 2.94E-04. This is half of the wavelength at 0.000588 mm (i.e., 0.000588/2 = 2.94E-04). Perform the ray trace once more (make sure you clear the detector first), and the result should be as shown in Fig. 3.147. Thus, the GLV double pixel image is now in its off-state, in 0th order imaging.

The "cat's ears" at the two ends of the distribution in Fig. 3.147 are a consequence of the absence of ribbon images on their opposite sides to overlap and interfere destructively. If you experiment a little with the magnitude of the gap (the X Position on object 1), you will see varying degrees of interference. For instance, making the gap "0" will improve the destructive interference because there is more overlap among ribbon images.

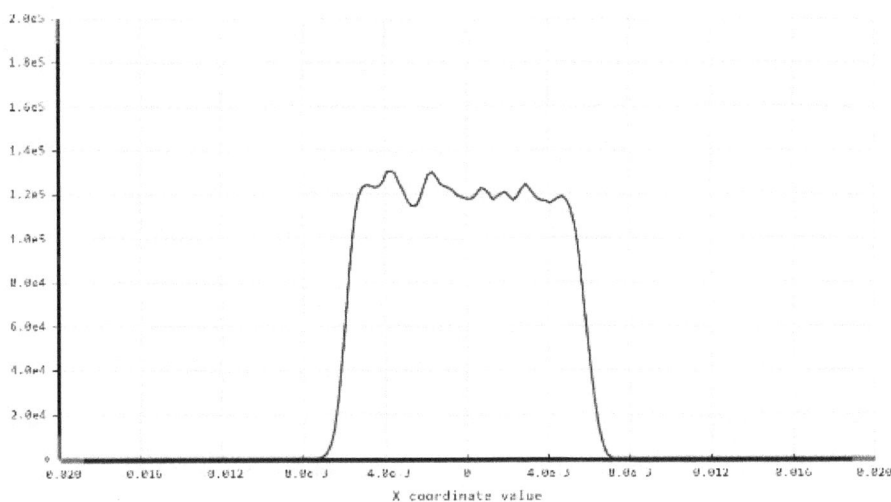

FIGURE 3.146 GLV double pixel coherent image irradiance distribution at zero ribbon deflection (on-state for 0th order imaging) for the model given by Figs. 3.140 – 3.145. The maximum scale is 200,000 (relative units).

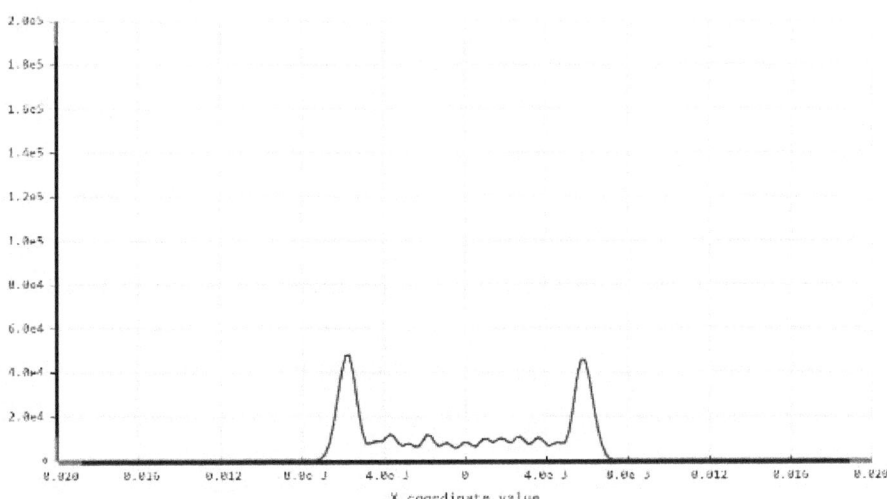

FIGURE 3.147 GLV double pixel coherent image irradiance distribution at $\lambda/2$ ribbon-source deflection (off-state for 0th order imaging) for the model given by Figs. 3.140 – 3.145. The maximum scale is 200,000 (relative units).

3.7 Fluorescence detection and DNA analysis

When light propagates through an absorbing medium, its irradiance in the medium is often described by Beer's law [110, 111]:

$$E(z) = E_o e^{-az}. \qquad (3.81)$$

In this equation, z is the propagation distance in the medium, and a is known as the absorption coefficient, with units of reciprocal length. If the medium converts the absorbed flux into heat, then no visible light is emitted, and atoms (and molecules) end up shaking about until conduction dissipates this energy. But if atoms and molecules shake, then so would their electrons. Therefore, some fraction of the incident flux would be re-emitted, perhaps in the infrared. But this is not fluorescence.

Certain molecules called *fluorophores* (i.e., dye stuff) re-emit visible light upon absorbing incident flux. The incident light is often called the *excitation* source, and the emission is the *fluorescence*. The fluorescence spectrum (aka *emission spectrum*) is at longer wavelengths than the incident excitation light's spectrum, with a peak whose wavelength is shifted from the fluorophore's *absorption* spectral peak by an amount often called the *Stokes Shift*. The magnitude of this shift is usually tens of nanometers (e.g., 20 – 30 nm). In a fluorescent medium, dyes are said to absorb incident flux in accordance with Eq. (3.81) as long as no other phenomena (such as Rayleigh or Mie scattering) serve to deviate the Beer law [112]. The fraction of the flux that is emitted as fluorescence is often quantified by the *quantum yield* [113], which is the ratio of the number of emitted photons to the number of absorbed photons.

A typical optical system layout to observe fluorescence is to assemble a light source, beamsplitter, excitation and emission filters, and a camera with some lens and sensor [114] such as shown in Fig. 3.148. Such a system of optical components forms a *fluorescence detection system*. The purpose of the excitation filter is to allow only a finite bandwidth of light from the source to illuminate the sample within the sample's absorption spectrum. Even if the sample's absorption spectrum is broad, you don't want to have the incident excitation spectrum to cover the entire spectral range of absorption, because otherwise some part of the emission spectrum will contain the incident flux's light, which is usually significantly higher in intensity than the emitted fluorescence and would therefore increase the background (which we clearly do not want). The purpose of the emission filter is to prevent any light from the source (whether from back-scatter at the sample or through stray scattering) to strike the sensor, and to instead allow only the emitted fluorescence to pass through and strike the sensor. The spectral bandpass of the emission filter is therefore relatively narrow (approximately in the range of 10 − 20 nm or more) within the fluorescence emission spectrum.

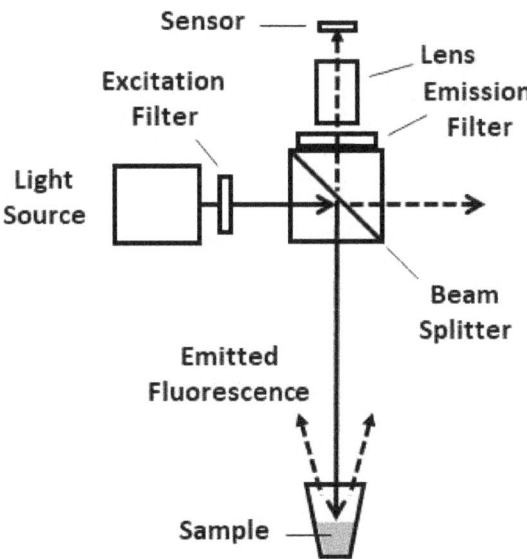

FIGURE 3.148 Typical arrangement of optical components to detect and observe emitted fluorescence from a fluorescent sample. Such a system is known as a *fluorescence detection system*.

Fluorescence detection is a major application in the life sciences industry. Many biological samples are fluorescent. Do you go clubbing and enjoy parties? If you do, then you must have experienced dancing in "black light", which is essentially near ultraviolet light with a deep blue color that excites your skin and teeth into emitting fluorescence. You don't need to wear emission filters over your eyes to see that fluorescence because it has a rather broad spectrum, and is quite bright.

In the life sciences and biomedical industry, *analytical instruments* that apply optical systems to examine biological samples usually make use of fluorescence. One significant area of interest is in the analysis of DNA, such as DNA sequencing and the detection of pathogens, cancer, and other illnesses. DNA does not possess fluorescence properties, but if specially designed fluorophores are connected or "tagged" to specific DNA, then we can detect the presence or absence of those specific DNA. This tagging is also often said to be "labelling". A very important class of instruments that make use of DNA labelled with dyes is called *quantitative polymerase chain reaction* (qPCR) instruments or qPCR systems. This is the subject of the next section.

3.7.1 The polymerase chain reaction (PCR)

In 1993, the Nobel Prize in chemistry was awarded to Karry B. Mullis and Michael Smith [115] for their work that led to the development of a method to essentially clone DNA molecules. This method is known as the polymerase chain reaction (PCR) [116]. Since then, the technique of PCR has been widely applied to many areas of biology and medicine. In qPCR, for example, a biological test or *assay* often concerns a test for the presence or absence of a specific gene in a DNA sample. Such an assay would comprise of a specially designed short single-stranded segment of DNA called a *primer** that could potentially bind with a complimentary single strand of the target sample DNA of interest. One places this primer into a chemical mixture containing billions of the four component molecules of DNA

---

* Short single-stranded segments of DNA are also called *oligonucleotides*.

called **nucleotides**. The nucleotides are known as adenine (A), thymine (T), guanine (G), and cytosine (C). You may have seen the first letters of their names mentioned in, say, tv shows, books, and articles. Specific sequences of nucleotides (aka **genes**) in DNA determine what biological entity the gene – through bio-chemical and physical processes – could end up "creating". Roughly speaking, the whole process that starts from genes (and reagents) to having some fully developed biological-entity is called **gene expression**.

Within a test tube, nucleotides, primers, an initial concentration of the DNA sample of interest, and a molecule called **DNA polymerase** are put together (including some other reagents) and made to undergo **thermal cycling**, which is a repetitive heating and cooling process of about 40 cycles. The maximum temperature is usually about 90 degrees Celsius or higher, and the minimum temperature is usually around 60 degrees Celsius. Whenever the DNA samples are heated to near the maximum, they split (recall from your biology textbook that DNA is a long double-stranded molecule, each strand consisting of sequences of nucleotides). This splitting process is called **denaturing**. Once denaturing has occurred, the mixture is cooled down to the **annealing** temperature of around 60 degrees Celsius. At approximately this temperature, if the sequence of nucleotides in a designed primer matches the specific gene of interest at a segment of the single-stranded DNA sample, a primer would attach itself onto the split DNA, a process called **hybridization**. The primer is said to be hybridized to the single-stranded DNA. The DNA polymerase then "stitches" nucleotides from the chemical mixture to the single-stranded DNA to generate a copy or "clone" of the original double-stranded sample DNA. The stitching process is also known as **extension**.

As long as the sequence of nucleotides in the designed primers match the gene sequence at the single-stranded DNA sample, this process of denaturing, annealing, and extension occurs at every thermal cycle (but the number of nucleotides decrease, so the efficiency at which PCR occurs also reduces). Thus, after the first thermal cycle, there would be double the number of original DNA samples. After the second thermal cycle, there would be double of the number of DNA samples from the prior cycle. And so on. This cyclic process of DNA cloning is the polymerase chain reaction. The cloning process is also referred to as *DNA amplification*, and the products (i.e., the clones) are often called the *amplicons*.

The *significance of PCR* to biology and medicine is this: if the gene sequence of, say, a virus has been studied and identified, then designing a primer with a sequence that matches that of a virus is a way to identify the presence or absence of that virus from a sample of DNA. Since PCR would only occur if the designed primer matches the target gene of interest, *the occurrence of PCR in a chemical mixture containing a sample of the gene from a virus is an indication of the presence of that virus*. It is no wonder why PCR has been and continue to be of significant value to medical researchers and biologists (and the life sciences market).

3.7.2 Real-time PCR

At the end of 40 cycles of PCR, one needs to know if the amplification reaction has indeed taken place. One needs to know if amplicons have been generated, and to do this, one must essentially detect and "count" the amplicons. This is done through the design of special dyes known as *probes* that would bind to the amplicons. There are usually two ways to do this. A common way is to design a probe that binds to double-stranded DNA without specificity (i.e., the probes would bind to any double-stranded DNA, even if they are not the target sample of interest). If PCR has taken place, there would be plenty more double-stranded DNA at the end of PCR than at the beginning, so one need only check this by placing probes into the mixture of amplicons and watch for fluorescence using a fluorescence detection system such as the optical system illustrated in Fig. 3.148. This approach to fluorescence detection in qPCR is known as *end-point analysis*.

Another approach is to check for fluorescence in real-time, which is a technique known as *real-time PCR* [117 – 119]. To do this, the mixture of reagents (i.e., target sample DNA, nucleotides, DNA polymerase, and primers) would include target-specific probes. These probes are specially designed oligonucleotide strands attached with a fluorophore and a *quencher*, which is a molecule that inhibits or "quenches" the fluorophore from emitting any fluorescence even when excitation light is present. At the annealing stage of PCR, primers and probes both attach to the single-stranded sample DNA (if the primers match in sequence to the target gene sequence of interest on the sample DNA). When DNA polymerase extends the single-stranded DNA, it would detach or *cleave* the probe from the single-stranded DNA sample. This also detaches the quencher from the probe, resulting in fluorescence if the excitation source illumination is present (which it would be in a *real-time PCR instrument*). As PCR continues, more and more probes get cleaved, and the intensity of fluorescence grows. To a reasonable approximation, the emitted fluorescence is proportional to the number of amplicons.[*] Thus, detection of this fluorescence is an indication of the occurrence of PCR, which in turn is an indication of the presence of the target gene sample of interest.

Real-time qPCR instruments apply the use of fluorescence detection systems for observing and quantifying amplicons. An example of such an optical system is depicted by the illustration shown in Figs. 1.3 and 2.19 (Secs. 1 and 2, respectively), based on a patent by Boege *et al.* [121]. The most challenging task in designing fluorescence detection systems is relating subsystem variables to system performance. Much work also involves ensuring that stray light does not leak into the sensor. As such, when modelling fluorescence detection systems, proper coatings must be applied onto element surfaces, especially on the filter and beamsplitter models. It is not a matter of which software you use. Ultimately, empirical

---

[*] Because of Beer's law, the absorption of the excitation light is a nonlinear process. Hence, if the absorption coefficient of a fluorescent medium is large, one would not expect the emitted fluorescence to be linearly proportional to the fluorophores present. Moreover, radiative transfer in an absorbing medium is generally rather complex [120].

measurements take precedence over and above flawed models. In Zemax OS, fluorescence from dye samples may be modelled [e.g., 122, 123]. In the next section, we will do something different, yet relevant to fluorescence detection. In particular, we will apply the results from Sec. 2.3.5.1 to simulate the detection of the *total* fluorescence from replicate dye-labelled DNA samples in qPCR.

3.7.3 Zemax OS Example: Optimizing detection from replicate wells using a lens' POP characteristic

Replication is the act of making provisions for having an ensemble of (hopefully) identical samples from which measurements can be taken for performing statistically sound analysis. In qPCR, therefore, samples and reagents are often placed into multiple test tubes or *wells* across a microtiter plate*. Fluorescence detection systems applied to qPCR often record fluorescence from wells by either scanning a detector from well to well [125, 126] across a microtiter plate, or by imaging the entire array of wells [121] on the microtiter plate. If replicate wells are scanned a well at a time, then one obtains fluorescence data on a per well basis. If replicate wells are imaged, then one obtains multiple well images on an image sensor. In either case, the recording of fluorescence emission from each replicate well is separated into individual data points (we're considering an image of a well as a single data point).

Once all fluorescence intensities have been recorded from each replicate well, it is reasonable to assume that one would compute the average intensity as a representation of the data from replicate wells. Of course, it is useful to also be able to examine the fluorescence from each well to check for anomalies. However, as an exercise, it seems rather interesting to consider an optical detection concept that eliminates the process of computing the average fluorescence from replicate wells. In other words, what if there is an efficient and cost-

---

* A microtiter plate is a tray (often made of plastic) where an array of wells (essentially, test tubes) hold reagent mixtures [e.g., 124] for analysis. Often, microtiter plates are used for PCR.

effective way to record the total fluorescence from all replicate wells? Such a system would not use an image sensor, for that could be somewhat costly. But it should also not scan a single sensor (such as a photodiode or photomultiplier sensor) across replicate wells, for it would be costly to have an electro-mechanical scanning system.

One approach could be to apply the results discussed in Sec. 2.3.5.1, where a small sensor, say, a photodiode, is mounted at the position of peak irradiance (POP) behind a lens, which is a position of high flux concentration. In prior work [127], I had shown that a POP exists behind a focusing lens if an array of small sources is mounted symmetrically about the optic axis at some finite distance from the lens. Such a POP would not be as pronounced as would be if the source is not an array, but it could still possess high flux density. Although some light is spread out beyond the POP area, the POP enables the use of small-size photodetectors for recording concentrated light. Let us try this idea in Zemax OS non-sequential (NSC) mode.

Set up Zemax OS in NSC mode, using a monochromatic wavelength setting at 588 nm, and with the NSC Component Editor objects shown in Fig. 3.149. Note that the "squeezed" column "Ins" is not used, so just ignore it. The other squeezed column is the "Reference Object" column, and only objects 12 and 13 have reference objects. Both are referenced to object 11. The following are the rest of the object parameters in the NSC Component Editor:

Objects 1 – 9: These are circular Lambertian disk sources, each with diameter 12.5 mm. They represent 9 fluorescent wells distributed in a rectangular grid array of 3 x 3 wells. Well 1 is at the top left corner. Well 2 is at the top. Well 3 is at the top right corner. Well 4 is to the left. Well 5 is centered to the optic axis of the lens. Well 6 is to the right…etc. Use a value of "1" for the "Cosine Exponent" for all objects 1 – 9. This is to model them as Lambertian surface emitters. Note that these wells are all 150 mm to the left of the front vertex of the lens. Leave the Power for each object to 1 Watt (default value).

Object 10: This is the lens' aperture stop. It is modelled as an annulus whose inner diameter is 30 mm, and outer diameter is 88 mm. It is located at the front vertex of the lens (object 11).

Object 11: This is a simple bi-convex lens with 50 mm radius of curvature, and 12 mm center thickness. Note that this is the same lens from Figs. 2.46 and 2.51 (Secs. 2.3.4.1 and 2.3.5.1), and the 3 x 3 well total array area is approximately equivalent to the circular Lambertian disk source in Fig. 2.51. Ensure that its front and back surfaces are given ideal coatings with transmission of near 100% (i.e., use I.99999999 for the coating under the Coat/Scatter tab of the object properties settings). Ensure that the lens edge is absorbing (use "Absorbing" under the "Face is" pull-down menu for the lens's side faces of the object properties settings).

Objects 12 – 13: These are 12 mm x 12 mm square detectors. Object 12 is located at the POP, while object 13 is located at the image plane at optimum focus. Set the number of X and Y pixels to 100 each.

	Object Type	Comment	Ref	Ins	X Position	Y Position	Z Position
1	Source Ellipse ▾	Well 1	0	0	-18.750	18.750	-150.000
2	Source Ellipse ▾	Well 2	0	0	0.000	18.750	-150.000
3	Source Ellipse ▾	Well 3	0	0	18.750	18.750	-150.000
4	Source Ellipse ▾	Well 4	0	0	-18.750	0.000	-150.000
5	Source Ellipse ▾	Well 5	0	0	0.000	0.000	-150.000
6	Source Ellipse ▾	Well 6	0	0	18.750	0.000	-150.000
7	Source Ellipse ▾	Well 7	0	0	-18.750	-18.750	-150.000
8	Source Ellipse ▾	Well 8	0	0	0.000	-18.750	-150.000
9	Source Ellipse ▾	Well 9	0	0	18.750	-18.750	-150.000
10	Annulus ▾	Stop	0	0	0.000	0.000	0.000
11	Standard Lens ▾	Lens	0	0	0.000	0.000	0.000
12	Detector Rectangle ▾	Det POP	11	0	0.000	0.000	42.000
13	Detector Rectangle ▾	Det Img	11	0	0.000	0.000	73.000

FIGURE 3.149 Zemax OS NSC Component Editor with objects to simulate fluorescence detection from replicate wells.

If the setup is right, you should see a NSC Shaded Model as shown in Fig. 3.150 when 1000 Layout Rays are applied to Well 5 (object 5).

The x, y, z display settings for Fig. 3.149 are 2.96935, -19.5009, and 13.9445, respectively.

FIGURE 3.150 Zemax OS NSC 3D Shaded Model for the system of objects given by Fig. 3.149. 1000 Monte-Carlo Layout Rays are traced from the central well (Well 5, object 5).

Now, perform a ray trace with a setting of 1,000,000 rays for each of the Wells (objects 1 – 9). With a smoothing value of 4 (and setting the Maximum Plot Scale at 0.053 Watts per cm^2, the detector outputs at the POP (object 12) and at the image plane (object 13) should appear as shown in Figs. 3.151 and 3.152, respectively (using "Inverse Grey Scale" as the display). Note that the POP has a slightly higher irradiance peak than the wells. If the "Show As" display setting is set to "Cross Section Row" for both detectors, one obtains the one-dimensional irradiance distribution plots shown in Figs. 3.153 and 3.154 for the POP detector and image plane detector, respectively. The result in Figs. 3.151 and 3.153 show that a single sensor with size 5 mm x 5 mm may be used to integrate flux from all 3 x 3 wells. Even if the integrated flux is not the total flux from all wells, its magnitude should be sufficient to record fluorescence, for it would be equivalent to the total flux from a single well image on the 12 mm x 12 mm detector. The values used for these detector sizes are just examples for a simple exercise. However, they do serve to demonstrate that one may consider smaller sized sensors for detecting fluorescence emission from replicate wells in qPCR systems.

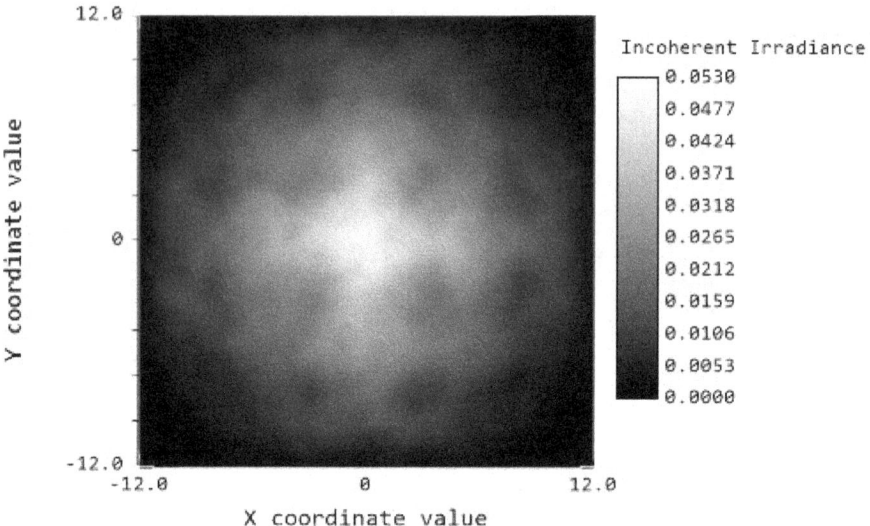

FIGURE 3.151　Inverse grey scale display for the detector at the POP.

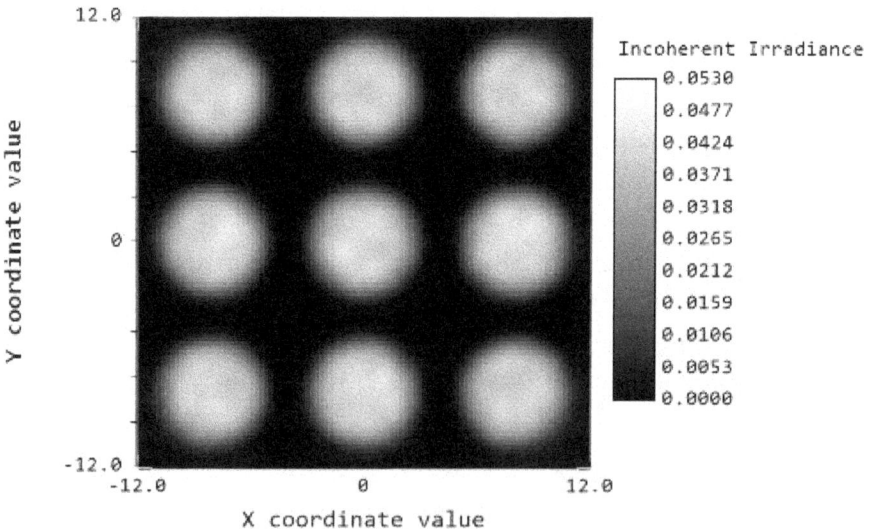

FIGURE 3.152　Inverse grey scale display for the detector at the image.

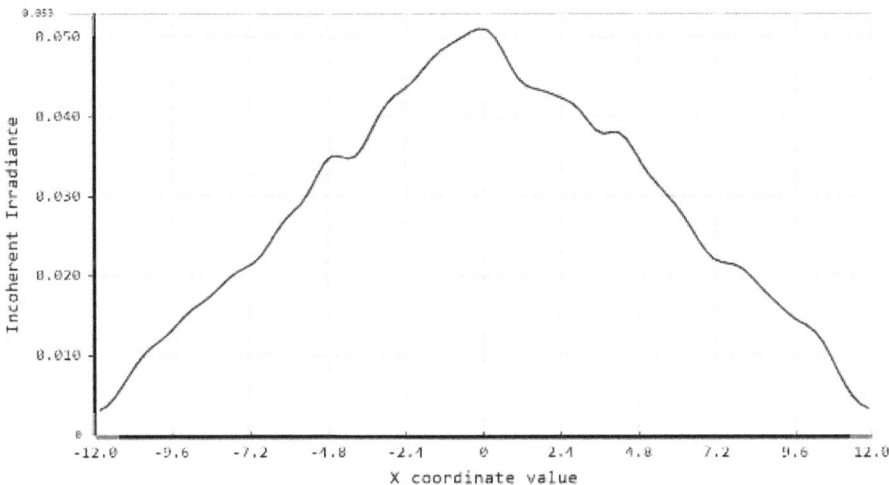

FIGURE 3.153 Cross section irradiance distribution at the POP for the central row of pixels for the detector in Fig. 3.151. The maximum scale is at 0.053 W/cm². The horizontal axis is in mm units.

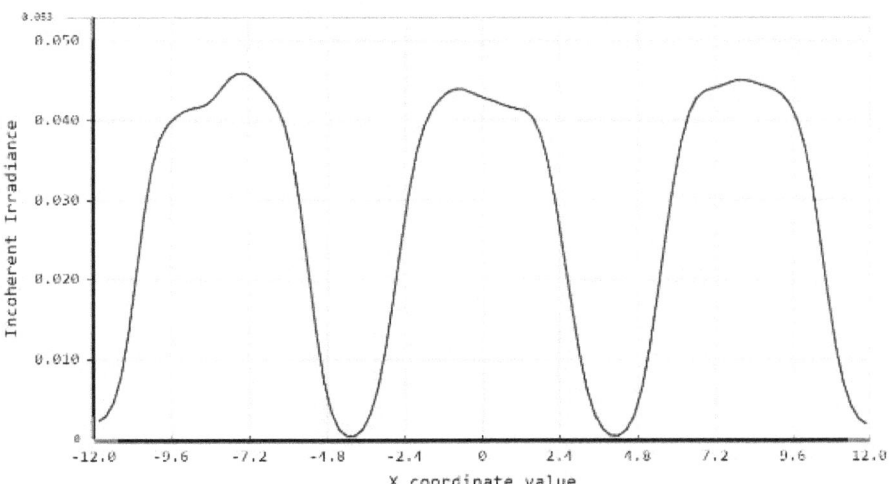

FIGURE 3.154 Cross section irradiance distribution at the image plane for the central row of pixels for the detector in Fig. 3.152. The maximum scale is at 0.053 W/cm². The horizontal axis is in mm units.

There are several other possible optical arrangements. For example, suppose an optical system is of the double-Gauss type, where the aperture stop is between a front and a rear lens group. If one optimizes the lens such that the POP of the first lens group is at the stop, then one may place a small photodetector at the first lens group's POP whilst letting the rest of the light pass through the stop and propagate towards the image plane of the total lens system (Fig. 3.155). In this way, one can use a "cheap" image sensor to record the image of replicate wells, and at the same time, record total flux at the POP of the first lens group. This would allow one to check for false positives and negatives across replicate wells. It may seem redundant for an optical system to possess both a POP detector and an image sensor, but if the image sensor is only used for imaging without consideration to noise, then it could perhaps be a simple and relatively inexpensive image sensor. But there is one clear disadvantage to optical arrangements given by Fig. 3.155 (and Fig. 3.150). The POP of any lens is highly dependent on source symmetry. Therefore, if, for some reason, one of the replicate wells is not emitting fluorescence (due, perhaps, to errors in pipetting reagents and probes into the well), then the POP's location shifts to some other location. Thus, when considering the use of lens POPs for optical detection, one should account for conditions where source asymmetries may arise.

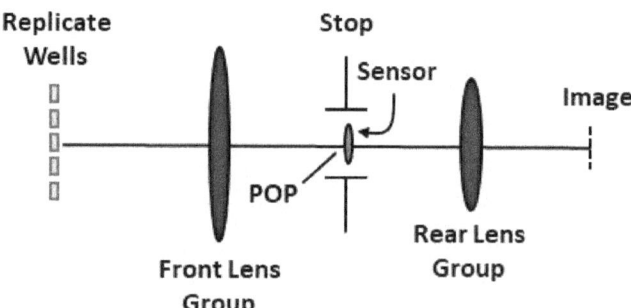

FIGURE 3.155 Optical arrangement for simultaneous detection of flux at the POP of a front lens group, and recording the system's final image.

Another possible optical arrangement is to return to having a well-to-well scanning system where a simple lens scans over each well, but the sensor (say, a photodiode) is mounted at the lens' POP rather than at its image plane. This of course assumes that each well appears as a circular Lambertian disk emitter, and they may not. One therefore has to test all of the assumptions made, and to compare empirical results with theoretical concepts.

References

1. T. V. Galstian (editor), *Smart Mini-Cameras*, (CRC Press, 2014).
2. P. P. Clark, "Lens Design and Advanced Function for Mobile Cameras," in *Smart Mini-Cameras*, edited by T. V. Galstian, (CRC Press, 2014), pp. 4 – 5.
3. See Ref. 1, p. 12.
4. J. Linkemann and B. Weber, "Global Shutter, Rolling Shutter – Functionality and Characteristics of Two Exposure Methods (Shutter Variants)," Basler White Paper available at www.baslerweb.com.
5. A. Theuwissen, "Digital Cameras: Smaller Pixels for Higher Resolution," available online from the Photonics Handbook at www.photonics.com.
6. V. Bolhouse, "Machine Vision: Costs Are Dropping and Capabilities Rising," in *The Photonics Design & Applications Handbook*, (Laurin Publishing, 2002), pp. H-147 – H-149.
7. A. W. Lohmann, "Scaling laws for lens systems," *App. Opt.* **28**(23), pp. 4996 – 4998 (1989).
8. R. Kingslake and R. B. Johnson, *Lens Design Fundamentals*, 2nd ed., (SPIE Press, 2010), p. 109.
9. M. Herzberger, "Color Correction in Optical Systems and a New Dispersion Formula," Opt. Acta **6**, p. 197 (1959).
10. M. Herzberger and N. R. McClure, "The Design of Superachromatic Lenses," *App. Opt.* **2**(6), pp. 553 – 560.
11. P. Robb, "Selection of optical glasses. 1: Two materials," *App. Opt.* **24**(12), pp. 1864 – 1877 (1985).
12. R. D. Sigler, "Glass selection for airspaced apochromats using the Buchdahl dispersion equation," *App. Opt.* **25**(23), pp. 4311 – 4320 (1986).
13. A. Yang, X. Gao, and M. Li, "Design of apochromatic lens with large field and high definition for machine vision," *App. Opt.* **55**(22), pp. 5977 – 5985 (2016).

14. R. Siew, "Practical automated glass selection and the design of apochromats with large field of view," *App. Opt.* **55**(32), pp. 9232 – 9236 (2016).

15. See, for example, Ref. 7, pp. 149 – 161.

16. S-B Rim, P. B. Catrysse, R. Dinyari, K. Huang, and P. Peumans, "The optical advantages of curved focal plane arrays," *Opt. Express* **16**(7), pp. 4965 – 4971 (2008).

17. D. Reshidko and J. Sasian, "Optical analysis of miniature lenses with curved imaging surfaces," *App. Opt.* **54**(28), pp. E216 – E223 (2015).

18. B. Guenter, N. Joshi, R. Stoakley, A. Keefe, K. Geary, R. Freeman, J. Hundley, P. Patterson, D. Hammon, G. Herrera, E. Sherman, A. Nowak, R. Schubert, P. Brewer, L. Yang, R. Mott, and G. McKnight, "Highly curved image sensors: a practical approach for improved optical performance," *Opt. Express* **25**, 13010-13023 (2017).

19. W. T. Welford, *Aberrations of Optical Systems*, (IOP Publishing, 1986), pp. 139 – 140.

20. http://www.varioptic.com

21. http://www.optotune.com

22. B. Berge, "Liquid Lens: A Key Adaptive Component for Cameras," in *Smart Mini-Cameras*, edited by T. V. Galstian, (CRC Press, 2014), pp. 149 – 179.

23. E. Simon, B. Berge; F. Fillit; H. Gaton; M. Guillet; O. Jacques-Sermet; F. Laune; J. Legrand; M. Maillard; N. Tallaron, "Optical design rules of a camera module with a liquid lens and principle of command for AF and OIS functions," in *Optical Design and Testing IV*, Y. Wang, J. Bentley, C. Du, K. Tatsumo, and H. P. Urbach (editors), *Proc. SPIE* **7849** (2010).

24. R. Siew and W. H. Kuek, "Minimizing thermal defocus effects in liquid lens based autofocus imaging systems using partial passive optical athermalization (PPOA)," Advanced Products Corporation Pte. Ltd. White Paper, (2017). Available at http://apc-vest.com/minimizing-thermal-defocus-effects-liquid-lens-based-autofcous-imaging-systems-using-partial-passive-optical-athermalization-ppoa/.

25. G. Berkovic and E. Shafir, "Optical methods for distance and displacement measurements," *Adv. Opt. Photon.* **4**(4), pp. 441 – 471 (2012).

26. J. Geng, "Structured-light 3D surface imaging: a tutorial," *Adv. Opt. Photon.* **3**(2), pp. 128 – 160 (2011).

27. A. A. Wajs, "Flash System for Multi-Aperture Imaging," United States Patent Application Publication No. **US 2013/0113988 A1**, (May 9, 2013).

28. See, for example, the Xbox® Kinect® sensor from Microsoft® at www.microsoft.com.

29. See Ref. 24, pp. 447 – 450.

30. W. J. Smith, *Modern Optical Engineering*, 2nd ed., (McGraw-Hill, 1990) pp. 270 – 273.

31. L. Baldwin, "Depth Determination Using Camera Focus," United States Patent Application Publication No. **US 2016/0248968 A1**, (Aug. 25, 2016).

32. D. Bakin, "Extended Depth of Field Technology in Camera Systems," in *Smart Mini-Cameras*, edited by T. Galstian, (CRC Press, 2014), pp. 99 – 147.

33. E. R. Dowski, Jr., and W. T. Cathey, "Extended depth of field through wave-front coding," *App. Opt.* **34**(11), pp. 1859 – 1866 (1995).

34. J. W. Goodman, *Introduction to Fourier Optics*, 2nd ed., (McGraw-Hill, 1996), pp. 148 – 151.

35. J. M. Geary, *Introduction to Lens Design With Practical Zemax*® *Examples*, (William-Bell, Inc., 2002), pp. 123 – 125.

36. E. H. Adelson and J. Y. A. Yang, "Single Lens Stereo with a Plenoptic Camera," *IEEE Transactions on Pattern Analysis and Machine Intelligence* **14**(2), pp. 99 – 106 (1992).

37. R. Ng, M. Levoy, M. Bredif, G. Duval, M. Horowitz, and P. Hanrahan, "Light Field Photography with a Hand-held Plenoptic Camera," *Stanford Tech Report* CTSR 2005-02.

38. See Ref. 1, pp. vii – xii.

39. D. Reshidko and J. Sasian, "Current trends in miniature camera lens technology," *SPIE Newsroom* (19, February, 2016). DOI: 10.1117/2.1201602.006327.

40. O. Cossairt, D. Miau, and S. K. Nayar, "Scaling law for computational imaging using spherical optics," *J. Opt. Soc. Am. A.* **28**(12), pp. 2540 – 2553 (2011).

41. A. Bruckner, "Multiaperture Cameras," in *Smart Mini-Cameras*, edited by T. Galstian, (CRC Press, 2014), pp. 247 – 252.

42. Y. Huo, C. C. Fesenmaier, and P. B. Catrysse, "Microlens performance limits in sub-2μm pixel CMOS image sensors," *Opt. Express* **18**(6), pp. 5861 – 5872 (2010).

43. M-S Kim, T. Scharf, S. Muhlig, M. Fruhnert, C. Rockstuhl, R. Bitterli, W. Noell, R. Voelkel, and H. P. Herzig, "Refraction limit of miniaturized optical systems: a ball-lens example," *Opt. Express* **24**(7), pp. 6996 – 7005 (2016).

44. Z. Chen and A. Taflove, "Photonic nanojet enhancement of backscattering of light by nanoparticles: a potential novel visible-light ultramicroscopy technique," *Opt. Express* **12**(7), pp. 1214 – 1220 (2004).

45. B. H. Walker, *Optical Design for Visual Systems*, (SPIE Press, 2000), p. 9.

46. F. L. Pedrotti, S. J., and L. S. Pedrotti, *Introduction to Optics*, 2nd ed., (Prentice Hall, 1993), p. 157.

47. See, for example, Ref. 46, p. 136.

48. R. Liang, *Optical Design for Biomedical Imaging*, (SPIE Press, 2010), pp. 103 – 104.

49. www.dynaoptics.com/oowa.html

50. www.dynaoptics.com

51. M. P. Schaub, *The Design of Plastic Optical Systems*, (SPIE Press, 2009), pp. 164 – 167.

52. ImageJ image processing tool, U. S. National Institutes of Health, Bethesda, Maryland, USA, https://imagej.nih.gov/ij/index.html, (1997-2016).

53. K. D. Long, H. Yu, and B. T. Cunningham, "Smartphone instrument for portable enzyme-linked immunosorbent assays," *Biomed. Opt. Express* **5**(11), pp. 3792 – 3806 (2014).

54. V. Lakshminarayanan, J. Zelek, and A. McBride, "'Smartphone Science' in Eye Care and Medicine," *Optics & Photonics News*, January 2015, pp. 44 – 51.

55. Md. A. Hossain, J. Canning, S. Ast, K. Cook, P. J. Rutledge, and A. Jamalipour, "Combined 'dual' absorption and fluorescence smartphone spectrometers," *Opt. Lett.* **40**(8), pp. 1737 – 1740 (2015).

56. Md. A. Hossain, J. Canning, S. Ast, K. Cook, and A. Jamalipour, "Optical fiber smartphone spectrometer," *Opt. Lett.* **41**(10), pp. 2237 – 2240 (2015).

57. S. Kim, D. Cho, J. Kim, M. Kim, S. Youn, J. E. Jang, M. Je, D. H. Lee, B. Lee, D. L. Farkas, and J. Y. Hwang, "Smartphone-based multispectral imaging: system development and potential for mobile skin diagnosis," *Biomed. Opt. Express* **7**(12), pp. 5294 – 5307 (2016).

58. J. Schlett, "Smartphones Poised to Shake Up Spectroscope," *Biophotonics*, January 2017, pp. 34 – 39.

59. https://www.gospectro-usa.com/

60. M. Born and E. Wolf, *Principles of Optics*, 6th ed., (Cambridge University Press, 1980), pp. 401 – 406.

61. See, for example, Ref. 46, pp. 251 – 252.

62. C. Palmer and E. Loewen, *Diffraction Grating Handbook*, 6th ed., (Newport Corp., 2005), p. 37.

63. Y. Feng and Y. Xiang, "Mitigation of spectral mis-registration effects in imaging spectrometers via cubic spline interpolation," *Opt. Express* **16**(20), pp. 15366 – 15374 (2008).

64. L. Guanter, K. Segl, B. Sang, L. Alonso, H. Kaufmann, and J. Moreno, "Scene-based spectral calibration assessment of high spectral resolution imaging spectrometers," *Opt. Express* **17**(14), pp. 11594 – 11606 (2009).

65. T. Skauli, "An upper-bound metric for characterizing spectral and spatial coregistration errors in spectral imaging," *Opt. Express* **20**(2), pp. 918 – 933 (2012).

66. M. Born and E. Wolf, *Principles of Optics*, 6th ed., (Cambridge University Press, 1980), pp. 513 – 516.

67. H. Hua and B. Javidi, "Augmented Reality: Easy on the Eyes," *Optics & Photonics News*, February 2015, pp. 26 – 33.

68. M. Wheeler, "HUD Systems: Augmented Reality is Coming to Your Windshield," *Photonics Spectra*, February 2016, pp. 34 – 37.

69. U. Vogel, "OLED Microdisplays: Advancing Virtual and Augmented Reality Smart Glasses," *Photonics Spectra*, April 2016, pp. 56 – 61.

70. "Bringing AR to the Masses," in *Optics & Photonics News*, June 2017, pp. 24 – 25.

71. https://www.x.company/glass/

72. X. Wu, J. He, J. Ellis, W. Choi, P. Wang, and K. Peng, "Which is a Better In-Vehicle Information Display? A Comparison of Google Glass and Smartphones," *J. of Display Technol.* **12**(11), pp. 1364 – 1371 (2016).

73. S. Feng, R. Caire, B. Cortazar, M. Turan, A. Wong, and A. Ozcan, "Immunochromatographic Diagnostic Test Analysis Using Google Glass," *ACS Nano* **8**(3), pp. 3069 – 3079 (2014).

74. A. Wang, A. Christoff, D. L. Guyton, M. X. Repka, M. Rezaei, A. O. Eghrari, "Google Glass Indirect Ophthalmoscopy," *Journal of Mobile Technology in Medicine* **4**(1), pp. 15 – 19 (2015).

75. E. Dehoog and J. Schwiegerling, "Optimal parameters for retinal illumination and imaging in fundus cameras," *App. Opt.* **47**(36), pp. 6769 – 6777.

76. E. Dehoog and J. Schwiegerling, "Fundus camera systems: a comparative analysis," *App. Opt.* **48**(2), pp. 221 – 228.

77. A. Gupta, B. Amirparviz, and S. Sharma, "Lightweight Eyepiece for Head Mounted Display," United States Patent Application Publication No. **US 2013/0070338 A1**, (Mar. 21, 2013).

78. A. Gupta, B. Amirparviz, and S. Sharma, "Lightweight Eyepiece for Head Mounted Display," United States Patent No. **US 9,013,793 B2**, (Apr. 21, 2015).

79. H. S. Raffle, and C-J. Wang, "Heads-up Display Including Eye Tracking," United States Patent Application Publication No. **US 2013/0207887 A1**, (Aug. 15, 2013).

80. R. W. Boyd, *Radiometry and the Detection of Optical Radiation*, (Wiley, 1983), pp. 91 – 93.

81. See Ref. 45, pp. 3 – 4.

82. P. Artal, "Optics of the eye and its impact in vision: a tutorial," *Adv. Opt. Photon.* **6**(3), pp. 340 – 367 (2014).

83. J. Carroll, D. C. Gray, A. Roorda, and D. R. Williams, "Recent Advances in Retinal Imaging With Adaptive Optics," *Optics & Photonics News*, January 2005, pp. 36 – 42.

84. J. Geng, "Three-dimensional display technologies," *Adv. Opt. Photon.* **5**(4), pp. 456 – 535 (2013).

85. A. R. L. Travis, "Autostereoscopic 3-D display," *App. Opt.* **29**(29), pp. 4341 – 4342 (1990).

86. J. B. Eichenlaub, "Developments in autostereoscopic technology at Dimension Technologies, Inc.," in *Stereoscopic Displays and Applications IV*, J. O. Merritt and S. S. Fisher (editors), *Proc. SPIE* **1915** (1993).

87. P. St. Hilaire, P-A. Blanche, C. Christenson, R. Voorakaranam, L. LaComb, B. Lynn, and N. Peyghambarian, "Are stereograms holograms? A human perception analysis of sampled perspective holography," *J. Phys. Conf. Ser.* **415** 012035 (2013).

88. J. T. McCrickerd and N. George, "Holographic stereogram from sequential component photographs," *App. Phys. Lett.* **12**(10), pp. 10 – 12.

89. T. N. Mitchell, "Method and apparatus for generating three-dimensional image formation using a single imaging path," United States Patent Application No. **US 2012/0188 A1**, (Jul. 26, 2012).

90. T. N. Mitchell and I. Shinkoda, "Method and apparatus for generating three-dimensional image formation," United States Patent Application No. **US 2013/0038690 A1**, (Feb. 14, 2013).

91. K. Iizuka, "Three-dimensional camera phone," *App. Opt.* **43**(34), pp. 6285 – 6292 (2004).

92. M-C. Park, S. J. Park, and J-Y Son, "Stereoscopic imaging and display for a 3-D mobile phone," *App. Opt.* **48**(34), pp. H238 – H243 (2009).

93. See, for example, the phone from the company Sharp: http://www.sharp-phone.com/en/index.html

94. See, for example, the phones from the company JSDigitech: www.j23d.cc

95. https://www.linkedin.com/pulse/glasses-free-3d-visual-treat-our-eyes-debesh-choudhury-ph-d-

96. K. Li, B. Robertson, M. Pivnenko, Y. Deng, D. Chu, J. Zhou, and J. Yao, "High quality micro liquid crystal phase lenses for full resolution image steering in auto-stereoscopic displays," *Opt. Express* **22**(18), pp. 21679 – 21689 (2014).

97. K. Li, A. O. Yontem, Y. Deng, P. Shrestha, D. Chu, J. Zhou, and J. Yao, "Full resolution auto-stereoscopic mobile display based on large scale uniform switchable crystal micro-lens array," *Opt. Express* **25**(9), pp. 9654 – 9675 (2017).

98. R. Siew, "Analysis of Biocular Images using 3D Glasses and Zemax: A Plausible Approach," available online at: http://customers.zemax.com/os/resources/learn/knowledgebase/analysis-of-biocular-images-using-3d-glasses-and-z

99. D. M. Bloom, F. S. A. Sandejas, and O. Solgaard, "Method and Apparatus for Modulating a Light Beam," United States Patent No. **5,311,360**, (May 10, 1994).

100. www.siliconlight.com

101. http://www.siliconlight.com/en/company/history.html

102. http://www.siliconlight.com/en/technology/glv.html

103. E. Tamaki, Y. Hashimoto, and O. Leung, "Computer-to-plate printing using the Grating Light Valve™ device," in *MOEMS Display and Imaging Systems II*, H. Urey, D. L. Dickensheets (editors), *Proc. SPIE* **5348** (2004).

104. See Ref. 34, p. 74.

105. See Ref. 46, pp. 341 – 346.

106. http://www.siliconlight.com/en/technology/imaging.html

107. Y. Reznichenko and H. A. Kelley, "Optical Imaging Head Having a Multiple Writing Beam Source," United States Patent No. **6,229,650 B1**, (May 8, 2001).

108. R. H. Siew, "Partial coherent imaging using the Grating Light Valve," in *Optical Scanning 2002*, S. F. Sagan, G. F. Marshall, and L. Beiser (editors), *Proc. SPIE* **4773**, pp. 92 – 101 (2002).

109. R. Thoma, V. Melzer, R. Siew, T. Damm, and P. Mueller, "System for Correction of Spatial Cross-talk and Pattern Frame Effects in Imaging Systems," United States Patent No. **6,897,994 B2**, (May 24, 2005).
110. D. C. O'Shea, W. R. Callen, and W. T. Rhodes, *Introduction to Lasers and Their Applications*, (Addison-Wesley, 1977), pp. 59 – 60.
111. See, for example, Ref. 48, pp. 3 – 4.
112. J. R. Lackowicz, *Principles of Fluorescence Spectroscopy*, 3rd ed., (Springer, 2006), pp. 58 – 59.
113. See Ref. 112, pp. 8 – 9.
114. See Ref. 112, p. 41.
115. https://www.nobelprize.org/nobel_prizes/chemistry/laureates/1993/
116. https://www.thermofisher.com/ca/en/home/life-science/cloning/cloning-learning-center/invitrogen-school-of-molecular-biology/pcr-education/pcr-reagents-enzymes/pcr-basics.html
117. https://www.thermofisher.com/ca/en/home/life-science/pcr/real-time-pcr/qpcr-education.html
118. https://www.thermofisher.com/ca/en/home/life-science/pcr/real-time-pcr/qpcr-education/essentials-of-real-time-pcr.html
119. K. Edwards, J. Logan, and N. Baunders, *Real-Time PCR: An Essential Guide*, (Horizon Bioscience, 2004).
120. R. W. Boyd, *Radiometry and the Detection of Optical Radiation*, (Wiley, 1983), pp. 63 – 66.
121. S. J. Boege, J. A. Hoshizaki, M. F. Oldham, and L. Ilkova, "Fluorescence Detector with Automatic Changing Filters," United States Patent No. **7,295,316 B2** (Nov. 13, 2007).
122. http://customers.zemax.com/os/resources/learn/knowledgebase/how-to-model-fluorescence-using-bulk-scattering
123. http://customers.zemax.com/os/resources/learn/knowledgebase/how-to-design-a-confocal-fluorescent-microscope-in
124. https://www.thermofisher.com/order/catalog/product/2101
125. J. Hoshizaki, H. G. King, J. P. Sluis, S. J. Boege, and M. F. Oldham, "Optical Scanning Configurations, Systems, and Methods," United States Patent No. **7,423,750 B2**, (Sep, 9, 2008).
126. S. J. Boege, "Concentrators for Luminescent Emission," United States Patent Application Publication No. **US 2007/0190642 A1**, (Aug 16, 2007).
127. R. Siew, "Axial nonimaging characteristics of imaging lenses: discussion," *J. Opt. Soc. Am. A* **33**(5), pp. 970 – 977 (2016).

Index

3D glasses, 276

3D imaging, 275

aberration-limited, 59

absorption coefficient, 310

absorption spectrum, 311

adaptive optics, 274

afocal attachments, 243

Airy disk

    in diffractive depth of focus, 50

amplitude spread function, 71

analytical instruments, 312

angular dispersion, 252

annealing, 313

anti-reflection

    (coating), 41

aperture stop

    defined, 33

aplantism

    independence of image irradiance
        from, 103

Apochromat

    multi-element lens system, 190

    thin lens doublet, 183

apodization

    defined, 67

Apodization, 66–74

Apparitions

    (ghost images) of optical systems,
        41

ASF. *See* amplitude spread function

Astigmatism

    defined, 40

augmented reality, 262

Autostereoscopic 3D imaging

    in mobile phone displays, 283

autostereoscopic displays, 276

Beer's law, 310

Blazing, 62–64

capability roadmap, 14

chief ray, 26, 129

chief ray angle

    in image space, 144

    in object space, 129

Chromatic Aberration, 178–202

coherence, 75–76

coherent transfer function, 75

coma

    defined, 40

complex degree of coherence, 298

conjugate, 25

convolution, 49, 59, 60

convolution theorem, 50

cosine^4th effect, 153

curved image sensors

    best focus position, 202

    interpretation of image distortion
        on, 205

customization

    for machine vision lenses, 155

Defocus

    aberration, defined, 40

denaturing, 313

density function. *See* Radiance,
defined

depth of field

    defined, 44

Depth of Field, 44–47

depth of focus

   defined, 47

   diffractive, 50

   geometric, 51

Depth of Focus, 47–49

Depth Sensing

   in machine vision, 218

differential distortion, 120

diffraction and aberrations

   impact on image irradiance, 90

diffraction-limited, 56

Diffractive Optics, 61–66

Distortion

   defined, 40

DNA, 312

   polymerase, 313

Effective F-number, 112–16

effective focal length

   defined, 29

embedded vision, 128

emission spectrum, 310

end point analysis

   in qPCR, 314

entrance and exit pupils

   influence on depth sensing, 219

   significance to phone attachments, 239

entrance pupil, 120

   defined, 34

etch depth, 65

etendue, 78–82

exit pupil, 121

   defined, 34

exitance, 84

extended depth of field imaging, 228

extension

   of DNA strand, 313

eyepiece, 238

Fermat's principle, 1

Field Curvature

   defined, 40

field lens, 26

field of view, 129

field stop, 146

first order

   properties of a lens, 29

fixed focus lenses, 52

fluorescence, 310

fluorescence detection system, 10, 310–12

fluorophores, 310

flux

   defined, 77

F-number

   of a lens with non-circular stop, 112

focusing

   a lens in machine vision, 206

Fourier optics, 4

Fourier Optics, 54–73

Fourier plane, 296

Fourier transform spectrometer, 16, 261

free spectral range, 253

freeform surface, 245

Fresnel factor and Fraunhofer factor, 55

f-stop, 35

fundus cameras, 262

fundus imager

using Google's Glass wearable computing device as, 264

Gaussian apodization, 70

Gaussian apodization factor

in Zemax OS, 71

Geometrical Optics, 23–53

glare radius, 152

glossy surfaces, 152

GLV. *See* Grating Light Valve

Google

Glass Wearable Computing Device, 262

Google Glass. *See* Google Glass Wearable Computing Device

Grating Light Valve

imaging, 292
operating principle, 292

hippo-focal distance, 53

holograms, 277

hyperfocal distance

defined, 52

Illumination

in machine vision, 150

illumination engineering, 111

image brightness, 86–90

image circle, 133

Image Circles, 137–42

image irradiance

axial, 88

Image Irradiance, 90–97

Imaging Principle, 23

interferometer, 16

irradiance

defined, 77

irradiance transport, 104

isoplanatic patches, 60

kinoform, 66

label-free detection, 14

Lambertian source, 82

laser speckle, 76

LED, 82

lens designers, 8–11

lens maker's formula, 33

liquid lenses, 210

luminous intensity, 82

magnification

in phone attachments, 235

magnification of the irradiance, 104

magnifier, 235

matte surfaces, 153

microscope objective, 237

miniaturization

limits of, 231
of spectrometers, 15–18

mobile phones

imaging attachments, 234

modern classical imaging, 5

modulation transfer function

effect of depth of field and focus, 44–46

MTF. *See* modulation transfer function

of a diffraction limited lens, 67

Near and far points

(relation to depth of field and focus), 46

(relation to imaging attachments), 234

Nonimaging

characteristics of imaging lenses, 105–12

nonimaging optics, 111

nucleotides, 313

numerical aperture

defined, 28

OOWA lens, 245

OPD. *See* optical path difference

optical coherence tomography, 14

optical designer. *See* lens designers

optical engineers, 8–11

optical path difference

defined, 37

optical path length

defined, 58

paraxial approximation, 29

paraxial image magnification

defined, 31

paraxial lens laws, 30

Paraxial Thin Lens Model, 31–32

partial coherence

in radiometry, 76

PCR. *See* polymerase chain reaction

real-time. *See* Real-Time PCR

Phase Terms, 56–58

photons, 2–6

physical etendue, 82

Physical Optics, 54–60

point spread function, 49, 56

polymerase chain reaction, 312

POP. *See* position of peak irradiance

position of peak irradiance, 105

in qPCR, 316

primer/s, 312

principal planes, 28

probes

in qPCR, 314

PTLM. *See* Paraxial Thin Lens Model

qPCR. *See* quantitative polymerase chain reaction

quantitative polymerase chain reaction

instruments, 312

quantum nonlinear optics, 15

quantum optical engineering, 15–18

quantum optical engineers, 15–18

quantum yield, 310

quencher

in qPCR, 315

radiance

defined, 77

radiance theorem, 78

radiance theorem for paraxial thin lens models, 86

radiant intensity, 82

Radiometry, 76–122

Raman imaging, 14

Real-Time PCR, 314

region of interest

defined, 53

relative illumination

defined, 36

Relative Illumination

effect of differential distortion, 163–78

effect of distortion and pupils, 116–22

relative partial dispersion, 182

resolution, 145

increase by apodization, 67

resolving power

of a spectrometer, 251

ROI

see region of interest, 53

RTPTLM. *See* radiance theorem for paraxial thin lens models

sensor formats, 145

shift-invariant, 60

signal to noise ratio, 147

slow light, 15–18

SNR. *See* signal to noise ratio

space-invariant, 60

spatial coherence, 75

    illumination on the GLV, 298

    partial, 299

spectral curvatures, 253

spectral resolution, 251

    geometric, 259

    in a miniaturized spectrometer, 16

Spectrometer

    phone attachment, 249

Spherical Aberration

    defined, 40

Stereo Imaging Systems, 275

stereoscopy, 276

stock lenses, 129

Stoke shift, 310

stray light, 41

    analysis, 43

sub-etendue, 81

symmetry (of lenses)

    influence on depth sensing, 221

technology roadmap, 14

telecentricity, 161

telescope, 238

Theorem of Malus and Dupin, 23

thermal imaging cameras, 13

t-number, 116

triangulation

    in depth sensing, 218

t-stop. *See* t-number

tube lens, 238

vignetting, 36

wavefront aberration function, 41

wavefronts, 4, 23

working distance, 132, 208

working f-number

    defined, 34

ABOUT THE AUTHOR

RONIAN SIEW became fascinated with optics and imaging at the age of eleven, and began constructing simple imaging systems using scrap lenses salvaged from junk. Today, Ronian has had more than twenty years of professional experience in optical engineering. His designs have included objectives, eyepieces, relays, fluorescence detection systems for DNA analysis, and more. Ronian has authored a number of technical articles, including scientific peer-reviewed and conference papers, and his work has been cited in Zemax's optical design program user's manual. Ronian holds a Master of Science in Optics and a degree in Physics from the University of Rochester. He also enjoys drawing cartoons, the piano, and is a physics bookworm. He lives with his wife and books in Vancouver, BC.

CPSIA information can be obtained
at www.ICGtesting.com
Printed in the USA
LVHW080549170420
653829LV00020B/2760